POSTHARVEST MANAGEMENT OF HORTICULTURAL CROPS
Practices for Quality Preservation

Postharvest Biology and Technology

POSTHARVEST MANAGEMENT OF HORTICULTURAL CROPS
Practices for Quality Preservation

Edited by
Mohammed Wasim Siddiqui, PhD
Asgar Ali, PhD

Apple Academic Press Inc. | Apple Academic Press Inc.
3333 Mistwell Crescent | 9 Spinnaker Way
Oakville, ON L6L 0A2 | Waretown, NJ 08758
Canada | USA

©2017 by Apple Academic Press, Inc.
Exclusive worldwide distribution by CRC Press, a member of Taylor & Francis Group
No claim to original U.S. Government works
Printed in the United States of America on acid-free paper
International Standard Book Number-13: 978-1-77188-334-4 (Hardcover)
International Standard Book Number-13: 978-1-77188-335-1 (eBook)

All rights reserved. No part of this work may be reprinted or reproduced or utilized in any form or by any electric, mechanical or other means, now known or hereafter invented, including photocopying and recording, or in any information storage or retrieval system, without permission in writing from the publisher or its distributor, except in the case of brief excerpts or quotations for use in reviews or critical articles.

This book contains information obtained from authentic and highly regarded sources. Reprinted material is quoted with permission and sources are indicated. Copyright for individual articles remains with the authors as indicated. A wide variety of references are listed. Reasonable efforts have been made to publish reliable data and information, but the authors, editors, and the publisher cannot assume responsibility for the validity of all materials or the consequences of their use. The authors, editors, and the publisher have attempted to trace the copyright holders of all material reproduced in this publication and apologize to copyright holders if permission to publish in this form has not been obtained. If any copyright material has not been acknowledged, please write and let us know so we may rectify in any future reprint.

Trademark Notice: Registered trademark of products or corporate names are used only for explanation and identification without intent to infringe.

Library and Archives Canada Cataloguing in Publication

Postharvest management of horticultural crops : practices for quality preservation / edited by Mohammed Wasim Siddiqui, PhD.

(Postharvest biology and technology book series)
Includes bibliographical references and index.
Issued in print and electronic formats.
ISBN 978-1-77188-334-4 (hardcover).--ISBN 978-1-77188-335-1 (pdf)

1. Horticultural crops--Postharvest technology. 2. Food crops--Postharvest technology. 3. Food industry and trade--Quality control. I. Siddiqui, Mohammed Wasim, author, editor II. Series: Postharvest biology and technology book series

SB318.P66 2016   635'.046   C2016-902517-9   C2016-902518-7

Library of Congress Cataloging-in-Publication Data

Names: Siddiqui, Mohammed Wasim, editor.
Title: Postharvest management of horticultural crops : practices for quality preservation / editor: Mohammed Wasim Siddiqui.
Description: Oakville, ON ; Waretown, NJ : Apple Academic Press, [2016] |
Includes bibliographical references and index.
Identifiers: LCCN 2016017384 (print) | LCCN 2016021532 (ebook) | ISBN 9781771883344 (hardcover : alk. paper) | ISBN 9781771883351 ()
Subjects: LCSH: Horticultural crops--Postharvest technology.
Classification: LCC SB319.7 .P683 2016 (print) | LCC SB319.7 (ebook) | DDC 635--dc23
LC record available at https://lccn.loc.gov/2016017384

Apple Academic Press also publishes its books in a variety of electronic formats. Some content that appears in print may not be available in electronic format. For information about Apple Academic Press products, visit our website at **www.appleacademicpress.com** and the CRC Press website at **www.crcpress.com**

This Book Is
Affectionately Dedicated
to
The World Food Preservation Centre, LLC
for
Education, Innovation, and Advocacy to Reducing Postharvest Food
Losses in Developing Countries

CONTENTS

Dedication .... *v*
Acknowledgments .... *ix*
List of Contributors .... *xi*
List of Abbreviations .... *xiii*
About the Book Series: Postharvest Biology and Technology ...*xv*
Books in the Postharvest Biology and Technology Series .... *xvii*
About the Editors .... *xix*
Preface .... *xxiii*

1. **Recent Advances in Postharvest Cooling of Horticultural Produce** 1
   Atef M. Elansari and Mohammed Wasim Siddiqui

2. **Postharvest Handling and Storage of Root and Tubers** 69
   Munir Abba Dandago

3. **Postharvest Management of Commercial Flowers** 91
   Sunil Kumar, Kalyan Barman, and Swati Sharma

4. **Postharvest Management and Processing Technology of Mushrooms** 151
   M. K. Yadav, Santosh Kumar, Ram Chandra, S. K. Biswas, P. K. Dhakad, and Mohammed Wasim Siddiqui

5. **Gibberellins: The Roles in Pre- and Postharvest Quality of Horticultural Produce** 179
   Venkata Satish Kuchi, J. Kabir, and Mohammed Wasim Siddiqui

6. **Advances in Packaging of Fresh Fruits and Vegetables** 231
   Alemwati Pongener and B. V. C. Mahajan

7. **Fresh-Cut Produce: Advances in Preserving Quality and Ensuring Safety** 265
   Ovais Shafiq Qadri, Basharat Yousuf, and Abhaya Kumar Srivastava

8. **Postharvest Pathology, Deterioration, and Spoilage of Horticultural Produce** 291

   S. M. Yahaya

9. **Natural Antimicrobials in Postharvest Storage and Minimal Processing of Fruits and Vegetables** 311

   Munir Abba Dandago

10. **ENHANCE: Breakthrough Technology to Preserve and Enhance Food** 325

    Charles L. Wilson

    Index *335*

ACKNOWLEDGMENTS

It was almost impossible to reveal the deepest sense of veneration to all without whose precious exhortation this book project could not be completed. At the onset of the acknowledgment, we ascribe all glory to the Gracious "Almighty Allah" from whom all blessings come. We would like to thank Him for His blessing to prepare this book.

With a profound and unfading sense of gratitude, we convey special thanks to our colleagues and other research team members for their support and encouragement for helping us in every step to accomplish this venture.

We are grateful to Mr. Ashish Kumar, President, Apple Academic Press, for publishing this book series, titled *Postharvest Biology and Technology*. We would also like to thank Ms. Sandra Jones Sickels and Mr. Rakesh Kumar of Apple Academic Press for their continuous support to complete the project.

In omega, our vocabulary will remain insufficient to express our indebtedness to our adored parents and family members for their infinitive love, cordial affection, and incessant inspiration.

LIST OF CONTRIBUTORS

Kalyan Barman
Department of Horticulture (Fruit and Fruit Technology), Bihar Agricultural University, Sabour, Bhagalpur – 813210, Bihar, India

S. K. Biswas
Department of Plant Pathology, C.S. Azad University of Agriculture and Technology, Kanpur, 208002, Uttar Pradesh, India

Ram Chandra
Department of Mycology and Plant Pathology, Institute of Agricultural Sciences, Banaras Hindu University, Varanasi, 221005, Uttar Pradesh, India

Munir Abba Dandago
Department of Food Science and Technology, Faculty of Agriculture and Agricultural Technology, Kano University of Science and Technology, Wudil, Kano State, Nigeria

P. K. Dhakad
Department of Mycology and Plant Pathology, Institute of Agricultural Sciences, Banaras Hindu University, Varanasi, 221005, Uttar Pradesh, India

Atef M. Elansari
Agricultural and Bio-Engineering Department, Alexandria University, Alexandria, Egypt; E-mail: aansari1962@yahoo.com

J. Kabir
Department of Postharvest Technology of Horticultural Crops, Bidhan Chandra Krishi Viswavidyalaya, Mohanpur, Nadia, West Bengal–741252, India

Venkata Satish Kuchi
Department of Postharvest Technology of Horticultural Crops, Bidhan Chandra Krishi Viswavidyalaya, Mohanpur, Nadia, West Bengal–741252, India

Santosh Kumar
Department of Plant Pathology, Bihar Agricultural University, Sabour, Bhagalpur, 813210, Bihar, India, E-mail: santosh35433@gmail.com

Sunil Kumar
Department of Horticulture, North Eastern Hill University, Tura Campus, West Garo Hills District, Tura – 794002, Meghalaya, India, E-mail: sunu159@yahoo.co.in

B. V. C. Mahajan
Punjab Horticultural Postharvest Technology Centre, PAU, Ludhiana, 141004, Punjab, India

Alemwati Pongener
ICAR-National Research Centre on Litchi, Mushahari, Muzaffarpur, 842002, Bihar, India, E-mail: alemwati@gmail.com

Ovais Shafiq Qadri
Department of Postharvest Engineering and Technology, Aligarh Muslim University, India,
E-mail: osqonline@gmail.com, Tel.: +91-9419041070

Swati Sharma
ICAR-National Research Centre on Litchi, Mushahari Farm, Mushahari, Muzaffarpur – 842002, Bihar, India

Mohammed Wasim Siddiqui
Department of Food Science and Post-Harvest Technology, BAC Bihar Agricultural University, India;
E-mail: wasim_serene@yahoo.com

Abhaya Kumar Srivastava
Department of Postharvest Engineering and Technology, Aligarh Muslim University, India,
E-mail: osqonline@gmail.com, Tel.: +91-9419041070

Charles L. Wilson
Founder/Chairman and CEO, World Food Preservation Center LLC, E-mail: worldfoodpreservationcenter@frontier.com

M. K. Yadav
Department of Mycology and Plant Pathology, Institute of Agricultural Sciences, Banaras Hindu University, Varanasi, 221005, Uttar Pradesh, India

S. M. Yahaya
Department of Biology, Kano University of Science and Technology, Wudil, P.M.B. 3244, Nigeria,
E-mail: sanimyahya@yahoo.com

Basharat Yousuf
Department of Postharvest Engineering and Technology, Aligarh Muslim University, India,
E-mail: osqonline@gmail.com, Tel.: +91-9419041070

LIST OF ABBREVIATIONS

ABA	abscisic acid
ACS	agriculturae conspectus scientificus
AVG	aminoethoxyvinylglycine
CA	controlled atmosphere
CAP	controlled atmosphere packaging
CFB	corrugated fiberboard boxes
CO	carbon monoxide
DACP	diazo-cyclopentadiene
DPCA	N-dipropyl (1-cyclopropenylmethyl) amine
DPSS	dimethyl-4-(phenylsulfonyl) semicarbazide
DX	direct expansion
EASP	electronic aroma signature pattern
ETH	etheophon
FAB	food, agriculture and biology
FEFO	first-expired-first-out
GA	gibberellic acid
GAP's	good agricultural practices
GGPP	geranylgeranyl pyrophosphate
GMP's	good manufacturing practices
HACCP	hazard analysis critical control point
HCFCs	halogenated hydrocarbons
HOCl	hypochlorite
HPTS	hydroxypyrene-1,3,6-trisulfonicacid
IPENZ	the Institution of Professional Engineers New Zealand
IPRH	in-package-relative-humidity
KMS	potassiummeta bisulphite
LAB	lactic acid bacteria
LDPE	low-density polyethylene
LPS	low pressure storage
LR-WPANs	low-rate wireless personal area networks
MA	modified atmosphere
MAC	medium access control

MAP	modified atmosphere packaging
MENA	Middle East and North Africa
MIR	mid infra-red
MJ	methyl jasmonate
MRI	magnetic resonance imaging
MRR	magnetic resonance relaxometry
MRS	magnetic resonance spectroscopy
NENA	Near East and North Africa
NIR	nuclear infra-red
NMR	nuclear magnetic resonance
OTR	oxygen transmission rate
PAA	peroxyacetic acid
PAL	phenylalanine ammonia lyase
PG	polygalacturonase
PHY	physical layer
PLC	programmable logic controllers
PLW	physiological weight loss
PME	pectin methyl esterase
PP	polypropylene
PPE	pomegranate peel extract
PVC	poly vinyl chloride
RFID	radio frequency identification
RH	relative humidity
RSCCS	refined smart cold chain system
RTE	ready-to-eat
SA	simulated annealing
SAF	Society of American Florist
SCADA	supervisory control and data acquisition systems
SCCS	smart cold chain system
STS	silver thiosulfate
TAL	tyrosine ammonia lyase
TDZ	thidiazuron
TTIs	time-temperature indicators
VFD	variable frequency drive
VSDs	variable speed drive
WHO	World Health Organization
WSN	wireless sensor network

ABOUT THE BOOK SERIES: POSTHARVEST BIOLOGY AND TECHNOLOGY

As we know, preserving the quality of fresh produce has long been a challenging task. In the past, several approaches were in use for the postharvest management of fresh produce, but due to continuous advancement in technology, the increased health consciousness of consumers, and environmental concerns, these approaches have been modified and enhanced to address these issues and concerns.

The **Postharvest Biology and Technology** series presents edited books that addressmany important aspects related to postharvest technology of fresh produce. The series presents existing and novel management systems that are in use today or that have great potential to maintain the postharvest quality of fresh produce in terms of microbiological safety, nutrition, and sensory quality.

The books are aimed at professionals, postharvest scientists, academicians researching postharvest problems, and graduate-level students.This series is intended to be a comprehensive venture that provides up-to-date scientific and technical information focusing on postharvest management for fresh produce.

Books in the series will address the following themes:

- Nutritional composition and antioxidant properties of fresh produce
- Postharvest physiology and biochemistry
- Biotic and abiotic factors affecting maturity and quality
- Preharvest treatments affecting postharvest quality
- Maturity and harvesting issues
- Nondestructive quality assessment
- Physiological and biochemical changes during ripening
- Postharvest treatments and their effects on shelf life and quality
- Postharvest operations such as sorting, grading, ripening, de-greening, curing etc
- Storage and shelf-life studies

- Packaging, transportation, and marketing
- Vase life improvement of flowers and foliage
- Postharvest management of spice, medicinal, and plantation crops
- Fruit and vegetable processing waste/byproducts: management and utilization
- Postharvest diseases and physiological disorders
- Minimal processing of fruits and vegetables
- Quarantine and phytosanitary treatments for fresh produce
- Conventional and modern breeding approaches to improve the postharvest quality
- Biotechnological approaches to improve postharvest quality of horticultural crops

We are seeking editors to edit volumes in different postharvest areas for the series. Interested editors may also propose other relevant subjects within their field of expertise, which may not be mentioned in the list above. We can only publish a limited number of volumes each year, so if you are interested, please email your proposal wasim@appleacademicpress.com at your earliest convenience.

We look forward to hearing from you soon.

Editor-in-Chief:
Mohammed Wasim Siddiqui, PhD
Scientist-cum-Assistant Professor | Bihar Agricultural University
Department of Food Science and Technology | Sabour | Bhagalpur | Bihar | INDIA
AAP Sr. Acquisitions Editor, Horticultural Science
Founding/Managing Editor, *Journal of Postharvest Technology*
Email: wasim@appleacademicpress.com
wasim_serene@yahoo.com

BOOKS IN THE POSTHARVEST BIOLOGY AND TECHNOLOGY SERIES

Postharvest Biology and Technology of Horticultural Crops: Principles and Practices for Quality Maintenance
Editor: Mohammed Wasim Siddiqui, PhD

Postharvest Management of Horticultural Crops: Practices for Quality Preservation
Editor: Mohammed Wasim Siddiqui, PhD, Asgar Ali, PhD

Insect Pests of Stored Grain: Biology, Behavior, and Management Strategies
Editor: Ranjeet Kumar, PhD

ABOUT THE EDITORS

Mohammed Wasim Siddiqui, PhD

Dr. Mohammed Wasim Siddiqui is an Assistant Professor and Scientist in the Department of Food Science and Post-Harvest Technology, Bihar Agricultural University, Sabour, India. His contribution as an author and editor in the field of postharvest biotechnology has been well recognized. He is an author or co-author of 34 peers reviewed research articles, 26 book chapters, two manuals, and 18 conference papers. He has 11 edited volumes and one authored book to his credit, published by Elsevier, USA; CRC Press, USA; Springer, USA; and Apple Academic Press, USA. Dr. Siddiqui has established an international peer-reviewed journal *Journal of Postharvest Technology*.

He has been honored to be the Editor-in-Chief of two book series titled "Postharvest Biology and Technology" and "Innovations in Horticultural Science" being published from Apple Academic Press, New Jersey, USA. Dr. Siddiqui is a Senior Acquisitions Editor in Apple Academic Press, New Jersey, USA for Horticultural Science. He has been serving as an editorial board member and active reviewer of several international journals such as *PLoS ONE* (PLOS), *LWT-Food Science and Technology* (Elsevier), *Food Science and Nutrition* (Wiley), *Acta Physiologiae Plantarum* (Springer), *Journal of Food Science and Technology* (Springer), and the *Indian Journal of Agricultural Science* (ICAR), etc.

Recently, Dr. Siddiqui was conferred with the Best Citizen of India Award (2016), Bharat Jyoti Award (2016), Outstanding Researcher Award (2016) by Aufau Periodicals, India, Best Young Researcher Award (2015) by GRABS Educational Trust, Chennai, India and the Young Scientist Award (2015) by Venus International Foundation, Chennai, India. He was also a recipient of the Young Achiever Award (2014) for outstanding research work by the Society for Advancement of Human and Nature (SADHNA), Nauni, Himachal Pradesh, India, where he is an Honorary

Board Member and Life Time Author. He has been an active member of the organizing committees of several national and international seminars/conferences/summits. He is one of key members in establishing the World Food Preservation Center (WFPC), LLC, USA. Presently, he is an active associate and supporter of WFPC, LLC, USA. Considering his outstanding contribution in science and technology, his biography has been published in "*Asia Pacific Who's Who*" and "*The Honored Best Citizens of India.*"

Dr. Siddiqui acquired a BSc (Agriculture) degree from Jawaharlal Nehru Krishi Vishwa Vidyalaya, Jabalpur, India. He received MSc (Horticulture) and PhD (Horticulture) degrees from Bidhan Chandra Krishi Viswa Vidyalaya, Mohanpur, Nadia, India, with specialization in the Postharvest Technology. He was awarded the Maulana Azad National Fellowship Award from the University Grants Commission, New Delhi, India. He is a member of Core Research Group at the Bihar Agricultural University (BAU) where he is providing appropriate direction and assistance to sensitize priority of the research. He has received several grants from various funding agencies to carry out his research projects. Dr. Siddiqui has been associated with postharvest technology and processing aspects of horticultural crops. He is dynamically involved in teaching (graduate and doctorate students) and research, and he has proved himself as an active scientist in the area of Postharvest Technology.

Asgar Ali, PhD

Prof. Asgar Ali is the Founding Director and Professor of Postharvest Biotechnology at the Centre of Excellence for Postharvest Biotechnology (CEPB), University of Nottingham Malaysia Campus. The CEPB is a global centre for postharvest research with the mandate of reducing postharvest losses. His research on postharvest biology and technology is internationally acknowledged, being notable on the development of natural edible coatings for the extension of shelf-life of perishable fruits and vegetables. This has resulted in a high frequency of publications in the top journals in the field, including the *Top 25 Hottest Articles by Science Direct* and publicly disseminated through popular media outlets such as *National Geographic* and EarthSky Net, from Hungary and USA, respectively, the Biotechnology and Biological Sciences Research Council (BBSRC) and Farming UK. Prof. Asgar also contributes to the field as a regular peer reviewer for high impact journals, and has been appointed Associate Editor for the *Journal of Horticultural Science and Biotechnology*, a leading peer-reviewed journal publishing high-quality research in horticulture since 1919.

Collaborations with government, academic and industrial bodies within and beyond Malaysia have been forged by Prof. Asgar. This has resulted in the expansion of his research area through direct funding into the Centre (CEPB). Dr. Asgar has served as chair, invited speaker, and keynote presenter for a number of international and national conferences and meetings in USA, UK, Malaysia, India, Turkey, and South Africa on recent advances in postharvest biotechnology fields. In addition, he was appointed as an international evaluator for proposals of international standards.

He was awarded a BSc Ag & AH and MSc (Horticulture) with first class from Chandra Shekhar Azad University of Agriculture and Technology, Kanpur India. In 2001, he was offered a Graduate Research Assistantship (GRA) in the Department of Crop Science, University Putra Malaysia for pursuing doctoral study in the area of Crop and Postharvest Physiology.

PREFACE

The eating quality of fresh horticultural products is mainly developed on the plant and can only be established at harvest. The harvesting of the commodity from the plant cuts off supply or accumulation of carbohydrates, water, and nutrients, owing to which the possibility for further improvement in the quality attributing components that is ceased. It is well established that these fresh horticultural products are living entities even after harvest. There are several pre- and postharvest strategies that have been developed to modify these physiological activities, resulting in increased shelf life. Therefore, it is very important to understand and apply the best technologies that positively influence quality attributes, including senescenal changes and, afterwards, the consumers' decision to purchase the product in the marketplace.

This book, titled *Postharvest Management of Horticultural Crops: Practices for Quality Preservation*, has been contributed to by experts of their fields, belonging both to developed and developing world. The book is consists of 10 chapters covering a thorough discussion on postharvest management strategies of fresh horticultural commodities.

Chapter 1 deals with the recent advances in postharvest cooling of horticultural produce. The chapter includes thorough coverage of different cooling systems of fresh commodities. Several photographs have been provided in the chapter to enhance understanding. The physiological factors, such as respiration and transpiration or water loss from the surface, affect the postharvest quality of the root or tuber crops in many ways. Chapter 2 discusses different postharvest handling and storage systems of root and tuber crops. The maximum potential vase life of cut flowers is very short due to high metabolic activities and other factors. For proper postharvest handling and prolonging vase life of cut flowers, several steps are needed to make the commercial floriculture venture a profitable trade are discussed in Chapter 3. Fresh mushrooms have very short self-life, and hence several technologies are recommended to increase their shelf-life. Chapter 4 describes in detail the postharvest management and processing technology of mushrooms. Research in the field of plant hormones is an interesting aspect of physiology in which new research findings have been established every year. Interaction of gibberellins with other plant hormones and its role in maintaining postharvest quality of horticultural produce has been highlighted in Chapter 5.

There is continuous demand for innovative and creative packaging from producers, processors, transporters, whole-sellers, retailers, and consumers to guarantee food quality, safety, and traceability. Chapter 6 deals with the advancement in packaging of fresh fruits and vegetables. There has been a continuous increase in consumer demand for convenient and minimally processed produce, including fresh-cut fruits and vegetables. In this curriculum, Chapter 7 precisely discusses the technological advances in preserving quality and ensuring safety of fresh-cut products. Chapter 8 covers important aspects of postharvest pathology, deterioration, and spoilage of horticultural produce. Fresh fruit and vegetables are perishable and susceptible to postharvest diseases, which limit the storage period and marketing life. The use of synthetic fungicides has many limitations and disadvantages. Chapter 9 describes the application of natural antimicrobials in postharvest storage and minimal processing of fruits and vegetables. Plants produce an array of phytochemicals in response to stress. Chapter 10 deals with a new concept of ENHANCE that has been supposed to be a breakthrough technology to preserve and enhance food.

The editors are confident that this book will prove to be a standard reference work, describing recent advancement in postharvest management of fresh horticultural commodities. The editors would appreciate receiving new information and comments to assist in the future development of the next edition.

CHAPTER 1

RECENT ADVANCES IN POSTHARVEST COOLING OF HORTICULTURAL PRODUCE

ATEF M. ELANSARI[1] and MOHAMMED WASIM SIDDIQUI[2]

[1]*Agricultural and Bio-Engineering Department, Alexandria University, Alexandria, Egypt, E-mail: aansari1962@yahoo.com*

[2]*Department of Food Science and Post-Harvest Technology, BAC Bihar Agricultural University, India, E-mail: wasim_serene@yahoo.com*

CONTENTS

Abstract 2
1.1 Introduction 3
1.2 The Importance of Precooling 5
1.3 Approaching the Optimum Precooling Method 6
    1.3.1 How Does the Fresh Produce get Precooled? 7
    1.3.2 Heat Load Calculations? 8
1.4 Types of Air Pre-Cooling Methods 9
    1.4.1 Natural Convection Air-Cooling (Room Cooling Method) 9
        1.4.1.1 Modified Room Cooling Method 12
    1.4.2 Forced Air-Cooling 13
1.5 Packaging 17
1.6 Capacity Design 18

1.7   System Classification 20
    1.7.1   Wet Cooling System (Ice Banks).................................... 20
    1.7.2   Dry System 24
1.8   Mobile Pre-Cooling Facilities.. 26
1.9   Hydrocooling 29
1.10 Vacuum Cooling... 34
1.11 Water Loss During Vacuum Cooling Determination 39
1.12 Mathematical Modeling of Vacuum Cooling Process.................. 40
1.13 Features and Benefits of Vacuum Cooling 42
1.14 Slurry Ice.. 42
    1.14.1 Direct Use of Slurry Ice 43
    1.14.2 Indirect Use of Slurry Ice.. 46
1.15 Control of the Cold Chain Projects.. 47
1.16 Variable Frequency Drive and Control Strategy 49
1.17 Temperature and Relative Humidity Control............................... 52
1.18 Energy Saving 56
1.19 Maintenance 59
1.20 Conclusion 60
Keywords 60
References.. 60

ABSTRACT

Temperature is the most single important factor that affects the quality and the shelf life and horticultural crops. The process of precooling is the removal of field heat as soon as possible after harvest since field heat arrest the deterioration and senescence process. The precooling process can be achieved via different methods forced air-cooling, hydrocooling, vacuum, slurry ice and evaporative cooling. Forced air precooling is the most common technique and is adapted to many commodities. The classification of the forced air precooling process includes wet deck system and the dry coil technique. Wet deck system is a mechanism, which provides air of low temperature and higher level of relative humidity, which minimizes the weight loss of produce during the process of cooling. Dry coil system uses a direct expansion or secondary coolant coil sized to operate at a small temperature difference, which will maintain a high relative humidity of the leaving air stream. An evaluation of precooling systems is presented through the current study that exhibits a description of the theory behind each system and its different components. Different control, management and monitoring requirement are discussed with the most recent advances. Maintenance to extend the lifetime of the hard ware and maximize system credibility is also presented. Through this chapter, it is aimed to promote interest in precooling and encourage its use on a more widespread basis via the illustration of the different systems details.

1.1 INTRODUCTION

Fresh produce (vegetables, fruits and cut-flowers) are living biological organisms that must stay alive and healthy even after harvest and during the handling chain until they are either processed or consumed. Highly perishable produce and because of their exposure to extremes of sun heat (field heat) and due to ambient temperature contain substantially more warmness at harvest than is normally acceptable during their subsequent marketable life chain or storage. Before harvest, the parent plant, compensate losses caused by respiration and transpiration by water, photosynthesis, and minerals. After separation of the parent plant (harvest), and if field heat is not properly and festally removed, it causes water loss, wilting and shriveling which leads to a serious damage in the appearance of produce (Siddiqui, 2016; Siddiqui et al., 2016; Elansari, 2009). Such heat also accelerates respiratory activity and degradation by enzymes. It encourages the growth of decay-producing microorganisms and increase the production of the natural ripening agent, ethylene. It is well documented that there is a correlation between food temperature and the rate of microbial growth. A rule of thumb is that a 1-h delay in cooling reduces a product's shelf-life by one day (Elansari and Yahia, 2012). This is not true for all crops, but especially for very highly perishable crops during hot weather.

Postharvest cooling was scientifically developed by US Department of Agriculture in 1904 (Ryall et al., 1982). The first commercial pre-cooling facility was built in Californian in 1955 and was used for cooling grapes destined to Florida market (Watkins, 1990). Several definitions for postharvest cooling can be found in the literature: the removal of field heat from freshly harvested produce in order to slow down metabolism and reduce deterioration prior to transport or storage; immediate lowering of commodity field heat following harvest; and the quick reduction in temperature of the product (Liberty et al., 2013).

The cold chain (Figure 1.1) is a shortened term encompassing all temperature management programs and other steps and processes that perishable must pass through to ensure they reach the end-consumer in a safe, wholesome and high-quality state. The cold chain should start immediately after harvest and contentious through the packing process, pre-storage, transportation and cool storage at the receiving market (Bharti, 2014). Another definition for the cold chain is the progressive removal of heat from the produce, starting as soon as possible after field harvest, in the shortest practical time period. The cold chain program should remove all field heat from the produce down to its lowest optimum storage and/ or shipping temperature.

FIGURE 1.1 Illustration of fresh produce cold chain.

1.2 THE IMPORTANCE OF PRECOOLING

Within the cold chain, temperature is the greatest determent and the most significant environmental factor that influences the deterioration rate of harvested commodities. The rate of respiration, and subsequently the rate of heat production depends on temperature, the higher the temperature, the higher the rate. Rapid precooling to the product's lowest safe temperature is most critical for Fresh produce with inherently high respiration rates. Rapid precooling is the first operation of the cold chain to be started from the instant of harvest, and considered the key element in modern marketing chain of fruits and vegetables. It removes field heat after harvesting, reduce breath function, retard ripening and control microbial processes (Siddiqui, 2015). It is also enhance keeping nutrition ingredients and fresh degree, improving cold-resisting ability, and avoiding chilling injury (Yahia and Smolak, 2014). Furthermore, precooling minimize the designed heat load needed for cold rooms and transport equipments. Investigations show that the postharvest losses of commercial fruit and vegetable is almost up to 25–30% without precooling in the whole storing and transporting chain while it is only 5–10% through precooling (Yang et al., 2007).

Postharvest cooling also provides marketing flexibility by allowing the grower to sell produce at the most appropriate time. Precooling also is applied as an important unit operation for post heat treatment for certain fruits (El-Ramady et al., 2015). The use of precooling after air-shipment can extend the shelf-life of certain fresh produce for considerable periods, by reducing the loss of moisture and maintaining a better firmness and texture and by limiting the increase of fiber content (Laurin et al., 2003, 2005). Precooling can be ranked as the most essential of the value added marketing services demanded by increasingly more sophisticated consumers.

The primary function of a well designed pre-cooling system is to be energy efficient and provide sufficient cooling capacity to ensure rapid pull down to desired temperature of a pallet load in certain conditions that are required by a product or process within a given space and time. A well-designed precooling system not only avoids wastage of electrical energy but also restricts the moisture loss within permissible limit. An accurate estimation of refrigeration load is the basis of designing and operating any type of precooling system. Refrigeration load is the rate of heat removal required to keep both the space and the product at the desired condition. The product-cooling load is one of the most important components of the refrigeration load, which contributes about two-thirds toward the total refrigeration load during the transient cooling period (Mukhopadhyay and Maity, 2015). To perform this function, equipment of the proper capacity and type must be selected, installed and controlled on a 24-h basis. The equipment capacity is determined by the actual instantaneous peak load requirements.

Thus, the refrigeration capacity in addition to cooling medium movement and operation control of the precooling process makes it different than just storing products in a conventional cold storage room. Therefore, pre-cooling must be considered independently from the cold storage and is typically a separate operation that requires specially designed equipment (Elansari, 2009).

1.3 APPROACHING THE OPTIMUM PRECOOLING METHOD

The capital investment and the running costs vary significantly among different pre-cooling methods. As an added value service, the expense of the selected technique must be covered through selling prices or other economic benefits. Various possible trades-offs can occur concerning the selection of certain method. Such practices may be based on certain conditions, such as amount and mix of produce handled, duration of pre-cooling season and its regional location, physical characteristics of the produce and its tolerance, specific market requirements, allowable pull-down time and the final desired temperature, sanitation level required to reduce decay organisms, packaging applied, further storage and shipping conditions, energy cost and availability, labor requirement, interest rates, building and equipment capital cost and its maintenance (Becker and Brian, 2006). These factors if not properly optimized, can lead to pre-cooling systems that do not achieve the required objectives or the cost benefit associated with the whole process is not feasible.

The process of heat removal from fresh produce can be achieved by several different methods; all involve the rapid transfer of heat from the product to a cooling medium, such as water, air, or ice. Such methods include as natural air-cooling or room cooling method, forced air-cooling, hydrocooling, ice cooling, slurry ice, vacuum cooling, and evaporative cooling, liquid nitrogen, mobile pre-cooler and in line pre-cooling (opti-flow cooling tunnels); each one is differing in heat removal efficiency and processing cost. One of the main advantages of hydrocooling is that, unlike air precooling, it removes no water from the product and may even revive slightly wilted products (Elansari, 2008). However, not all kinds of products tolerate hydrocooling (Tokarskyy et al., 2015). The most common method being utilized for precooling of fresh product is forced air-cooling. It is one of the few fast-cooling methods used with a wide range of commodities (Defraeye et al., 2015).

Fresh produce is usually cooled down to its maximum shelf life temperature with various techniques. Forced air-cooling is the most common method adapted for many types of vegetables fruits and cut flowers. Hydrocooling uses water as the cooling medium and therefore one of its advantages is that it removes no water from the produce and may even revive slightly wilted product. Vacuum cooling has been traditionally used as a precooling treatment for leafy vegetables that release water vapor rapidly allowing them to be quickly cooled. Precooling with top icing is a common practice with green onions and broccoli, where the flaks of ice are placed on top of packed containers. Table 1.1 indicates the optimum precooling methods for selected types of vegetables and fruits.

1.3.1 HOW DOES THE FRESH PRODUCE GET PRECOOLED?

- The temperature of the air inside the cooling facility is lower than the load of fresh produce, so the heat is moving out of the fruit to the surrounding air.
- During rapid heat transfer, a temperature gradient develops within the product, with faster cooling causing larger gradients. This gradient is a function of product properties, surface heat transfer parameters, and cooling rate.
- The evaporator contains refrigerant boiling at low pressure and temperature. As the refrigerant boils or evaporates it absorbs a lot of heat. This heat is removed from whatever surrounds the evaporator, usually air or secondary refrigerant.

- As the refrigerant flows inside the evaporator, it makes it always colder than the air in the cooling facility, thus the refrigerant is absorbing the heat carried out by the air that is drawn over the evaporator through the fans. The refrigerant is transferred from the liquid state to the vapor state.
- As the time goes, heat contained on the air is absorbed and the temperature of the produce is getting lowered.
- The refrigerant is sucked from the evaporator as superheated low-pressure gas and is compressed to a higher pressure passing through the compressor of the refrigeration system. Compressing the refrigerant gas increases it temperature and heat content and does not remove any of the heat transferred from the cooling facility.
- The high-pressure superheated vapor flows into the condenser where it changes from a gas to a liquid and heat is released. The process is the reverse to what is taking place in the evaporator. The cooling of this process is achieved by using ambient air (air-cooled condenser) or water (evaporative condenser). Even on a 45°C day, the outside ambient air or cooling water temperature from the cooling tower is lower than the condenser pipes and fins temperature and so the heat is transferred from the refrigerant through the pipes and fins of the condenser to the ambient or water.
- Now the heat load is pulled out of fruit and released to the atmosphere outside the cooling facility and cooling of the fruit has accomplished.

1.3.2 HEAT LOAD CALCULATIONS?

The most common term used to quantify refrigeration capacity or heat load is the refrigeration ton. One ton of refrigeration is defined as the energy removed from one ton of water so it freezes in 24 h. It is equivalent to 3.5 kW. The refrigeration capacity needed for pre-cooling is much greater than that required for holding a product at a constant temperature (Elansari, 2009). Therefore, the efficient design of pre-cooling systems that pull-down the heat load requires accurate estimation of the pre-cooling times of fruits and vegetables as well as the corresponding refrigeration capacity. However, it is uneconomical to have more refrigerating capacity available than is needed. The total heat load comes from the product, surroundings, air infiltration, containers, and heat-producing devices, such as motors, lights, fans, and pumps. Product heat accounts for the major portion of total heat load on a pre-cooling system. Product heat load depends on product initial and desired final temperature, cooling rate, weight of product cooled in a given time, and specific heat of the product. Heat from respiration is part of the product heat load, but it is generally small. No rule of thumb can be followed in that regard although that some figures are available in the literatures (Thompson, 2006). It is has been a usual practice to have a safety margin to overcome the peak load of the theoretical calculation. Nowadays and with modern numerical techniques a more practical cold store operation, the safety margin can be reduced to a more realistic level (Nahora et al., 2005; Chourasia and Goswami, 2007).

1.4 TYPES OF AIR PRE-COOLING METHODS

1.4.1 NATURAL CONVECTION AIR-COOLING (ROOM COOLING METHOD)

Conventional refrigerated storage facility is any building or section of a building that achieves controlled storage conditions using thermal insulation and refrigeration equipment. Such facilities are classified as coolers with commodities stored at temperatures usually above 0°C. They can be also classified into small, intermediate and large storage rooms, ranging from small rooms utilizing prepackaged refrigerator units to massive cold storage cooler warehouses. This method is the simplest and the slowest cooling method, in which the bulk or containerized commodity is placed in a refrigerated room for several hours or days. Typically cooling rates are not as good as other methods, though this can be enhanced by the use of forced ventilation via a letterbox wall or velum sheet. In this way some soft fruits may be cooled in less than 2.5 h, however other crops, such as Brussels or Cauliflower may take 24 h or may be longer. Air is circulated by the existing fans from the evaporator coil in the room where produce is cooled by exposure to cold air around the produce package. Air within the room is cooled with a direct expansion (DX) refrigeration system. The use of this type of cooling enables the produce to be both cooled and stored. Typically the produce being placed on the ventilation wall until cooled and then being moved to another part of the store for holding or dispatching, making space warm produce on the wall. This decrease the handling steps required and it eliminates the capital investment needed for fast cooling. Also this system tends to be less heavy on power consumption.

Room cooling (Figure 1.2) is used for produce sensitive to free moisture or surface moisture and for very small amounts of produce or produce that does not deteriorate rapidly. Exposing certain type of fruit to specific durations of cold storage has been shown to enhance ripening due to increased ethylene synthesis in the tissue (Mworia et al., 2012). For apple, the room cooling method is very common where it kept refrigerated in rectangular bins with lateral holes to let the cool air in and the temperature is usually maintained below 1°C (Russell, 2006). Citrus fruit is also used to be cooled using room cooling method (Defraeye et al., 2015).

In this method produce is loaded into a refrigerated space where cold air is circulated within the room and around the produce by the refrigeration fans. Cold air does not circulate readily through the packaged produce. Heat exchange is mainly by conduction through the container walls to the cold outer surface. The method to be effected needs a uniform air distribution, (at least 60 to 120 m.min air circulation), spaced stacking for airflow between containers and well ventilated containers.

FIGURE 1.2 Room cooling method.

Coolers of this type generally have less ability to remove heat from the product and lacks the air movement needed for rapid cooling compared with other pre-cooling methods. The half-cooling time may be 12–36 h so three half-cooling times (7/8 cooling time) will be 36–108 h (Ross, 1990). The efficiency of a forced-air-cooling system compared to a cooling room for grapefruits resulted in a reduction of 6.7°C in 1 h and 14.6°C after 2.5 h, compared to 2°C and 3.5°C for 1 h and 2.5 h, respectively, for the cooling room (Barbin et al., 2012). Due to its slow cooling rate, the produce takes long time to reach the desired final temperature. Unless the room is designed to deliver high level of relative humidity, the cooling systems will have sufficient time to remove moisture from the air, and subsequently the dry air will draw moisture out of the product, which will progressively dehydrate. Produce is largely composed of water where the loss of this natural moisture will reduce quality, taste, texture and shelf life. Most of these rooms especially in developing countries are equipped with direct expansion commercial refrigeration system (DX), which is not recommending for perishable storage. The installed evaporators usually have small surface area and large ΔT (temperature difference between room air and coil) that increase the water loss from the produce. Another disadvantage is that air velocity decreases with increasing distance from the source, causing produce stacked further from the fans to have less air passage over it.

Defrost is another problem for this kind of cold room. In a typical cool store, fans circulate air over the refrigerator coils. To maintain a storage temperature of 0°C the temperature of the coils will have to be appreciably below 0°C. Moisture is therefore removed from the air and this accumulates as ice on the coils. This why a defrost system is a basic requirement since such cold rooms would sometimes run as low as −2°C for certain products like grapes. Electrical defrost introduce extra heat load to the system and cause great fluctuation in room temperature.

The nature of the DX cooler has the negative effect of removing moisture from the air as it passes over the evaporator, this can be minimized by the careful design of the cooler surface, however some moisture and hence weight loss is inevitable. Humidification systems are recommended to reduce the losses by the introduction of water into the air. Systems, such as ultrasonic nozzles have been applied, though care must be taken to avoid excessive frosting on the evaporative coil face and water being lost from the produce being cooled. Evaporative humidification is a good alternative in which the water is transferred into the air in vapor form avoiding introducing this moisture as free water. Also care must also be taken to avoid freezing of the produce.

1.4.1.1 Modified Room Cooling Method

If the facilities are to be used for rapid pre-cooling, the capacity of the refrigeration system must be increased. The amount of increase will be determined by the rate of harvest, the desired cooling time and the required temperature drop. For the big and medium room, it is expected to have sufficient cooling capacity to pre-cool predetermined amount of produce according to its conditions. For a small room, an essential step is the determining of the capacity of the installed refrigeration system. Knowledge of the system control will be needed in addition to the produce initial temperature, final temperature, thermal properties, and the space requirements to place tunnel load. Based upon this data and the estimated cooling capacity of the storage space, the optimum amount of produce to be pre-cooled can be estimated. An auxiliary cooling fan is put in position after the pallets are placed in the room. Pallets are stacked in even numbers in set positions on the cool room floor. A tarp is rolled down over the bins to direct airflow (Figure 1.3). The forced air fan is wheeled in position against the pallets. The fan is turned on which draws air through the pallets. After forced air-cooling is completed the fan can simply be shut off and the pallets remain in position for room storage. Barbin et al. (2012) compared exhaustion and blowing air using an experimental portable forced-air tunnel built inside an existing cold store. The device was designed to improve cooling rates inside storage room without the need for a cooling tunnel. A heterogeneity factor was proposed for air circulation evaluation and compared with convective heat transfer coefficient (h) values. Lower modules of heterogeneity factor values represent smaller temperature differences among samples used. Comparing the two different airflow processes, heterogeneity factor values were similar for regions where the cooling air could flow without obstructions. However, larger differences were observed for regions with hampered air circulation. Results indicated that the air distribution, as well

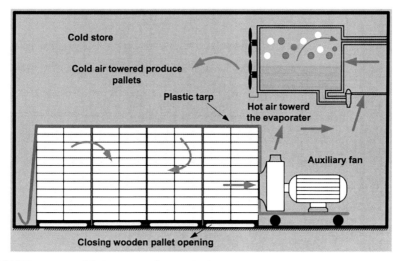

FIGURE 1.3 Modified room cooling method.

as the heat transfer, occurs more uniformly around the products in the exhausting process than in the blowing system.

1.4.2 FORCED AIR-COOLING

Forced-air-cooling is considered an improved technique compared by the room cooling method since the cold air is forced through produce packed in boxes or pallet bins via its venting areas. A number of airflow configurations are available, but the tunnel cooler is the most common (Figures 1.4–1.6). In the tunnel system, which is a patch type, pallets are lined up in front of a pressure fan and covered with a tarp to form a tunnel. Cold air is pulled through the tunnel of covered pallets so the air must go through the containers. The product is cooled in batches and cooling times range from 1 h for cut flowers to more than 6 h for larger fruit (Thompson, 2004).

Vertical Airflow forced air-cooling (Figures 1.7 and 1.8) use pallet racking, so that pallets can be stacked 2-high. If 12 pallets occupy a floor space footprint, with a trapped tunnel precooler system, the vertical airflow design allows 24 pallets to be cooled in that same space. One advantage of the vertical design system is that it eliminates the traditional

FIGURE 1.4 Forced air cooler tunnel type.

FIGURE 1.5 Forced air cooler tunnel type during operation.

Recent Advances in Postharvest Cooling of Horticultural Produce

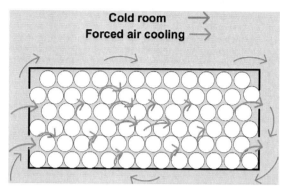

FIGURE 1.6 Concept of Forced air cooler compared by room cooling method.

FIGURE 1.7 Vertical Airflow forced air cooling

precooling problem of "last pallets to cool," which are typically those two-pallet positions furthest from the suction fan or fans. The system offer superior speeds cooling with a flow rates flow up to 2.35 L/s/kg compared by 1 L/s/kg for the trapped tunnel precooler. Through such system, Strawberry cooling time can be reduced from 1.5–2 h to about (Thompson

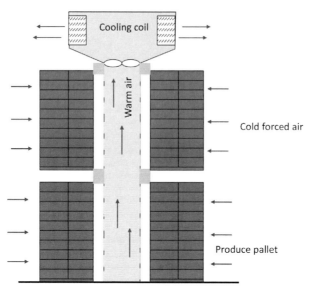

FIGURE 1.8 Concept of vertical Airflow forced air cooling

et al., 2010). So as the design precools faster, at the same time it physically doubles precooling pallet positions where the capacity can actually triple. The new design of such technology offer several advantages, such as faster cooling, increased capacity per unit area, potential for reduced cost per unit cooled, more uniform product temperature, some can be operated at field side and automated process control.

The disadvantages of high airflow method and technique are the high-pressure drop across pallets where doubling airflow increases pressure drop by a factor of about 4. This is also reflected in the high consumption of fan electricity where doubling airflow increases electricity demand fans by a factor of 7–8. Subsequently, an increased heat load because of the fan heat. The use of high venting area reduces the pressure drop across the pallet.

The continuous system where product is moved through a cooler on a conveyer has largely been abandoned in favor of batch cooling due to the high cost of conveyer systems. Some recent application for that type of configuration is reported for specific application, such as tying it in a production line for fresh-cut produce (Christie, 2007).

1.5 PACKAGING

Package design is a subject of ongoing and active research in the food industry due to its importance in the forced-convective cooling process and its complexity (Dehghannya et al., 2010–2012; Ferrua and Singh, 2009a, 2009b; Pathare et al., 2012). Optimum package design is very product-specific due to the large variety in size and shape and thermal–physical properties of different fresh produce. Often, a compromise has to be made between optimal ventilation (percentage and shape) and mechanical strength of the containers, which is required for stacking them and for protecting the fruit. The way of packaging and the packaging materials should be properly selected to avoid any blockage of air passage and allow good air-flow to achieve the cooling rate desired. Packed produce with airflow restricting materials should be taken in consideration when sizing the system airflow and static head pressure of the fans. Boxes should have about 5% sidewall vent area to accommodate airflow without excessive pressure drop across the box (Kader, 2002). Packing table grapes via sea shipments is an example; it needs a lot of packing materials that cannot be avoided, such as consumer bags and unvented liner (Luvisi et al., 1995). Crisosto et al. (2002) reported an air-flow rate of 9.35 m^3/h/kg that overcome the heavy internal package of table grapes boxes during the precooling process. Luvisi et al. (1995) reported a value of 216 min for the 7/8th cooling time of grapes that were bagged and packed in corrugated box. The corresponded initial and final temperatures were 21.1 and 1.7°C, respectively. For most systems, fans are being selected based on a maximum static pressure of 200 Pa (Hugh and Fraser, 1998).

The complex and chaotic structure within fresh produce ventilated packages during a forced air precooling process complicates the mathematical analysis of heat and mass transfer considering each individual produce. The complexity of the physical structure of the packed systems and the biological variability of the produce make both experimental and model-based studies of transport processes challenging time consuming. Ventilation of the produce packages should be designed in such a way that they can provide a uniform airflow distribution and consequently uniform produce cooling. Total opening area and opening size and position show a significant effect on pressure drop, air distribution uniformity and cooling efficiency (Pathare et al., 2012). Recent advances in measurement and mathematical modeling techniques, such as CFD, have provided powerful tools to develop detailed investigations of local airflow rate and heat and mass transfer processes within complex packaging structures.

Ferrua and Singh (2011) proposed a new packaging design capable of promoting a more uniform and energy-efficient performance during forced air-cooling using CFD modeling. It was reported that and for the same airflow conditions, the new design significantly improved the uniformity and energy-efficiency of the process, while replicating of the cooling rate of commercial designs. In particular, no significant differences were found among the cooling rates of individual clamshells, and the pressure drop across the system was decreased by 70%. Defraeye et al. (2013) analyzed the cooling performance of newly design pack, Supervent and Ecopack with citrus fruit during precooling process using CFD modeling. The best cooling performance was found for Ecopack where the uniformity of fruit cooling and the magnitude of the convective heat transfer coefficients, in a specific container and between different containers on the pallet, was the best for Ecopack container, followed by the Supervent and the standard container. The new container designs thus clearly showed significant improvements in cooling performance. A 3-D CFD model of ventilated packaging was applied to fresh produce where the cooling rate increased with an increase in vent area up to a limit. It was found that a vent area beyond 7% did not substantially increase cooling rate (Delele et al., 2013).

1.6 CAPACITY DESIGN

Forced air pre-cooling facility design involves a variety of tasks, including planning, site selection, architectural and structural design, refrigeration system design, equipment selection and installation, construction, supervision, inspection, maintenance and management. In addition, considerations of building and safety codes, efficient operation, and cost effectiveness make the design procedure more exhausting. The first step in a forced air pre-cooling facility design is for the designer to develop an exact set of specifications that meets all the interests of the facility owner. Specifications for the overall facility must consider the individual product specifications, forced air arrangements, environmental conditions, and other miscellaneous aspects of the design process.

One of the critical design parameter is the required capacity for a forced air precooling facility. The capacity mainly depends upon factors, such as the total production of the farm, nature of the produce and its thermophysical specifications, expected duration of production season and modes of material handling. An essential step in determining the capacity requirements of the prospective refrigerated storage facility is to acquire data concerning the traffic levels of the harvested produce. In addition, space requirements for loading docks, product handling and logistics must be estimated. Based upon this data and the estimated capital and operating costs per unit volume of the storage space, optimum dimensions may be determined.

For forced-air-cooling, the refrigeration capacity requirements (Figure 1.9) are much greater than just storing products in a typical cold storage room and might be as much as 5 or 6 times greater than the requirements for a standard cold room design (Elansari, 2009). Sufficient cooling capacity allows room air temperature to be stable throughout the cooling process and avoids temperature rising that slows cooling rates. Cooling time in forced-air-coolers is controlled by volumetric airflow rate and product diameter (Flockens and Meffert, 1972; Gan and Woods, 1989).

For the estimation of the refrigeration capacity needed, Wade (1984) developed an equation for the estimation of the load required in terms of

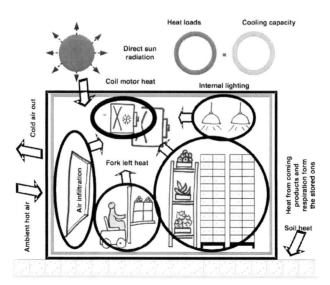

FIGURE 1.9 The refrigeration capacity requirements.

the rate of heat loss from the cooling produce. The developed model uses the seven eighth cooling times and the lag factor, which is an empirical measure of the thermal properties of the product. Thompson and et al. (1998) reported a calculation method for the estimation of the peak refrigeration tonnage associated with product cooling based on certain assumptions. Heat from miscellaneous source, such as fan motors was taken as a percentage of the product load. Watkins (1990) developed a cooling load calculation method and graphs were presented that show the relationship between the air-flow rate and the cooling rate required for different commodities based of a pallet load. The method was specified for the systems, which use an auxiliary fan with the existing cold stores. Elansari and Yahia (2012) charted the cooling capacity required for table grape, mango, melon, strawberries and green been as a function of precooling cycle designed and the initial temperature of the produce.

1.7 SYSTEM CLASSIFICATION

There are generally two designs of forced are precooler. They are: (i) wetted-coil or spry deck style; and (ii) dry-coil high humidity style. The two systems have significant differences in design concepts and philosophy. Each has advantages and disadvantages that should be considered to determine which is the best for a specific commodity.

1.7.1 WET COOLING SYSTEM (ICE BANKS)

The practice of precooling and cold storage fruits, vegetables and flowers in a high humidity atmosphere has been applied for many years in the U.S. and it is has been used commercially for some 25–30 years (James, 2013). Several systems are available for achieving this, such as the ice bank system and many other forms branded by various manufacturers. The wet deck system (Figures 1.10 and 1.11) was developed by the Institute of Agricultural Engineering in the 1970s (Farrimond et al., 1979; Geeson, 1989; Rule, 1995; Macleod-Smith et al., 1996; Tassou and Xiang, 1998). It is the common precooling systems installed in many pack-house facilities especially in developing countries where ice cold water is brought into intimate contact with the recalculating air within a cooler (Elansari,

Recent Advances in Postharvest Cooling of Horticultural Produce

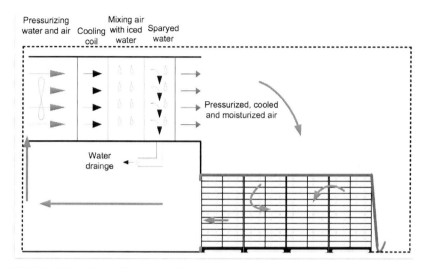

FIGURE 1.10 Wet cooling system (Ice banks).

FIGURE 1.11 Wet cooling system (Ice banks) ready for operation.

2003; Ahmad and Siddiqui, 2015). Wet Deck systems have the ability to maintain low temperatures and high relative humidities with lower running costs than conventional systems, making them suitable for long- and medium-term storage of a number of vegetable crops (Farrimond et al.,

1979). Wet air-cooling has been used successfully for the pre-cooling and/ or storage of grapes, mushrooms cucumbers, carrots, cauliflowers, tomatoes, strawberries, cut flowers white and red cabbage, Brussels, spinach potted plants and flowers, lettuce chicory potatoes, celery chicory roots cheese, leeks.

Warm produce is loaded in the precooling room in open crates stacked to allow forced circulation of air through the crates. The cooling unit is usually located near the end of the room. Air is circulated by the wet air-cooler to the opposite end of the room where it is drawn through the stacked produce pallets and returns to unit. A false wall or a plenum chamber (Tassou and Xiang, 1998) at the end of the room creates a positive pressure in the space to force cold air evenly through the produce and forms a return air passage to the cooling unit. Each cold room may have one or more unit operating in parallel based on the total capacity required. The circulation rate is typically 40 air changes per hour (Benz, 1989).

Wet cooling system is an alternative to simple direct expansion cooling where the refrigeration is supplied in the front of the water pumped from the ice water tank, which works as thermal storage unit at the top of the fill pack heat transfer surface (cooling tower), thus, cooling the air and warming the water. The formation of the ice on the surface of the evaporative coil occurs when the refrigeration load is light and melts when the load goes up. Air-cooler, which can cause damage to the produce, are stripped from the air stream by directional mist eliminators. The water is prevented from freezing completely through mechanical agitation, which also maintains good heat transfer rates between the refrigerated plates and the water (Tassou and Xiang, 1998). The air exits the cooler at temperatures as low as 1.5°C and relative humidities as high as 98%.

Wet cooling system is suitable for most crops other than those that require low humidity storage, such as dry bulb onions and produce that is required to be stored much lower than 1°C. When combined with a forced ventilation system, the precooling cycle maybe shortened to only 2 h, however bulkier and packaged produces will last longer (10–17 h). Due to the high relative humidity of the cooled forced air, the water losses from such systems are minimal. As with the DX system, the cooler provides both cooling and holding possibilities. Freezing of the crop is not possible, though care must be taken with crops that are sensitive to chill damage.

Since wet spray and wet deck systems are recirculated water system, the cooler must be designed to control disease organisms that enter the system via the coming produce. The water acts as an effective air scrubber and can be very efficient in removing air borne contaminations into the water stream. Chlorine is commonly used, and requires concentrations of 100–150 ppm available chlorine for water near 0°C. However, chlorine is corrosive to many common metals, thus care must be taken to determine if chlorine can be used with the cooling equipment installed.

Conventional commercial refrigeration or industrial system using either semi-hermetic compressors or screw working with ammonia or halocarbon refrigerant are used to supply the required refrigeration capacity to charge the ice chiller thermal storage unit. In order to reduce energy and capital cost, the ice also can be built at night or when they're no loads. An evaporative or air-cooled condenser rejects the heat from the refrigeration system.

Tator (1997) summaries the disadvantages of the wet deck precooling system where it is usually designed with a smaller coil surface. The coil must operate at a high temperature difference, usually 5–6°C delta t (Δt). That system can only cools the fruits to usually 2.5–3°C or above. Cross contamination can occur unless the recirculated water is chlorinated. Wet air produces wet product surfaces that may detract from the appearance, make handling difficult, or provide an enhanced environment for microbial growth. Due to the wet air used, packaging must be water resistant, hence waxed face packs or plastic trays are usually required.

Varszegi (2003) conducted an experiment to determine the relationship between the bacterial growth on mushroom cap and the wet forced air precooling methods (forced wet cooling and vacuum cooling) and found that vacuum cooling provided the longest period of time needed to reach the maximum value of microbial population and this method was found beneficial for the quality. However and with a view to reduce the weight loss during the conventional vacuum- cooling, ice bank cooling of mushrooms is now in vogue where a stack of mushrooms is passed through forced draft of chilled but humidified air from the ice bank (Rai and Arumuganathan, 2008).

Elansari et al. (2000) mentioned that the wet deck system is not the optimum precooling technique for sea shipment produce since it is not capable of reaching the lowest recommended temperature for certain product like table grapes and strawberries. Also the ice bank coolers required a larger space (James, 2013). However, the wet air-cooler offers some economic advantage in addition to reduce weight loss:

- Smaller refrigeration plant since peak heat loads are met by the reserve of ice. The plant therefore runs for longer periods at full capacity.
- Running a refrigeration plant at full load (as ice bank systems operate) is more feasible than running at part load and therefore the overall efficiency of the plant is greater.
- Energy saving since smaller plant consumes less power.
- Portion of the refrigeration capacity is utilized to accumulate a reserve of ice during the nighttime where electrical power is cheaper.

1.7.2 DRY SYSTEM

This system uses a direct expansion (as detailed on Figure 1.12) or secondary coolant coil sized to operate at a small temperature difference between

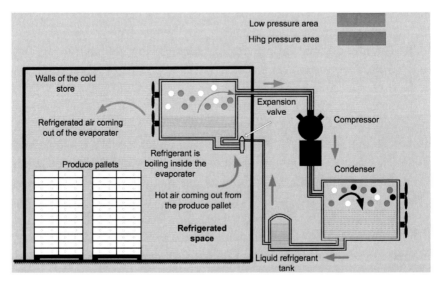

FIGURE 1.12 The details of the refrigeration DX system

room air and coil (ΔT), which will maintain a high relative humidity of the leaving air stream. The dry-coil system can maintain 85–90% relative humidity during the precooling process if properly designed and operated. The DX system is not recommended for high humidity fruit precooling. However, customers sometimes, due to economical reasons, buy according to the lowest price, and then they have to compromise. For a bigger size a flooded ammonia systems is an obvious choice for different reasons. A flooded ammonia system achieves less temperature fluctuation, which is especially critical for the precooling process. Another reason is mainly for its lack of oil separation problems and better efficiency, providing the plant with less cost for Kwh.

In case of DX system using commercial type style it must incorporated in its refrigeration loop different components that maintain high level of relative humidity to enhance its efficiency. Elansari (2009) indicated different details for the dry-coil concept that utilizes a semi-hermetic condensing unit working with R-134a (Figure 1.13). The refrigerant main loop for each tunnel includes a liquid receiver; a thermostatic expansion valve;

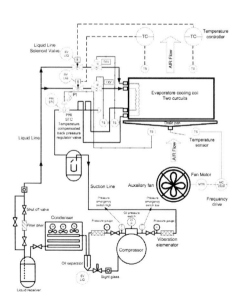

FIGURE 1.13 Dry-coil concept that utilizes a semi-hermetic condensing unit working with R-134a.

and a plate-finned tube evaporator coil. Each compressor is equipped with a capacity control that controls the delivered cooling capacity by 50%.

The evaporator coil should match the same capacity and conditions of the condensing unit with two circuits. A separate axial auxiliary fan is used to circulate the designed amount of air against 400 Pa static pressure.

A wide fin spacing evaporators is used (1.575 cm/fin) to guarantee a good supply of air through the precooling cycle and to avoid any blocking of the coil by dirt or frost. In order to maintain a relative humidity level not less than 85%, the coil is designed to have larger facing area. The installation includes a temperature compensated back-pressure regulator valve. Its function is to maintain the evaporating temperature at the required setup conditions and preventing it from falling down at the end of the precooling cycle. Therefore, the system minimizes the dehydration effect might happen due to the big difference in ΔT.

The air-flow rate supplied by the auxiliary fan in each precooling tunnel is controlled via a variable frequency drive (VFD). VFDs are an electronic motor controller that is used to reduce fan speed after the heat field has been pulled down to storage temperatures. By other word, as the precooling process nears its end, water loss from product should be avoided by minimizing air-flow which can be reduced as low as 50%. The VFDs offer very attractive energy savings. At half fan speed, the fans will consume only about 15% of full speed power (Morton and McDevitt, 2000). A safety cut-off arrangement is installed at the front of the air return channel to sense the return air temperature and stop the auxiliary fan if the temperature is less than 0°C. That is to prevent any freezing might happen for the produce being precooled.

1.8 MOBILE PRE-COOLING FACILITIES

A mobile precooler is one, which removes the field heat at the farm and during transit period. Commercial mobile precooling system had been previously designed, in which three-precooling unit container loads of product could be precooled simultaneously (Green, 1997). The capital investment and running cost of the system are very high due to its capacity, which exceeds the production of the average size facilities. It consumes about eight, of fuel per hour to run the ammonia screw compressors.

Talbot and Flitecher (1993) utilized a trial mounted cooling unit equipped with two 10.5 kW packaged air conditioner units, a high-pressure blower and a self constructed cooling chamber for cooling a pallet of containerized product as a mobile precooling facility. Boyette and Rohrbach (1990) promoted a similar idea that applies two to three tons refrigeration air conditioning with integrated fan unit to supply the cooled air through the length of insulated flexible duct, which holds the product being precooled. The cooling rate reported for previous units were slow and the product load exceeded the design load by 30% apart from the very limited capacity, which is only for one pallet. The water loss was a major concern for both units. The used air conditioning systems were to comfort the human body rather than the fresh product

Elansari et al. (2001) designed a portable forced air-precooling unit using 40" high cube bottom air delivery reefer container. The precooling unit was modified by using a bulkhead door, and the floor T-sections were blocked in order to short cycle the cooled air around the precooled pallets. The average pallets grapes temperature was lowered by 18°C in 8 h. The product load exceeded the available load for the unit by about 50%, which caused longer cooling time. The designed refrigeration capacity of the reefer container was to hold and maintain the temperature of the shipment and not to pull down the field heat of the shipment.

Elansari (2009) described the development and performance of a portable forced air-cooling unit exclusively designed to satisfy different precooling requirements (Figures 1.14–1.16). It took 150 min to cool down 2.3 tons of Strawberries from 22°C initial temperature to 1–4°C final temperature. The unit is simple and use on-shelf refrigeration components. The cooling system uses Scroll compressor that has proven to be efficient and reliable with respect to the precooling requirements. The unit is an insulated container (8590 × 2990 × 2940 mm) divided into three sections as shown in Figure 1.15, a machine room; a cooling chamber which represents the false wall and finally the main cooling area that holds the stacked produce pallets. The dimensions and weight of the unit were to accommodate highway regulations. The unit can run with a separate motor generator fueled with diesel/electrical portable power unit for keeping it running while off the road.

In this regards, Barbin et al. (2012) suggested a portable precooling tunnel that improve cooling rates inside storage room without the need for

FIGURE 1.14 Portable precooling unit loaded with strawberries.

FIGURE 1.15 The machine room of the Portable precooling wit maneurop hermetic compressor.

FIGURE 1.16 External view of the Portable precooling wit the electrical generator beside.

a the conventional forced precooling tunnel. As an alternative to forced-air precooling, warm loading of citrus fruit into refrigerated containers for cooling during marine transport was explored (Defraeye et al., 2015). Although a refrigerated container was theoretically able to cool the produce in less than 5 days, the experiment showed that these cooling rates are not currently achieved in practice, bearing in mind that step-down cooling was applied. Future improvements in the technique point towards an improved box design and better stacking on the pallet, and to reducing airflow short-circuits between pallets is still required.

1.9 HYDROCOOLING

Hydrocooling as shown in (Figure 1.17) is the process or technique of arresting the field heat of fruits and vegetables after harvesting by immersion in ice or cold water. Hydrocooling is one of the fastest precooling methods. One main advantage of hydrocooling is that it does not remove water from the produce and in contrary; it may even revive slightly wilted produce (Elansari, 2008). Hydrocooling is an effective method for rabidly precools a wide range of fruits and vegetables in containers or in bulk. It is normally only applicable to fresh fruits and vegetables that can withstand

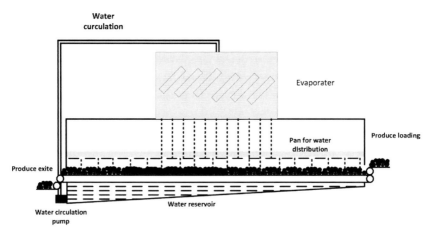

FIGURE 1.17 Continues type hydrocooler.

water immersion (peach, cherry, avocado, mango, sweet corn, and carrot), it is also may be applied to vacuum packs of prepared foodstuffs.

For the cherry industry as an example, the high efficiency of hydrocooling system is achievable due to the large heat capacity and high rate of heat transfer of agitated water. At typical flow rates and temperature differences, water removes heat about 15 times faster than air, resulting in threefold shorter cooling time in comparison with products cooled by forced air, or 10-fold, when products are placed in conventional or storage room (Manganaris et al., 2007).

Based on the hydrocooler type, hydrocooling process is achieved by immersing or flooding products in chilled water or spraying chilled water over the products. Water is an excellent heat transfer refrigerant compared to air where the convection resistance at the product surface is usually negligible. During the hydrocooling process, the main resistance to heat transfer is the internal resistance of the product, and internal heat is removed once it arrives at the surface. The temperature difference between the product surface and the cooling water is normally less than 0.5°C. Under idealized conditions, the convection heat transfer coefficient and the cooling rate per unit surface area can be 680 W/m2·°C and 300 W/m2, respectively (Cengel and Ghajar, 2013).

For efficient hydrocooling, water must be kept as cold as possible without endangering produce. In commercial practices, water temperature is usually kept around 0.5°C except for chilling sensitive commodities. The water in a hydrocooling system is cooled by passing it through cooling coils in which a refrigerant flows at about −2°C. Hydrocooler usually uses a plate-type heat exchanger to cool the recirculated water to 0.5°C, either placed directly over the belt conveyor or on the process floor near the hydrocooler belt. The plates are refrigerated using R-717 or R-22. Usually, the refrigerant is supplied from the central equipment room. The water is normally recirculated within a closed system to save both water and energy. However, recirculation can cause cross contamination for fruits and vegetables and this why chemicals, such as active chlorine (or ozone) are commonly added (usually at a rate of 50–100 mg/kg water) to reduce bacteria build-up in water (Suslow, 1997) to disinfect the water used in the process and therefore minimize the potential risk of spreading any contamination.

The variation of the mass-average product temperature with time is shown (Figure 1.18) for some fruits (ASHRAE, 1993). The typical seven-eighths cooling times are 10 min for small-diameter products like cherries and up to 1 h for large products, such as melons. It is clear that reducing the

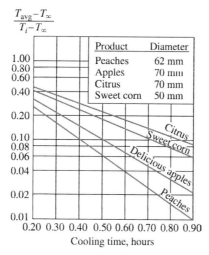

FIGURE 1.18 Cooling rates for different produce with varying diameter.

temperature difference between the fruit and the water to 10% of the initial value takes about 0.4 h for peaches while it take 0.7 h for citrus fruits. The size of the fruit is an important factor influencing the hydrocooling rate in addition to other factors, such as water temperature, produce orientation and water flow pattern. Hydrocoolers can be portable, extending the cooling season. Containers used in hydrocooling must be water-tolerant.

Container design and the stacking arrangement of the produce are critical to achieve efficient hydrocooling. Water distribution within the containers and the amount of water flowing out of the container through the side-walls influences the effectiveness of hydrocooling process. The containers should be designed to provide an efficient and uniform cooling throughout the entire volume of container and throughout an entire stack of containers. In terms of uniform water distribution, the width of the openings on the bottom of containers is also important (Pathare et al., 2012). Vigneault et al. (2004) investigated the non-uniform water supply inside plastic collapsible containers used for three types of produce during the hydrocooling process. The study recommended to use a container base opening that covers approximately 5.2% of the bottom surface which will allows a more uniform water distribution and insures the fastest cooling rate by obtaining higher minimum flow rate in each section of the container.

Forced-air-cooling is traditionally, the most common method applied for the fats cooling of strawberries in pack-house facilities where the typical cooling times for the pulp temperature to reach 3°C ranges from 60 to 90 min. However, the final strawberry pulp temperature can vary widely according to the location within the cooling tunnel, resulting in uneven cooling and a delay in achieving the desired final temperature. In addition, water loss has been associated with forced-air-cooling process, contributing to reduced shelf life and the quality of the strawberries. Recently, the application of hydrocooling was extended to strawberries leading to overall better quality than forced-air-cooled, with significant differences in epidermal color, weight loss, incidence and severity of decay (Ferreira et al., 2006; Jacomino et al., 2011). Hydrocooling did not affect the fruit quality during cold storage in terms of physical and chemical analyzes, freshness or decay. Use of this method resulted in fruit that were 2–3% heavier than those that were forced-air-cooled by the end of the storage time. For strawberries, hydrocooling is an alternate method that has several advantages compared to forced-air-cooling, including a faster cooling time (12–13 min), and reduced dirt/field debris, and overall microbial load (Jacomino et al., 2011). Based on the current practices, strawberries are unwashed and field-packed for fresh market, which increases the risk of microbial contamination during cultivation, harvest and postharvest, handling. Fresh and frozen strawberries have been associated with several reported foodborne illness outbreaks (FDA, 2011), which highlight a need for better sanitation and process control. The use of antimicrobials sodium hypochlorite (HOCl, 100 mL/L) and peroxyacetic acid (PAA, 80 mL/L) are both effective in reducing surface contamination on strawberries during hydrocooling (Tokarskyy et al., 2015). Sreedharan et al. (2015), reported that and compared to forced-air-cooling, hydrocooling significantly reduced salmonella survival on inoculated intact strawberries, with levels below the enumerable limit (1.5 log CFU/berry) by day 8. Hydrocooling reduced the initial salmonella levels by 1.9 log CFU/berry, while the addition of 100 or 200 ppm HOCl reduced levels by 3.5 and 4.4 log CFU/berry, respectively.

The immersion hydrocooling with sanitized water for strawberry shippers in which the fruit were uniformly cooled in approximately 13 min, potentially increasing throughput by 4- to 8-fold (Tokarskyy et al., 2015). Hydrocooling of strawberries in clamshells cooled at the same rate as those in bulk; after 14 days at 2°C, quality of hydrocooling fruit was equal to or better than forced air-cooling of the fruit.

Also for Blueberry, the current practices are forced-air-cooled for 60–90 min to 2–3°C pulp temperature. Carnelossi et al. (2014) compared the cooling efficiency and the effect of forced-air-cooling with hydrocooling and with hydrocooling plus forced-air-cooling on fruit (Emerald and Farthing varieties) quality. The results indicated that 'emerald' was more sensitive to hydrocooling than 'farthing,' where several fruit from the former showed skin breaks. Both cultivars had no decay during storage.

For sweet cherries, it was mentioned that hydrocooling shortly after harvest (4 h) and then transporting fruit in cold flume water during packing are used to maximize postharvest quality, but can cause fruit splitting (Wang and Long, 2015). In a simulated commercial procedure, hydrocooling cherry fruit in appropriate $CaCl_2$ solutions (i.e., 0.2–0.5%) for 5 min and then passing the fruit in cold flume water for 15 min increased fruit firmness, retarded losses in ascorbic acid, titratable acidity, and skin color, and reduced splitting and decay following four weeks of cold storage.

The effectiveness of hydrocooling treatment with a minimum delay after harvest to suppress decay and prolong the storage life of many produces is still being evaluated. Liang et al. (2013) investigate the influence of hydrocooling at 1, 2, 4, and 6 h post-harvest on the storage life and quality of the litchi cultivar, Feizixiao, by comparing litchi with and without hydrocooling treatment. The observed parameters included variations in temperature during hydrocooling, biochemical properties of the pericarp, and fluctuations in the content of soluble solids and titratable acids in the aril during storage. Hydrocooling for 30 min reduced the temperature of the pericarp by 6.2 ± 0.3°C. It also delayed the increase in electrolyte leakage and polyphenol oxidase and peroxidase activity in the pericarp.

Thorpe (2008) developed a commercial large scale, transportable hydrocooler with a capacity of 6 tons/h of broccoli from 30°C initial temperature to a final one of 12°C/h. The water in the hydrocooler is recirculated and the water consumption is estimated to be 35 L/ton of round fruit, such as apples and 75 L/ton of broccoli. If the water were not recirculated, the water consumption would be 60,000 L/ton of produce cooled. The electrical running cost is estimated to be 20 kWh and 16 kWh per ton of produce when the throughput is 4 ton/h and 6 ton/h, respectively. If the water were not recirculated the running cost would be about 300 kW per ton. This may be regarded as being infeasible.

It can concluded that hydrocooler and in case of custom designed hydrocooler to meet specific needs provide fast, reliable and efficient means of cooling many water tolerant fruits and vegetables, such as sweet corn, broccoli, artichokes, asparagus, avocados, green beans, beets, Brussels sprouts, cantaloupes, carrots, celery, cherries, strawberries, endive, greens, kale, leeks, nectarines, parsley, peaches, radishes, romaine lettuce, spinach, turnips, watercress and more.

1.10 VACUUM COOLING

In the mid-20th century, vacuum cooling was developed by the University of California (Tragethon, 2011). Vacuum cooling (Figure 1.19) is a batch process where the products are cooled by vaporizing some of the water content of the products under low-pressure conditions. By other words, vacuum precooling, which uses the principle of absorbing the latent heat of vaporization by water under vacuum pressure to rapidly cool down

FIGURE 1.19 Vacuum cooler.

freshly harvested produce, thus inhibiting respiration, is applied to maintain the storage quality of fruits and vegetables (Liu et al., 2014).

The newly discovered vacuum cooling technology at that time was commercial applied by several companies to meet the postharvest handling needs of their customers. The first commercial vacuum cooling industrial facility could precool five pallets of product as a batch. This is achieved by lowering the product temperature below 4.4°C as soon as possible after harvesting. Prior to vacuum cooling technology the precooling process used to last hours to reach the pulp temperature of the product below 4.4°C using other precooling techniques. Nowadays, precooling process using vacuum cooling can be shortened to as little as 35 minutes.

The concept behind the vacuum cooling is due to the thermodynamic properties of water, namely: the latent heat of vaporization which is absorbed from the products during evaporation process results in lowering its temperature. Water is considered a natural refrigerant with a commercial name of R718. Liquid water as a refrigerant will boil at 100°C where it is well established that boiling water at higher elevations, such as in the mountains, causes the water to boil at a temperature less than 100°C. Vacuums cooling of leafy vegetables (such as, lettuce) is based on lowering the pressure of the air-tight (sealed) cooling tube to the saturation pressure that meet the desired final low temperature, and evaporating some water from the products to be cooled. During this process, free water evaporates at the temperature corresponding to the boiling (flash) point and since the saturation pressure of water at 0°C is 0.61 kPa, the products can be cooled to 0°C by lowering the pressure to that level. The cooling rate can be increased by lowering the pressure below 0.61 kPa, but this is not desirable because of the danger of freezing and the added cost (Cengle and Boles, 2014). With the constant reduction in the pressure of the vacuum cooling system, the progressing evaporation of the product takes place. Practically and when a product is subjected to vacuum gradually, the flash point of the water decreases and some of the water boils until new equilibrium conditions is obtained (Alibas and Koksal, 2014; Reno et al., 2011; Rodrigues et al., 2012).

Looking at Figure 1.20 in a vacuum cooler, one can distinguish two stages. Primary, the produce with at initial temperature of 25°C for an example, are brought into the vacuum tube and the operation starts. The temperature in the tube remains unchanged until the saturation pressure is reached, which is 3.17 kPa at 25°C. In the second stage that follows, saturation conditions are maintained inside at progressively lower pressures and the corresponding lower temperatures until the desired final temperature is achieved which is usually slightly above 0°C.

Compared by other conventional precooling techniques, vacuum cooling is considered the most expensive choice. One of the reasons for that is

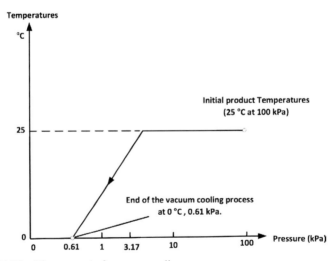

FIGURE 1.20 The concept of vacuum cooling.

its limited application, which is: much faster cooling. It is applied to any fresh produce, which has free water provided that its structure will not be affected by the removal of such water. The rate of cooling and effectiveness of vacuum cooling are mainly related to the ratio between its evaporation surface area and the mass of foods. Vegetables with large surface area per unit mass and a high tendency to liberate moisture, such as lettuce and spinach are good example for vacuum cooled products. In the contrary, produce with low ration of surface area to mass are not suited to vacuum cooling, especially those that have relatively water-resistant peels, such as tomatoes and cucumbers. Some products, such as mushrooms and green peas can be vacuum cooled successfully by wetting them first.

Precooling of mushrooms is a major traditional application of vacuum cooling. The porous structure and high moisture content of mushrooms have made this possible (Zheng and Sun, 2005). For mushrooms (He et al., 2013) reported a cooling time of 25 min from 25.1°C initial temperature to 2.4°C final temperature where the weight loss was 5.3%. The effects of vacuum cooling on the color, firmness, polyphenol oxidase and membrane permeability of mushroom after cooling and storage were determined. The results showed that vacuum cooling significantly reduced the polyphenol oxidase and membrane permeability. It has been seen shown that fresh produce can be cooled much more rapidly and efficiently with vacuum cooling than with conventional cooling (Ozturk et al., 2011).

Cauliflower heads, whose initial temperature was 23.5 ± 0.5°C, were cooled until the temperature reached at 1°C using different precooling methods (Alibas and Koksal, 2014). It was found that the most suitable cooling method to precool cauliflower in terms of cooling time and energy consumption was vacuum, followed by the high and low flow hydro and forced-air precooling methods, respectively. The highest weight loss was observed in the vacuum precooling method, followed by the forced-air method. However, there was an increase in the weight of the cauliflower heads in the high and low flow hydro precooling method. The best color and hardness values were found in the vacuum precooling method. Among all methods tested, the most suitable method to precool cauliflower in terms of cooling and quality parameters was the vacuum precooling method.

Rahi et al. (2013) vacuum cooled cabbage where the results indicated that pressure 0.7 kPa reduce the cooling time by 17% and 39% compared with 1 and 1.5 kPa, respectively. The optimum selected pressure was 0.7 kPa that will minimize weight loss. It has been also found that temperature distribution within the products during vacuum cooling despite the cabbage complex structure was homogeneous. Water loss during vacuum cooling is unavoidable due to the essence of vacuum process. Percent product yield, water loss and cooling time where significantly improved by regulation of pressure.

Liu et al. (2014) studied the effect of vacuum precooling process on leaf lettuce, which is a complex process of heat and mass transfers. Based on the properties of leaf lettuce in vacuum precooling process, an unsteady computation model was constructed to analyze the factors affecting vacuum precooling. Some factors, such as the precooling temperature, pressure and quantity of the spray-applied water were verified throughout the experiment. The study showed that the measured and simulated values were basically the same, and the overall trend was similar. The lower the vacuum pressure, the greater the cooling rate lettuce and water loss rate.

Garrido et al. (2015) compared four precooling systems including room cooling, forced-air-cooling, hydrocooling and vacuum cooling for their effects on quality and shelf-life of baby spinach. Leaf water content increased after cooling in hydrocooling and vacuum cooling but more significantly in winter while in spring, differences among treatments were not significant. The color measured as Chroma was more vivid in hydrocooling and vacuum cooling just after processing but after storage, no differences among pre-cooling treatments were observed. In winter, there were no significant differences in the respiration rate among precooling systems applied. However, in spring, hydrocooling and vacuum cooling decreased respiration rate and modified less the headspace gas composition of the packages. Surprisingly, visual quality was significantly lower in vacuum cooling compared with the rest of precooling treatments due to the higher degree and number of damaged leaves. In conclusion, selection of the precooling method is critical during warm weather due to the higher field temperature at harvest.

In recent years, vacuum-cooling technology has attracted much attention and its application has been extended to precooling of cut flowers. In 2013, FlowerForce, Netherlands, started to use the new loading cooler. They choose the vacuum cooler as the best solution due to the quickest way to cool their products. Using vacuum cooling significantly increases the shelf life of the flowers and reduces the health risk caused by organism growth.

Most of the existing precooling facilities and equipment has been designed to use halogenated hydrocarbons (CFCs and HCFCs) whose emissions to the environment are depleting the ozone layer and contributing significantly to global warming since the refrigerant leakage rates of the vapor-compression systems to the environment is about 15% of the total charge per annum (Elansari and Bekhit, 2015). Producing and handling of CFCs is banned in most of the world and many HCFC refrigerants are only short-term alternative and becoming more expensive and less efficient. With the phase-out of R22, which together with ammonia was a popular refrigerant in food processing. Vacuum cooling machines have a large potential market, in food cooling and processing industry. As its refrigerant is water, which is more environment-friendly, it can be widely used and has no limit.

The removal of water vapor from a product during vacuum cooling results in the loss of heat, approximately equivalent to the latent heat of vaporization of water. An advantage of vacuum cooling is that it is possible to stop the cooling process at a predetermined pressure and temperature. Water loss can be minimized by spraying the produce with water before cooling. Some coolers are equipped with water spray systems that are activated during the cooling cycle, such systems is called hydro-vacuum methods. Like hydrocooling water, this water must be disinfected if it is recirculated.

1.11 WATER LOSS DURING VACUUM COOLING DETERMINATION

The amount of cooling (heat removed from the product) is proportional to the weight of water evaporated, w_v, and the latent heat of vaporization of water at the average temperature, h_{fg}, and is determined from:

$$Q_{vacuum} = w_v \, h_{fg} \text{ (kJ)}$$

Since:

$$Q_{vacuum} = m_p \, C_p \, \Delta T \text{ (kJ)}$$

where, m_p is the product mass, C_p is the specific heat of the product (kJ/kg·°C) and ΔT is the temperature difference between the product initial temperate and the final desired temperature (°C). Therefore, during vacuum cooling, the amount of water vapor generated (also cooling loss) can be calculated by:

$$w_v = m_p\, C_p\, \Delta T / h_{fg}\ (kg)$$

If the initial temperature of the product to be vacuum-cooled is 25°C and the desired final temperature is 0°C, the average heat of vaporization can be taken to be 2472 kJ/kg, which corresponds to the average temperature of 12.5°C. This is adapted from the properties of saturated water tables by interpolation (Cengel and Ghajar, 2011) as shown. Assuming that the specific heat of products is about 4.12 kJ/kg·°C, the evaporation of 0.01 kg of water will cool down 1 kg of product by 24.72/4.12 = 6°C. by other words, the vacuum-cooled products will lose 1% moisture for each 6°C drop in their temperature. This means the products will experience a weight loss of 4% for a temperature drop of about 24°C. To minimize the product moisture loss and enhance the effectiveness of vacuum cooling, the products are often wetted prior to cooling.

1.12 MATHEMATICAL MODELING OF VACUUM COOLING PROCESS

Modeling of vacuum cooling process is useful. It can lead not only to a better design of vacuum cooling equipment, but also to a better understanding of the effects of the process on the physical, chemical and sensory properties of the products.

Isik (2007) tested and compared the results of thermodynamically analysis implemented to the vacuum cooling of lettuce. According to the findings of the trial and results of the thermodynamically model, it is possible to predict the weight loss within an error of 2.12%, close to the other parameters to be used in the design of vacuum precooling system, such as temperature, pressure, enthalpy and entropy on specified points using the mathematical model prepared from thermodynamically equations. Moreover, the fact that the power need, the most important parameter in the design of the system, could be predicted with a minimal error (0.162%)

which reveals that the thermodynamically model could be applied to the design of a vacuum precooling system.

Liu et al. (2014) and through an unsteady computation constructed model verified some factors that influence the vacuum cooling process such as the precooling temperature, pressure and quantity of the spray-applied water. The study showed that the measured and simulated values were basically the same, and the overall trend was similar. Their work discovered that the quantities of leaf lettuce covered with water were equal to 4.211–5.977% of the total sample mass and the mass loss of the sample was 1.987–2.873%. Under precooling pressure of 600, 1000, and 1500 Pa, the mass loss was 2.758, 2.701, and 1.929%. After that, the results of calculation indicated that the quantities of capture water of the water-catcher was 1.607–2.567 g, and the cooling capacity of the total sample was 3.722–5.946 W in vacuum precooling process. The results reveal that the model of leaf lettuce was fitted and it was confirmed by the experimental data.

Zhang et al. (2014) constructed a coupled model for the porous food vacuum cooling process based on the theory of heat and mass transfer. Sensitivity analyzes of the process to food density, thermal conductivity, specific heat, latent heat of evaporation, diameter of pores, mass transfer coefficient, viscosity of gas, and porosity were investigated. The results indicated that the food density would affect the vacuum cooling process but not the vacuum cooling end temperature. The surface temperature of food was slightly affected and the core temperature is not influenced by the changed thermal conductivity. Change of the Specific heat as well as latent heat of evaporation affected both, the core temperature and surface temperature. The core temperature is affected by the diameter of pores while the surface temperature is not affected obviously.

As indicated before, water-spraying is regarded as an effective method to reduce the weight loss of product during vacuum cooling process. Tian et al. (2014) investigated the effect of vacuum cooling factors on the weight loss of broccoli, and attempted to optimize the treatment conditions by simulated annealing (SA) technique. An algorithm based on simulated annealing meta-heuristic technique was established to identify the optimum condition for vacuum cooling treatment of broccoli. Results indicated that the simulated annealing algorithm could adjust well with the simulation of the broccoli vacuum cooling process. The optimum condition was at 200 Pa of pressure, 274 g broccoli, 6% of water volume and 40 min processing time. Under this circumstances, a product with only 0.35% of weight loss and 1.48°C final temperature was obtained. The developed method may used to effectively control the weight loss during vacuum cooling process and reducing the economic loss.

1.13 FEATURES AND BENEFITS OF VACUUM COOLING

The features and benefits of vacuum cooling can be summarized as follows:

1. Vacuum cooling machines have a large potential market, in food cooling and processing industry as its refrigerant is water, which is more environment-friendly where it can be widely used and has no limit.
2. Solve the internal field heat problem in 20–30 min, inhibiting organism growth of their own.
3. Cool down the temperature inside and outside the products in a consistent steady and uniform way.
4. Applicable to fruits and vegetables harvested on rainy days where it can quickly take away surplus moisture on their surface, achieving cooling effect.
5. Hydro-vacuum cooler designed with additional water circuit meets the rapid cooling while avoid excessive moisture loss.
6. Safe and stable since electric components are imported from famous suppliers to ensure safe working and long service life.

1.14 SLURRY ICE

In the past, most of ice forms involve a certain level of manual handling for transportation from one place to another. It also has sharp edges and quite coarse that may injury the fresh product's surface in case of using it as a direct contact chiller. Such ice performance to lower the heat load of the product is poor due to the limited heat transfer performance when releasing their latent heat of fusion. The ice slurry overcomes most of these disadvantages since it has a high-energy storage density because of the latent heat of fusion of its crystals. Due to its large heat transfer surface area created by its numerous particles, it results in fast cooling rate. The principal advantage of liquid icing is the much greater contact between the ice and product afforded by this method in addition of being economical and environmental nature.

Ice slurry as shown in (Figure 1.21) is a homogenous mixture of small ice particles and carrier liquid which can be either pure freshwater or a binary solution consisting of water and a freezing point depressant. The most commonly used freezing point depressants in industry are Sodium chloride, ethanol, ethylene glycol and propylene glycol. Over the last two decades interest in using phase-change ice slurry coolants has grown significantly (Kauffeld et al., 2011). Slurry ice can be used in two ways; directly for the rapid chilling of fresh produce or indirectly as a secondary refrigerant within a refrigeration loop for the cold chain elements applied to the fresh produce industry such as cold storage and refrigerated transportation. With such refrigeration system and if one allows a temperature change of −12°C upto −8°C, the enthalpy content for ice slurry will be approximately eight times higher than for any conventional heat transfer fluid (secondary refrigerant) based on water such as propylene glycol (Rhiemeier et al., 2009).

1.14.1 DIRECT USE OF SLURRY ICE

Slurry ice can be used directly for rapid chilling or precooling for vegetables that can tolerate water such as asparagus, cauliflower, broccoli,

FIGURE 1.21 Handling of slurry ice.

green onions, cantaloupes, leafy greens, carrots and sweet corn, spinach, parsley, and Brussels sprouts (El-Ramady et al., 2015; Kitinoja, 2013). For broccoli as an example, rapid chilling by ice slurry to the field-packed, waxed broccoli cartons, immediately after harvesting prevents wilting, suppresses enzymatic degradation and respiratory activity; slows or inhibits the growth of decay-producing micro- organisms, and reduces ethylene production. The use if slurry ice with broccoli ensures that broccoli heads are retained in a fresh and attractive condition throughout the cold chain, right to the consumer.

Package ice can be used only with water tolerant packages such as waxed fiberboard, plastic or wood (Kitinoja and Thompson, 2010). The ice remaining after cooling protects the produce from warming and dehydration during transport (Vigneault et al., 2009). There are several ways to inject slurry ice in the carton packed with a variety of produce. The easiest icing method is to add a measured amount of ice manually to the top of each carton although the method usually resulted in uneven cooling of produce. The method is low efficiency, since it takes 5 min for two dedicated workers to ice a pallet of 30 cases (Boyette and Estes, 2000), this is only marginally justified for small-scale operations.

Kauffeld et al. (2010) described the use of an automatic pallet icing chamber design that can greatly improve the icing efficiency. The design incorporates a stainless steel enclosure capable of handling a pallet of 48 cases (9 kg broccoli per case) during each icing cycle where only a single operator is required to move the produce pallet into the chamber. Once a locally positioned, icing machine is switched on and the two front doors are closed automatically. A pump begins to circulate the ice slurry in the mixing tank located right underneath the chamber to the top of the enclosure, where it is distributed to four vertical slots built on the side walls. Then, ice slurry is forced to flow through the hand openings and fill the voids throughout the produce within the cases in a about 90 seconds. As water drains off, ice particles are tightly packed with the produce. The pallet is then moved out of the chamber. Therefore, slurry ice can be beneficiary in both small and large operations despite the fact that the produce is wet during the process, liquid icing is an excellent cooling method (Kanlayanarat et al., 2009).

The liquid ice slurries range in water to ice ratio from 1:1 to 1:4. The liquid nature of the slurry allows the ice to move throughout the box, filling all of the void volume of the container, reaching all the crevices and holes around the individual units of the product. The slurry maintains a constant low temperature level during the cooling process, and provides a higher heat transfer coefficient than water or other single-phase liquids. These features of ice slurry make it valuable in many applications among them is fresh produce handling. For example, the ice slurry based thermal storage system within the fresh produce packhorse and cold stores in the form of dense ice slurry during nighttime hours when power is cheap. Latter on and during the daytime working hours the cold energy can then be quickly released by melting the ice slurry for produce chilling when electricity might be several times more expensive.

In a recent study, Rawung et al. (2014) used a simple tropical cooler with fresh cabbage in order to analyze cooler air circulation, cooling rate, storage duration, and cabbage loss using ice. Results showed a highest cooling rate of ice at room temperature of 0.64°C/h and weight loss of cabbage was reduced to only an average of 0.83%.

For Broccoli, four cooling methods were tested, room cooling, forced air-cooling, hydrocooling and package icing (Kochhar and Kumar, 2015). The temperatures of all four cooling mediums were in the range of 0–1°C. Based on the obtained results, it was concluded that package icing and hydrocooling were better methods of cooling compared to forced air-cooling and hydrocooling.

A Canadian manufacturer (SUNWELL, 2015) has recently developed an ice slurry system for the preservation of fresh products, which are traditionally stored and preserved in ice such as broccoli, green onions, corn, herbs and other delicate produce. In this system, the slurry ice formed inside an ice generator are pumped into an insulated storage-dispenser insulate tank, where they remain suspended in water. Dry ice crystals from the top of the tank are then mixed with a small amount of water and the mixture is pumped with a positive displacement pump via a pipe system to the display cases where it is spread over the display surface with a flexible hose. Boxes containing produce are stacked on a pallet. Ice slurry is then injected into the boxes. The entire pallet is rapidly chilled in 36 sec. The excess water is drained away, leaving the cartons and uniformly packed in slurry ice and ready for storage and distribution. Unlike other icing systems, this system reduces the shipping weight by selecting the amount of slurry ice packed in to each box. The amount of water with the ice slurry can varies according the temperature wanted to be attained and it could be ranged from 65 to 80% where the diameter of ice crystals is as low as 50–500 μ micron.

Slurry ice may also be mixed with other additives, such as ozone to inhibit microbial growth and to increase shelf life and maintain sensory quality, as demonstrated (Keys, 2015; Lu et al., 2012), but there are few studies comparing ozonized flake ice to ozonized slurry ice.

1.14.2 INDIRECT USE OF SLURRY ICE

Elansari and Yahia (2012) presented a design for the fresh produce cold store using slurry ice as secondary refrigerant in which the relative humidity can be maintained at higher level leading to a minimum weight loss during cold storage (Figure 1.22). The system is a thermal storage one and can provide two temperature range, the first is for banana ripening rooms along with other cold storage products, 10°C and above, while the other is

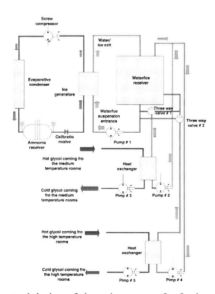

FIGURE 1.22 Conceptual design of slurry ice system for fresh produce.

for potato cold store, 4°C. The system consists of a main circuit in which in an ice generator is producing slurry ice via a heat exchanger using ammonia as the primary refrigerant. The harvested slurry ice is accumulated and stored in an insulated tank with an optimum capacity. The tank is manufactured from a plastic material or a stainless steel and to be positioned in a shaded area. The slurry ice within the tank is always agitated to keep its homogeneity. The heat generated within the agitator is to be considered during the design stage as a heat source. The first pump is located in lower level with respect to the tank; it pulls the melted water due to the heat added by the agitator and return it continuously to the generator. This keep the slurry ice quality as per required. Over the last fifteen years there have been a large number of installations completed in over 40 countries for direct contact cooling of various food products (Kauffeld, et al., 2010; Matsumoto et al., 2010).

Slurry ice is also used in refrigerated trucks transporting fresh produce. It was found that ice slurry refrigeration system operates at a higher efficiency than the standard on-board truck cooling system where it is in operation in Japan (Kato and Kando, 2008). Ice slurry is produced at a central plant and is charged into special heat exchangers in the insulated boxes fitted into the truck. Carbon dioxide emissions associated with the refrigeration system could therefore be reduced by 20–30% (Kato and Kando, 2008). In addition, the engine in the ice slurry cooled trucks can be switched off completely at the points of goods pick-up and delivery, therefore reducing noise and exhaust emissions, a feature which is especially valuable in large cities with air quality problems (Kauffeld, et al., 2010).

Slurry ice is undoubtedly a promising technology the postharvest refrigeration of the horticultural crop that should be encouraged because of its numerous advantages, in particularly energy savings and for being environmentally friendly. Further research and improving work need to conducted particularly on its effect and performance in keeping produce quality and extending its shelf life for different products and under various circumstances.

1.15 CONTROL OF THE COLD CHAIN PROJECTS

In industrial installations that use refrigeration systems that are associated with the fresh produce industry; the optimization of energy consumption associated with the achievement of high quality standard and extended shelf life is one of the main objectives of the modern innovation. The two sides on any cold chain element is the mechanical side and the air side. The mechanical side is where the hardware of the refrigeration cycle is located, the machine room. The control algorithms for the mechanical side has no direct relation to the quality, maturity, storage, etc., of the produce in the cold room (Brettl, 2001). The airside is the insulated cooling chamber where we cool the produce. Cold stores (chamber) or refrigerated warehouses are defined as these facilities where perishable foodstuffs are handled and stored under controlled temperatures with the aim of maintaining quality. The main stages in controlling the process and in assessing potential energy saving opportunities are audit existing refrigerating equipment, check controls and set points, reduce heat loads, improve defrosting, reduce temperature lifts in refrigerating plant, optimize compressor and system operation, institute planned maintenance.

For some products, other conditions related to the postharvest stage, besides temperature and relative humidity, control might be required such as; the moisture content and/or the composition of the surrounding atmosphere has to be changed like in the case of potato storage or for controlled-atmosphere (CA) storage or ultra-low-oxygen (ULO) storage.

Accurate control of temperature, relative humidity (RH), and airflow significantly affects grape metabolism in terms of volatile compounds. Temperature plays a key role in accelerating or delaying the desired water loss during the handling of grapes but it is mainly important for the modulation of volatile compound metabolism and the formation of volatile acidity (Chkaiban et al., 2007; Silva and Teruel, 2011).

As an example and for CA room and for long-term storage, the quality is maintained by controlling certain parameters (Brettl, 2001). These parameters include:

- Temperature differences in the storage room (for example: −1 to −2°C).
- Temperature fluctuation (0.1 to 0.2°C).
- Relative humidity (92–95%RH).
- RH differences (5 or 10%).
- Temperature fluctuation in the space.
- RH fluctuation in the space.
- O_2 measurement accuracy.

- CO_2 measurement accuracy.
- Nitrogen pull down.
- Low limit of O_2
- Low limit of CO_2
- Air tightness of the chamber.
- Air circulation figure rate.

We will limit our discussion in this section to the airside of the refrigeration process only where we concern about temperature, relative humidity and controlling the air pattern in the forced air precooling process.

1.16 VARIABLE FREQUENCY DRIVE AND CONTROL STRATEGY

Another aspects of the control of different cold chain element is its strategies which should be generally intend to reduce the fluctuation of the controlled environment temperature and often minimize energy consumption associated with the system operation. A failure in cold chain causes lower durable produce and uneconomical use of energy for cooling and storage. According to Meneghetti et al. (2013), when cold chain is interrupted, it can create gaps for deterioration due to water condensation on the product, providing an excellent environment for fungi growth and other microorganisms.

Therefore, during the short- or long-term cold storage period, it is essential to maintain the steady temperature in a narrow range and no major variation or fluctuations, in spite of the existence of disturbances resulted from different heat sources. Such heat sources included heat generation due: the biological activity, the operation of electric motors of the evaporator; presence of operators, the heat loss through walls, floor and roof and heat losses due to frequent opening of the chamber, in addition other factors.

In the forced air precooling process, the effect of airflow blockage and guide technology applied on energy consumption is vey much related the type of control strategy implemented. The velocities and temperatures of the air in the cold zone for different designs of airflow blockage and guide boards should be very carefully planned and evaluated since the airflow pattern plays a key role on energy efficiency, precooling time, and productivity (Akdemir, 2012).

In potato cold store, the inadequate and poorly manage airflow distribution through stacks of bagged potatoes could also result in non uniform humidity which might lead to condensation of moisture where relative humidity reaches saturation or excessive dehydration where the relative humidity remains below 80%. The prevailing conditions in potato cold stores lead to storage losses up to 10%, against the prescribed maximum limit of 5% during the storage period of 8 months (Chourasia and Goswami, 2009). Therefore, one of the main aims in designing a storage system is to ensure a uniform targeted airflow, which leads to better temperature and humidity control.

Most stores are designed to provide an airflow of 0.3 m^3/min. per ton of product, based on the maximum amount of fresh produce that can be stored in the chamber (Akdemir and Arin, 2006; Cold Chain Development Center, 2010). This is needed to cool product to storage temperature and also may be needed if the produce has a high respiration rate. This high airflow rate can cause excessive water loss from products, and fans are a considerable source of heat, so the system should be designed to reduce airflow to 0.06 m^3/min to 0.12 m^3/min. of airflow per ton. Motor speed control systems, such as variable rate –frequency control controllers (Figure 1.23) for alternating current motors, are used to control fan speed at the lowest possible

FIGURE 1.23 New set of FVD being installed for a precooling station.

speed that will prevent unacceptably warm product in the storage as well as minimizing weight loss.

In the forced air precooling process, the airflow rate supplied by the auxiliary fan in the precooling tunnel is controlled via a variable frequency drive (VFD). Also and for cold store, the flow air supplied by fan evaporators can be controlled by VFD (Elansari, 2009). VFDs are an electronic motor controller that is used to reduce fan speed after the heat field has been pulled down to storage temperatures. For the forced air precooling and as the process nears its end, water loss from product should be avoided by minimizing air-flow which can be reduced as low as 50%. The VFDs offer very attractive energy savings. At half fan speed, the fans will consume only about 15% of full speed power (Morton and McDevitt, 2000). In a CA facility, fans are typically operated at full speed for several weeks following room seal. At that point, fan speed can be immediately reduced to 50%, or can be staged down over several weeks, again with a minimum of 50% speed (Becker et al., 2013). Therefore, benefits of evaporator fan VFD control include smooth temperature control and subsequently controlling relative humidity and weight loss.

In 2007, PG&E conducted a demonstration of variable frequency drives for a vacuum cooler (PG&E, 2008). They demonstrated a 29% electricity savings and a 29% reduction in demand compared with a conventional vacuum cooler. Research conducted in the Pacific Northwest for the Northwest Energy Efficiency Alliance (2008) reported an improved product quality and reduced mass loss in fruit stored in controlled atmosphere rooms with VFD controls on evaporator fans. VSDs applied also to evaporator fans in cold stores provided good temperature control. The report indicated that it is very feasible to use VFD to control motor speed in evaporators with fan motors greater than 1 hp. Thus for all motor sizes, the motor speed should be controlled based on targeted temperature, required with a provision for a minimum speed setting that can be defined by the operators of the refrigerated warehouse.

The other alternative to VFDs is fan cycling by the on-off method. Excessive fan cycling can cause an increase in shrinkage of fresh produce due to depressed humidity levels in the room; poor or irregular temperatures in the fruit and poor air circulation in parts of the room in addition to a permanent unwanted oscillations in the chamber temperature

(Meneghetti et al., 2013). In the other hand, implementing VSDs on evaporator fan motors, fan speed can be modified to match varying cooling loads. At low loads, reducing the speed of the fan decreases the power consumption of its motor significantly, as power is proportional to the cube of speed. For example, reducing fan speed by 20% will reduce power requirement by approximately 50% (NSW Government: Office of Environment and Heritage, 2011).

Therefore, it could be concluded that, the marketable life of most fresh vegetables can be extended by prompt storage in an environment that maintains product quality. The desired environment can be obtained in facilities where temperature, air circulation, relative humidity, and sometimes atmosphere composition can be controlled (El-Ramady et al., 2015).

There are a number of benefits to implementing variable speed drives on the evaporator fans motors including:

- Energy savings due to reduced operation speeds.
- Maintenance cost savings due to reduced operation hours.
- Labor savings due to reduced maintenance required.
- For cold storage, reduced fan speed may improve the storage of perishables such as potatoes and apples in a controlled atmosphere.
- The mass loss from fruit is reduced.
- Provides outstanding humidity control.
- Allows modification in air-flow.

1.17 TEMPERATURE AND RELATIVE HUMIDITY CONTROL

The production, storage, distribution and transport of fresh produce (vegetables, fruits and cut flowers), are taking place continuously and around the clock all over the world. The key success for such handling and supply chain is the control of temperature and relative humidity, which are very essential (Garcia et al., 2011; Melis et al., 2015). Mainly, the term "cold chain" defines the sequences of interdependent equipment and processes employed to grantee the temperature preservation of perishables and other temperature-controlled products from the harvesting to the consumption end in a safe, wholesome, and good quality state (Elansari and Yahia, 2012). For an example, the inadequacy of sufficient and efficient cold chain infrastructure is a major contributor to food losses and waste in NENA (Middle East and North Africa) as undeveloped countries, estimated to be 55% of fruits (FAO, 2011). This amounts to up to 215 kg/year per capita, which not only exacerbates the food insecurity in poor countries and the high reliance on imports, but is a waste of scarce natural resources (water and land, most acutely) and a source of economic losses and environmental problems (FAO, 2014). A reliable and efficient cold chain cannot only contribute to minimizing losses and waste in the quantity and quality of food, but can also improve the efficiency of food supply chains and compliance with food safety and quality standards, thus also reducing health problems and costs associated with the consumption of unsafe food. In addition, reducing food losses and waste will also minimize food secrecy and thus exposure to food price volatility for countries dependent on food imports. Cold chain development is, therefore, a necessary step in improving food and nutrition security worldwide (FAO, 2012).

The quality of fresh produce might change rapidly due to inadequate temperature and relative humidity conditions during different cold chain steps especially handling warehousing and transport. Inadequate temperature is second on the list of factors causing foodborne illnesses, surpassed only by the presence of initial microflora in foods (López and Daeyoung, 2008). Also, temperature is considered the most important single factor influencing the quality and shelf life of fresh produce in postharvest stage (Thompson et al., 2002). Water loss is one of the main causes of deterioration that reduces the marketability of fresh produce. Transpiration is the loss of moisture from living tissues. This process causes most weight loss of stored fruit. Temperature and relative humidity of the product, temperature of the surrounding atmosphere, and air velocity all affect the amount of water lost from perishable food products. In the contrary the use of poor controlled humidifiers to increase relative humidity leads to free water accumulation or condensation is also a problem as it encourages microbial infection and growth, and it can also reduce the strength of non-waxed cardboard boxes (Burg, 2014).

Also, there is an increasing pressure of traceability in the food chain, statutory requirements are up-warding stricter and there is increasing demand to develop standardized traceability systems. From the raw material to the sale of goods, more and more information needs to be gathered and made available. We should take in our consideration also the new concept of first-expired-first-out (FEFO). The basic idea is to apply stock rotation in such a way that the remaining shelf life of each item is best matched to the remaining transport duration options, to reduce product waste during transportation and provide product consistency at the store (Jedermann et al., 2014).

In refrigerated trucks or marine containers, temperatures rise very rapidly if a reefer unit fails. A recent study shows temperature-controlled shipment rise above the specified temperature in 30% of trips from the supplier to the distribution center, and in 15% of trips from the distribution center to the store. Lower-than required temperatures occur in 19% of trips from supplier to distribution center and in 36% of trips from the distribution center to the store (Garcia and Lunadei, 2011).

Thus, studying and analyzing both temperature gradient data and relative humidity inside precoolers, refrigeration rooms or warehouses, containers and trucks is a primary concern for the fresh produce industry. Any temperature disturbance can undermine the efforts of the whole chain (Mahajan et al., 2014). Maintaining appropriate conditions over the whole chain is a very challenging task where negligence or mishandling in the logistics of perishable food products is very familiar. A lot of reports in the literature give many cases where the inadequate management during temperature control usually leads to losses in the food chain (postharvest, distribution and at home). However, in reality less than 10% of such perishable foodstuffs are in fact currently refrigerated (Coulomb, 2008). Also it should be mentioned that, the production of food involves a significant carbon investment that is squandered if the food is then not utilized. In the planning phase for an element of the cold chain, the costs of a new refrigeration system can sometimes quickly be recovered in energy savings over an old system, which is achieved by precise and better control (Energy Efficiency Best Practice Guide Industrial Refrigeration, 2009).

The operation of the cold chain element of perishable produce requires both automatic and manual control of the equipment in order to properly pull down field heat in a precooler, optimum storage and relative humidity within a cold store as an example. The most successful cold chain elements are those whose owner and operators understand the need to continuously measure system performance and energy consumption. Such project cannot meet the ultimate goals of produce quality and optimum energy consumption and the most feasible running or operating cost without proper control. Thus and based on the above, studying and analyzing temperature gradient data inside precoolers, refrigeration rooms, containers and trucks is a primary concern for the industry where they are the main input for any system control.

Garcia et al. (2009) illustrated the great potential of a specific type of motes, providing information concerning several parameters such as temperature, relative humidity, door openings and truck stops. They also developed a Psychometric charts for improving the knowledge about water loss and condensation on the product during shipments.

Aung et al. (2012) discussed the application of radio frequency identification (RFID) systems in logistics applications to track and trace the location of produce throughout different points in the supply chain. RFID tags attached to produce are capable of providing real-time tracking information across the supply chain. Applying such technology can lead to a better decision with the fresh produce supply chain and could then be made based on information (temperature and relative humidity) and not only the location of an asset, but also its condition (Roussos et al., 2008).

Vandana et al. (2014) designed a low cost data logger prototype suitable for Cold Chain Logistics. The proposed data logger is capable of measuring levels of temperature (T), humidity (H), and carbon monoxide (CO). It is capable of alerting the user regarding the parameter changes using SMS, so that early precaution steps can be taken. The system also incorporates GPS module, which enables the live tracking capability of the cargo at any point of time.

Chandra and Lee (2014) presented a system comprising of Arduino wireless sensor network and Xively sensor which can be an ideal system to monitor temperature and humidity of cold chain logistics. The application is making use of the internet of things (IoT), which is a new evolution in technological advancement taking place in the world today. The combination of wireless sensor networks and cloud computing is becoming a popular strategy for the IoT era. The cold chain requires controlled environment for sensitive products in order for them to be fit for use. The monitoring process is the only assurance which tells if a certain process has been carried out successfully. Taking advantage of IoT and its benefits to monitor cold chain logistics will result in better management and product handling.

Melis et al. (2015) presented the results of a combination of RFID and wireless sensor network (WSN) devices in a set of studies performed in three commercial wholesale chambers of 1848 m3 with different set points and perishable produces. Up to 90 semi-passive RFID temperature loggers were installed simultaneously together with seven motes, during one week in each chamber. 3D temperature mapping charts were obtained and also the psychometric 32 data model was implemented for the calculation of enthalpy changes and the absolute water content of air. It was concluded that, the feedback of data, between RFID and WSN made it possible to estimate energy consumption in the cold room, water loss from the products and detect any condensation over the stored commodities.

1.18 ENERGY SAVING

Fresh produce industry includes facilities engaged in precooling and cold storage of fruits, vegetables and cut-flowers. Such industry consumes a considerable amount of fuels and electricity per year to run its refrigeration plants and its supporting systems. Apart from energy consumption, cold storage facilities are responsible for approximately 2.5% of global green house gas emissions through direct and indirect energy consumption (Reinholdt, 2012). Therefore, energy efficiency improvement is a vital goal to reduce these costs and to increased predictable earnings, especially in times of high-energy price volatility. There are a variety of opportunities available at individual plants that handle fruit and vegetable to reduce energy consumption in a cost-effective manner.

Many opportunities exist within fresh produce facilities to reduce energy consumption while maintaining or enhancing productivity and quality provided that it pursued in a coordinated fashion at multiple levels within a facility. At the hardware (component and equipment) level, energy efficiency can be enhanced through sustainable preventative and predictive maintenance programs, proper loading and operation, and upgrading of older components and equipment with higher efficiency models (e.g., high efficiency motors) whenever feasible. At the process stage and via process control and optimization, the operations can be pursued and run at maximum efficiency. At the facility level (precooling and cold store), the efficiency of space lighting and cooling can be improved while total facility energy inputs can be at the minimum level through process integration and combined heat and power systems, where feasible. Lastly, at the level of the organization, energy management strategies can be adapted to ensure a strong corporate framework exists for energy monitoring, target setting (temperature and pull down time), employee involvement, and continuous improvement and training.

Supervisory control and data acquisition systems (SCADA) can be very helpful in energy monitoring and metering of cold store for fresh produce project. SCADA is fast data-acquisition software for monitoring and control power availability of electrical distribution networks. The software gives operators exceptional knowledge and control of their network through an intuitive, interactive and customizable interface. With fast, consistent access to actionable information, SCADA system is more effective at protecting and optimizing their electrical distribution network, thereby improving both its efficiency and productivity.

All critical cold store rooms temperatures and relative humidity, plant temperatures and pressures, will be observable through the SCADA system. It will record and file all relevant data allowing subsequent viewing and reporting of all previous working conditions and plant operational parameters. Historical data should be backed up to provide a permanent record of product storage history as well as energy consumption and its circumstances. The SCADA computer and operating system software should be upgraded at least every five years to ensure the system remains currency with IT industry personnel skills (IPENZ, 2009). In the past, large food storage operations may have been staffed 24 h a day seven days a week, but in recent years, the advent of PLC and SCADA systems with monitoring and alarming features have provided a more economical alternative. For example, the SCADA system can send error messages on the status of the refrigeration system to the plant operator's pager.

Yu et al. (2013) used a programmable logic controllers (PLC) to control a SCADA system that was designed to keep fruits and vegetables fresh with ozone. To address the problem of system accuracy and real-time monitoring, Rockwell configuration software (CITECT) was used, the Cicode function, SWOPC-FXGP/WIN-C principles of programming and the information transfer between the PC and the control system to provide real-time monitoring of the ozone concentration and the temperature of the cold storage. It was reported that the accuracy of the system was improved. Experiments that have been conducted showed that the system was successfully used to preserve a crop of kiwi fruit. Long-range automatic control technology and agricultural technology were implemented to optimize the ozone treatment used to preserve fruits and vegetables.

Thompson et al. (2010) analyzed utility bills, facility equipment, operation and production records from seven forced-air-cooling operations. Also the range of electricity use for commercial forced-air-cooling facilities were documented to evaluate the electricity use and conservation options for the major system components, and to estimate annual electricity use for forced-air-cooled produce in California. It was confirmed that electricity use was the greatest for fruit cooling, with nearly as much for direct operation of fans plus field heat removing. Power demand for operating and cooling lights, removing heat gain through walls and operating and cooling lift trucks comprised the next largest energy consumption in decreasing order of use. Options for reducing electricity use of each system were suggested. Possible methods of reducing electricity use are to utilize produce containers with appropriate venting area and minimum amounts of internal packaging materials. Increasing product throughput per unit of refrigerated area has great potential to improve efficiency.

In refrigerated spaces, energy-efficient lighting can produce additional coincidence cooling savings of 30–40% (Raftery and Cummings, 2013). LEDs are substantially more energy-efficient and give off much less waste heat than HIDs, reducing cooling loads and maintenance costs for cooling equipment. LED lighting system can achieve a 90% reduction in lighting energy costs and generate 30–40% additional coincidence cooling savings.

Mulobe and Huan (2012) studied the impact of airflow efficiency or stacking style on the rate of energy consumption by evaporator fans motors where variable speed drive (VSDs) technology on evaporator fans motors for fresh produce cold store were applied. VSDs reduce motor electricity consumption by 30–60%; other benefits include prolonging equipment life through motor speed adjustments according to refrigeration load.

Hilton and Airah (2013) detailed how at one of the largest cold stores in Australia, energy efficiency was improved from 53.5 kWh/m^3 to 37.6 kWh/m^3. Over this period the total storage capacity increased by 34.5% from 106,270 to 142,970 pallets but the total electricity consumption (kWh) did not change. The major contributors to improving energy efficiency were:

1. Constructing the new buildings and refrigeration plants to high energy-efficiency standards.
2. Energy-efficiency benchmarking of the existing facility
3. Improved monitoring and control of chamber temperatures.
4. Improvements in door design to reduce infiltration.
5. Retrofitting energy efficient LED (light-emitting diode) lights.
6. Retrofitting VFDs to existing screw compressors, freezer and condenser fans.
7. Over-sizing evaporative condensers.
8. Power factor correction and Voltage optimization.
9. Rain water harvesting to substitute for potable condenser feed water.

1.19 MAINTENANCE

Maintenance can have many objectives. Long time ago maintenance was seen as repairing those items that have broke down for whatever reason, so called corrective maintenance. A step was set when preventive maintenance became more common, preventing breaking down of items or replacing the subject item before it broke down. Nowadays, the objective of maintenance is not only to have the refrigerating plant available at all time, but also to maintain its capacity, efficiency and the quality of the stored. Another, not less important objective is safety with regard to people, stored foodstuffs and environment.

As any piece of mechanical equipment, also a refrigerating plant needs maintenance in order to keep it operational with the original capacity and efficiency for a long period of time. Maintenance should not be restricted to equipment with moving parts only. Inspection and maintenance must comprise the complete installation from compressors, coolers, condensers and pumps to controls, piping and insulation and even the primary and secondary refrigerant. A maintenance comprehensive plan should be developed for all equipment, including the building itself.

As a minimum, the maintenance program should include periodic inspection and maintenance of the following items:

1. Building: vapor retarder, structure and piping insulation, doors, floors.
2. Material handling system: forklifts, conveyors, pallet racks.
3. Refrigeration equipment: compressors, heat exchangers, pumps, tanks and receivers, condensers, evaporators, fans, piping, valves, instrumentation, purgers, system oil management.
4. Safety apparatuses: fire detection devices and alarms, refrigerant leak detectors and alarms, fire extinguishing devices, relive valves.

1.20 CONCLUSION

Different precooling methods are presented along with its recent applications. To maximize the benefits of each system, careful design and selection is required in order to minimize capital investment as well as the running cost. Saving energy approaches should be considered and implemented during different stages. Investment on the management and controlling apparatus will be reflected on the performance as well as the quality of the produce.

KEYWORDS

- cold chain
- forced air
- hydrocooling
- precooling
- slurry ice
- vacuum

REFERENCES

Aaron Raftery & Kelsey Cummings (2013). Leveraging Energy-Efficient Lighting Technologies to Reduce Waste Heat and Operating Costs. 35th Annual Meeting International Institute of Ammonia Refrigeration March 17–20, 2013, The Broadmoor Colorado Springs, Colorado.

Ahmad, M.S., & Siddiqui, M.W. (2015). *Postharvest Quality Assurance of Fruits: Practical Approaches for Developing Countries*. Springer, New York. pp. 265.

Akdemir, S. (2012). Energy Consumption of An Experimental Cold Storage. *Bulgarian Journal of Agricultural Science, 18*(6), 991–996.

Al-Ansari, A.M. (2009). Design Aspects in the Pre-Cooling Process of Fresh Produce. In: Quality Retention During Postharvest Handling Chain, Sivakumar, D. (Ed.). 49–57, Global Science Book, UK.

Al-Ansari, A.M., & Yahia, E.M. (2012). Cold Chain for Perishable Foods (In Arabic). FAO Publication.

Alibas, I., & Koksal, N. (2014). Forced-air, vacuum, and hydro precooling of cauliflower (*Brassica oleracea* L. var. botrytis cv. Freemont), Part I. Determination of precooling parameters. *Food Science and Technology (Campinas), 34*(4), 730–737.

Aung, M.M., Chang, Y.S., & Won, J.U. (2012). Emerging RFID/USN applications and challenges. *International Journal of RFID Security and Cryptography, 1*, 3–8.

Badia-Melis, R., Ruiz-Garcia, L., Garcia-Hierro, J., & Villalba, J.I.R. (2015). Refrigerated Fruit Storage Monitoring Combining Two Different Wireless Sensing Technologies: RFID and WSN. *Sensors, 15*(3), 4781–4795.

Barbin, D.F., Neves Filho, L.C., & Silveira, V. (2012). Portable forced-air tunnel evaluation for cooling products inside cold storage rooms. *International Journal of Refrigeration, 35*(1), 202–208.

Becker, B. (2013). Green-guide for Sustainable Energy Efficient Refrigerated Storage Facilities. Energy Research and Development Division Final Project Report.

Becker, B.R., & Fricke, B.A. (2006). Best practices in the design, construction, and management of refrigerated storage facilities. International Institute of Ammonia Refrigeration Annual Meeting, *28*, 341–388.

Benz, S.M. (1989). Wet air-cooling. International Institute of Ammonia Refrigeration Annual Meeting, *11*, 85–94

Bharti, A. (2014). Examining market challenges pertaining to cold chain in the frozen food industry in Indian retail sector. *J. Manage. Sci Tech., 2*(1), 33–40.

Boyette, M.D., & Estes, E.A. (2000). Postharvest Technology Series AG-414-5. Carolina Cooperative Extension Service. Crushed and Liquid Ice Cooling.

Boyette, M.D., & Rohrbach, R.P. (1990). A low-cost, portable, forced-air pallet cooling system. *Applied Engineering in Agriculture, 9*, 97–104.

Burg, S.P. (2014). Capillary Condensation in Non-Waxed Cardboard Boxes. In Hypobaric Storage Food Industry: Advances in Application and Theory. Academic Press: London, UK.

California Energy Commission (2008). Final Report Refrigerated Warehouses. Refrigerated Warehouse CASE Report. Pacific Gas and Electric Company, USA.

Carlos Teles Ribeiro da Silva, & Bárbara J. Teruel, M. (2011). Automatic Integrated Control for the Drying Process of Wine Grapes. João Proceedings of the 6th CIGR Section VI International Symposium "Towards a Sustainable Food Chain" Food Process, Bioprocessing and Food Quality Management Nantes, France – April 18–20, 2011.

Chandra, A.A., & Lee, S.R. (2014). A Method of WSN and Sensor Cloud System to Monitor Cold Chain Logistics as Part of the IoT Technology. *International Journal of Multimedia and Ubiquitous Engineering, 9*(10), 145–152.

Chkaiban, L., Botondi, R., Bellincontro, A., Santis, D., Kefalas, P., & Mencarelli, F. (2007). Influence of postharvest water stress on lipoxygenase and alcohol dehydrogenase activities, and on the composition of some volatile compounds of Gewürztraminer grapes dehydrated under controlled and uncontrolled thermohygrometric conditions. *Australian Journal of Grape and Wine Research, 13*(3), 142–149.

Chourasia, M.K., & Goswami, K. (2009). Efficient Design, Operation, Maintenance and Management of Cold Storage. *Journal of Biological Sciences, 1*(1), 70–93.

Chourasia, M.K., & Goswami, T.K. (2007). Steady state CFD modeling of airflow, heat transfer and moisture loss in a commercial potato cold store. *International Journal of Refrigeration, 30*(4), 672–689.

Christie, S. (2007). Pre-Cooling Fresh-Cuts; Cold Chain Begins Before Processing Starts. Fresh Cut. Great American Publishing.

Coulomb, D. (2008). Refrigeration and cold chain serving the global food industry and creating a better future: two key IIR challenges for improved health and environment. *Trends in Food Science and Technology, 19*(8), 413–417.

Crisosto, C.H., Thompson, J.F., & Garner, D. (2002). Table grapes cooling. Central Valley Postharvest Newsletter. Cooperative Extension, University of California, Kearney Agricultural Center, *11*, 5–13.

Defraeye, T., Lambrecht, R., Tsige, A.A., Delele, M.A., Opara, U.L., Cronjé, P., Verboven & Nicolai, B. (2013). Forced-convective cooling of citrus fruit: Package design. *Journal of Food Engineering, 118*(1), 8–18.

Defraeye, T., Verboven, P., Opara, U.L., Nicolai, B., & Cronjé, P. (2015). Feasibility of ambient loading of citrus fruit into refrigerated containers for cooling during marine transport. *Biosystems Engineering, 134*, 20–30.

Dehghannya, J., Ngadi, M., & Vigneault, C. (2010). Mathematical modeling procedures for airflow, heat and mass transfer during forced convection cooling of produce: a review. *Food Engineering Reviews, 2*(4), 227–243.

Dehghannya, J., Ngadi, M., & Vigneault, C. (2011). Mathematical modeling of airflow and heat transfer during forced convection cooling of produce considering various package vent areas. *Food Control, 22*(8), 1393–1399.

Dehghannya, J., Ngadi, M., & Vigneault, C. (2012). Transport phenomena modeling during produce cooling for optimal package design: Thermal sensitivity analysis. *Biosystems Engineering, 111*(3), 315–324.

Delele, M.A., Ngcobo, M.E.K., Getahun, S.T., Chen, L., Mellmann, J., & Opara, U.L. (2013). Studying airflow and heat transfer characteristics of a horticultural produce packaging system using a 3-D CFD model. Part I: model development and validation. *Postharvest Biology and Technology, 86*, 536–545.

Don Tragethon (2011). Vacuum Cooling – The Science and Practice. 2011 Industrial Refrigeration Conference and Heavy Equipment Show Caribe Royale Orlando, Florida.

Dustin R. Keys (2015). Cooling Characterizations and Practical Utilization of Sub-micron Slurry Ice for the Chilling of Fresh Seafood. MSc Thesis. Oregon State University.

Edition, A.H.F.S. (1993). American Society of Heating, Refrigerating and Air Conditioning Engineers. Inc., Atlanta, GA.

Elansari, A.M. (2003). Forced air fast cooling system of Egyptian fresh Strawberries. *Misr Journal for Agriculture Engineering, 20*, 571–586.

Elansari, A.M. (2008a). Hydrocooling rates of Barhee dates at the Khalal stage. *Postharvest Biology and Technology, 48*(3), 402–407.

Elansari, A.M. (2009). Design of portable forced – air precooling system. *Journal of the Saudi Society of Agricultural Sciences*, *2*, 38–48.
Elansari, A.M., & Alaa El-din Bekhit (2015). Freezing/thawing technologies. In: Advances in Meat Processing, Alaa El-din Bekhit, (Ed.). CRC Press. In Press.
Elansari, A.M., Hussein, A.M., & Bishop, C.F. (2000). Performance of wet deck (ice bank) pre-cooling systems with export produce from Egypt. *Landwards*, *55*(4), 20–25.
Elansari, A.M., Shokr, A.Z., & Hussein, A.M. (2000). The Use of Sea-Shipment Container As a Portable Pre-Cooling Facility. *Misr J. Agric. Eng.*, *17*, 401–411.
Energy Efficiency Best Practice Guide Industrial Refrigeration. 2,9. Sustainability Victoria.
FAO (2011). Global Food Losses and Food Waste: Extent, Causes, and Prevention. In: Gustavsson, J., Cederberg, C., Sonesson, U., van Otterdijk, R., & Meybeck, A. FAO, Rome, Italy.
FAO (2012). Proceedings: Expert Consultation Meeting on the Status and Challenges of the Cold Chain for Food Handling in the Middle East and North Africa (MENA) Region. In: Kader, A., & Yahia, E. FAO/RNE, Cairo, Egypt.
FAO (2014). Strategic Framework to Reduce Food Losses and Waste in NENA. FAO/RNE, Cairo, Egypt.
Farrimond, A., Lindsay, R.T., & Neale, M.A. (1979). The ice bank cooling system with positive ventilation. *International Journal of Refrigeration*, *2*(4), 199–205.
Ferreira, M.D., Brecht, J.K., Sargent, S.A., & Chandler, C.K. (2006). Hydrocooling as an alternative to forced-air-cooling for maintaining fresh-market strawberry quality. *Hort Technology*, *16*, 659–666.
Ferrua, M.J., & Singh, R.P. (2009a). Modeling the forced-air-cooling process of fresh strawberry packages, Part I: Numerical model. *International Journal of Refrigeration*, *32*(2), 335–348.
Ferrua, M.J., & Singh, R.P. (2009b). Modeling the forced-air-cooling process of fresh strawberry packages, Part II: Experimental validation of the flow model. *International Journal of Refrigeration*, *32*(2), 349–358.
Ferrua, M.J., & Singh, R.P. (2011). Improved airflow method and packaging system for forced-air-cooling of strawberries. *International Journal of Refrigeration*, *34*(4), 1162–1173.
Fockens, F.H., & Meffert, H.F. (1972). Biophysical properties of horticultural products as related to loss of moisture during cooling down. *Journal of the Science of Food and Agriculture*, *23*(3), 285–298.
Fraser, H.W. (1998). *Tunnel Forced-Air-Coolers for Fresh Fruits and Vegetables*. Ottawa.
Gabor H., & Airah, M. (2013). Reducing energy use in the cold storage industry – A case study. *Ecolibrium*, 44–50.
Gan, G., Woods, J.L. (1989). A deep bed simulation of vegetable cooling. In: Dodd and Grace (Eds.). Land and Water Use. Rotterdam: Balkema, pp. 2301–2308.
Garrido, Y., Tudela, J.A., & Gil, M.I. (2015). Comparison of industrial precooling systems for minimally processed baby spinach. *Postharvest Biology and Technology*, *102*, 1–8.
Geeson, D.J. (1989). Cooling and storage of fruits and vegetables. In: Proceedings of the Institute of Refrigeration *85*, 65–76.
Green, T. (1997). Mobile forced Air-cooling service. 5436 North Sunrise Ave, Fresno, California. http://www.coolforce.com/index1024.htm.

Handry Rawung, Senia Ubis, Stella Kairupan, Hildy Wullur, & Dedie Tooy (2014). Analysis of a Cooling System for Cabbage in a Box Cooler. International Conference on Food, Agriculture and Biology (FAB, 2014) June 11–12, 2014, Kuala Lumpur (Malaysia).

Hassan R. El-Ramady, Éva Domokos-Szabolcsy, Neama A. Abdalla, Hussein S. Taha, & Miklós Fári (2015). Postharvest Management of Fruits and Vegetables Storage. In: Sustainable Agriculture Reviews. E. Lichtfouse (Ed.). Springer International Publishing, Switzerland.

He, S.Y., Yu, Y.Q., Zhang, G.C., & Yang, Q.R. (2013). Effects of Vacuum Pre-Cooling on Quality of Mushroom after Cooling and Storage. *Advanced Materials Research*, *699*, 189–193.

http://www.sustainability.vic.gov.au/~/media/resources/documents/services%20and%20 advice/business/srsb%20em/resources%20and%20tools/srsb%20em%20best%20 practice%20guide%20refrigeration%202009.pdf

Hugh, W., & Fraser, P. (1998). Tunnel forced – air-coolers. Canadian Plan Service 98–031, 1–10

IPENZ (2009). Cold Store Engineering in New Zealand. The Institution of Professional Engineers New Zealand (IPENZ).

Isik, E. (2007). Comparison of the Thermodynamically Analysis of Vacuum Cooling Method with the Experimental Model. *American Journal of Food Technology*, *2*(4), 217–227.

Jacomino, A.P., Sargent, S.A., Berry, A.D., & Brecht, J.K. (2011). Potential for grading, sanitizing, and hydrocooling fresh strawberries. *Proc. Fla. State Hort. Soc.*, *124*, 221–226.

James F. Thompson, F. Gordon Mitchell, & Robert F. Kasmire (2002). Cooling Horticultural Commodities. In: Postharvest technology of horticultural Crops. Kader, A. A. (Ed.). Coop. Ext. Service. University of California. Special Publication, 3311 Agr., & Nat. Resources Publication, Berkley, CA 94720.

James, S.J. (2013). Refrigeration Systems. In: Handbook of Food Factory Design. Christopher G.J. Baker (Ed.).

Jedermann, R., Nicometo, M., Uysal, I., & Lang, W. (2014). Reducing food losses by intelligent food logistics. *Philosophical Transactions of the Royal Society of London A: Mathematical, Physical and Engineering Sciences*. A *372*, 20130302.

Josef Brettl (2001). Refrigeration and Controlled Atmosphere. 200–1 IIAR Ammonia Refrigeration Conference Long Beach, CA. Technical Paper # 7, 268–282.

Kader, A.A. (2002). Postharvest Technology of Horticultural Crops. Coop. Ext. Service. University of California. Special Pubi. 3311. Agr., & Nat. Resources Pubi., Berkley; CA 94720.

Kanlayanarat, S., Rolle, R., & Acedo Jr., A. (2009). Horticultural Chain Management for Countries of Asia and the Pacific Region: A Training Package.

Kato, Y., & Kando, M. (2008). Development of thermal energy storage technologies for vehicle use. In: *IEA Annex 18 Workshop in Freiburg*, Germany.

Kauffeld, M., Wang, M.J., Goldstein, V., & Kasza, K.E. (2010). Ice slurry applications. *International Journal of Refrigeration*, *33*(8), 1491–1505.

Kitinoja, L., & Thompson, J.F. (2010). Pre-cooling systems for small-scale producers. *Stewart Postharvest Review*, *6*(2), 1–14.

Kochhar, V., & Kumar, S. (2015). Effect of Different Pre-Cooling Methods on the Quality and Shelf Life of Broccoli. *Journal of Food Processing and Technology*, 6(3).

Lars Reinholdt. (2012). Energy consumption storage facilities examined in ICE-E. Danish Technological Institute

Laurin, E., Nunes, M.C.N., & Emond, J.P. (2003). Forced-air-cooling after air-shipment delays asparagus deterioration. *Journal of Food Quality*, 26(1), 43–54.

Laurin, E., Nunes, M.C.N., & Emond, J.P. (2005). Re-cooling of strawberries after air shipment delays fruit senescence. *Acta Horticulture*, 682, 1745–1751

Liang, Y.S., Wongmetha, O., Wu, P.S., & Ke, L.S. (2013). Influence of hydrocooling on browning and quality of litchi cultivar Feizixiao during storage. *International Journal of Refrigeration*, 36(3), 1173–1179.

Liberty, J.T., Okonkwo, W.I., & Echiegu, S.A. (2013). Evaporative cooling: A postharvest Technology for fruits and vegetables preservation. *Int. J. Sci. Eng. Res*, 4(8), 2257–2266.

Liu, E., Hu, X., & Liu, S. (2014). Experimental study on effect of vacuum pre-cooling for post-harvest leaf lettuce. *Research on Crops*, 15(4), 907–911.

Liu, E., Hu, X., & Liu, S. (2014). Theoretical Simulation and Experimental Study on Effect of Vacuum Pre-Cooling for Postharvest Leaf Lettuce. *Journal of Food and Nutrition Research*, 2(8), 443–449.

López, T.S., & Kim, D. (2008). Wireless sensor networks and RFID integration for context aware services. *Information and Communications University 119 Yuseong-gu*, 305–714.

Lu, F., Liu, S.L., Liu, R., Ding, Y.C., & Ding, Y.T. (2012). Combined effect of ozonized water pretreatment and ozonized flake ice on maintaining quality of Japanese sea bass (*Lateolabrax japonicus*). *Journal of Aquatic Food Product Technology*, 21(2), 168–180.

Luvisi, D., Shorey, H., Thompson, J.F., Hinsch, T., & Slaughter, D. (1995). Packaging California grapes. University of California, DANR, Publication 1934.

Macleod-Smith, R.I., & Espen, V.J. (1996). Modern practices in wet air-cooling for pre-cooling and storage of fresh producer. *The Australian Institute of Refrigeration Air Conditioning and Heating Journal*, 50, 31–40.

Mahajan, P.V., Caleb, O.J., Singh, Z., Watkins, C.B., & Geyer, M. (2014). Postharvest treatments of fresh produce. *Philosophical Transactions of the Royal Society of London A: Mathematical, Physical and Engineering Sciences*, A372.

Manganaris, G.A., Ilias, I.F., Vasilakakis, M., & Mignani, I. (2007). The effect of hydrocooling on ripening related quality attributes and cell wall physicochemical properties of sweet cherry fruit (*Prunus avium* L.). *International Journal of Refrigeration*, 30(8), 1386–1392.

Marcelo A.G. Carnelossi, Steven A. Sargent, & Adrian D. Berry. (2014). 2014 ASHS Annual Conference/Hydrocooling, Forced-air-cooling and Hydrocooling Plus Forced-air-cooling. American Society for Horticultural Science. Salon.

Matsumoto, K., Kaneko, A., Teraoka, Y., & Igarashi, Y. (2011). Development of ice slurry for cold storage of foods in wide temperature range. *Transactions of the Japan Society of Refrigerating and Air Conditioning Engineers*, 27, 281–291.

Meneghetti, C.R., Tizzei, A., Cappelli, N.L., Umezu, C.K., & Bezzon, G. (2013). A Mathematical model for the cold storage of agricultural products. *Revista Ciência Agronômica*, 44(2), 286–293.

Morton, R.D., & McDevitt, M.L. (2000). Evaporator Fan Variable Frequency Drive Effects on Energy and Fruit Quality. In: 16th Annual Postharvest Conference, Yakima, WA, Mar (pp. 14–15).

Mulobe, N.J., & Huan, Z. (2012). Energy efficient technologies and energy saving potential for cold rooms. In: *Industrial and Commercial Use of Energy Conference (ICUE), 2012 Proceedings of the 9th IEEE.*, pp. 1–7.

Mworia, E.G., Yoshikawa, T., Salikon, N., Oda, C., Asiche, W.O., Yokotani, N., Abe, D., Ushijima K., Nakono R., & Kubo, Y. (2012). Low-temperature-modulated fruit ripening is independent of ethylene in 'Sanuki Gold' kiwi fruit. *Journal of Experimental Botany*, *63*(2), 963–971.

Nahor, H.B., Hoang, M.L., Verboven, P., Baelmans, M., & Nicolai, B.M. (2005). CFD model of the airflow, heat and mass transfer in cool stores. *International Journal of Refrigeration*, *28*(3), 368–380.

Nelson, K.E., & Ahmedullah, M. (1976). Packaging and decay-control systems for storage and transit of table grapes for export. *American Journal of Enology and Viticulture*, *27*(2), 74–79.

Nimai Mukhopadhyay, N., & Bodhisattwa Maity (2015). A Theoretical Comparative Study of Heat Load. Distribution Model of a Cold Storage. *International Journal of Scientific and Engineering Research, 6*(2).

NSW Government: Office of Environment and Heritage (2011). Energy Saver: Technology Report – Industrial refrigeration and chilled glycol and water applications. [Online] Available at: www.environment.nsw.gov.au/resources/sustainbus/110302ESRefrigRprtLowRes.pdf

Ozturk, H.M., Ozturk, H.K., & Kocar, G. (2011). Comparison of vacuum cooling with conventional cooling for purslane. *International Journal of Food Engineering*, *7*(6), 2.

Pacific Northwest for the Northwest Energy Efficiency Alliance report. (2008). [Online] Available at: http://aceee.org/files/proceedings/2003/data/papers/SS03_Panel4_Paper_09.pdf

Pathare, P.B., Opara, U.L., Vigneault, C., Delele, M.A., & Al-Said, F.A.J. (2012). Design of packaging vents for cooling fresh horticultural produce. *Food and Bioprocess Technology*, *5*(6), 2031–2045.

Rai, R.D., & Arumuganathan, T. (2008). *Postharvest Technology of Mushrooms*. National Research Centre for Mushroom, Indian Council of Agricultural Research.

Rawung, H., Ubis, S., Kairupan, S., Wullur, H., & Tooy, D. (2014). Analysis of a Cooling System for Cabbage in a Box Cooler. International Conference on Food, Agriculture and Biology (FAB-2014) June 11–12, Kuala Lumpur, Malaysia, pp. 20–23.

Reno, M.J., Prado, M.E.T., & Resende, J.V.D. (2011). Microstructural changes of frozen strawberries submitted to pre-treatments with additives and vacuum impregnation. *Food Science and Technology (Campinas)*, *31*(1), 247–256.

Rhiemeier, J.M., Harnisch, J., Ters, C., Kauffeld, M., & Leisewitz, A. (2009). Comparative assessment of the climate relevance of supermarket refrigeration systems and equipment. *Environmental Research of the Federal Ministry of the Environment, Nature Conservation and Nuclear Safety Research Report*, *206*(44), 300.

Rodrigues, L.G.G., Cavalheiro, D., Schmidt, F.C., & Laurindo, J.B. (2012). Integration of cooking and vacuum cooling of carrots in a same vessel. *Food Science and Technology (Campinas)*, *32*(1), 187–195.

Ross, D.S. (1990) Postharvest cooling basics. Facts Agricultural Engineering/University of Maryland, Cooperative Extension Service. *178*, 1–8.

Roussos, G. (2008). *Networked RFID: Systems, Software and Services*. Springer Science & Business Media.
Ruiz-Garcia, L., & Lunadei, L. (2011). The role of RFID in agriculture: Applications, limitations and challenges. *Computers and Electronics in Agriculture*, 79(1), 42–50.
Rule, J. (1995). Wet air-cooling using ice storage. *The Australian Institute of Refrigeration Air Conditioning and Heating Journal*, 49, 19–22.
Russell, K. (2006). Refrigeration for controlled atmosphere storage of apples in the 21st century. *International Institute of Ammonia Refrigeration Annual Meeting 28*, 275–314.
Ryall, A.L., Lipton, W.J., & Pentzer, W.T. (1982). Handling, Transportation and Storage of Fruits and Vegetables. Vol. 1. AVI Pub. Co., Westport CT.
Ryall, Albert Lloyd, & Wilbur Tibbils Pentzer. *Handling, Transportation and Storage of Fruits and Vegetables. Volume 2. Fruits and Tree Nuts*. AVI Publishing Co. Inc., 1982.
Siddiqui, M.W. (2015). *Postharvest Biology and Technology of Horticultural Crops: Principles and Practices for Quality Maintenance*. CRC Press, Boca Raton, Florida, USA. pp. 550.
Siddiqui, M.W. (2016). *Eco-friendly Technology for Postharvest Produce Quality*. Academic Press, Elsevier Science, USA. pp. 324.
Siddiqui, M.W., Ayala-Zavala, J.F., & Hwang, C.A. (2016). *Postharvest Management Approaches for Maintaining Quality of Fresh Produce*. Springer, New York. pp. 222.
Sreedharan, A., Tokarskyy, O., Sargent, S., & Schneider, K.R. (2015). Survival of *Salmonella* spp. on surface-inoculated forced-air-cooled and hydrocooled intact strawberries, and in strawberry puree. *Food Control*, 51, 244–250.
Sun, D.W., & Zheng, L. (2006). Vacuum cooling technology for the agri-food industry: Past, present and future. *Journal of Food Engineering*, 77(2), 203–214.
Sunwell Technologies Inc. http://www.sunwell.com.
Suslow, T. (1997). *Postharvest Chlorination: Basic Properties and Key Points for Effective disinfection*. University of California, Division of Agriculture and Natural Resources.
Talbot, M.T., & Fletcher, J.H. (1993). Design and development of a portable forced-air-cooler. *Proceedings-Florida State Horticultural Society, 106*, 249–249.
Tassou, S.A., & Xiang, W. (1998). Modeling the environment within a wet air-cooled vegetable store. *Journal of Food Engineering*, 38(2), 169–187.
Tator, R. (1997). Developing the cold chain for horticultural exports. Ministry of Agricultural and Reclamation, Egypt. ATUT technical report. ATUT technical report.
Thompson, J.F. (2004). The commercial storage of fruits, vegetables, and florist and nursery stocks. Agriculture Handbook Number 66, USDA, ARS.
Thompson, J.F. (2006). Requirements for successful forced-air-cooling. Washington Tree Fruit Postharvest Conference, Yakima, WA.
Thompson, J.F., Gordon, M.F., Rumsey, T.R., Kasmire, R.F., & Crisosto, C. (1998). Commercial cooling of fruits, vegetables and flowers. University of California Division of Agricultural and Natural Resources. Publication No. 21567.
Thompson, J.F., Mejia, D.C., & Singh, R.P. (2010). Energy use of commercial forced-air-coolers for fruit. *Applied Engineering in Agriculture*, 26(5), 919–924.
Thorpe, G.R. (2008). *The Design and Operation of Hydrocoolers: a Smart Water Funded Project* (Doctoral dissertation, Victoria University).

Tian, D., Alves, W.A.L., Araújo, S.A., Santana, J.C.C., Jiangang, L., Jianchu, C., & Donghong, L. (2014). Simulation approach for optimal design of vacuum cooling on broccoli by simulated annealing technique. *International Journal of Agricultural and Biological Engineering, 7*(5), 111–115.

Tokarskyy, O., Schneider, K.R., Berry, A., Sargent, S.A., & Sreedharan, A. (2015). Sanitizer applicability in a laboratory model strawberry hydrocooling system. *Postharvest Biology and Technology, 101*, 103–106.

Vandana, Madiwale, W.G., & Awasthi, N. (2014). An Efficient Data Logger System for Continuous Monitoring and Traceability of Cargo: Application of GPS and GSM Technology. *International Journal of Research in Engineering and Technology, 3*(6), 569–572.

Varszegi, T. (2003). Bacterial growth on the cap surface of *Agaricus bisporus. Acta Horticulturae. 599*, 705–710.

Vigneault, C., Markarian, N.R., Da Silva, A., & Goyette, B. (2004). Pressure drop during forced-air ventilation of various horticultural produce in containers with different opening configurations. *Transactions of the ASAE, 47*(3), 807–814.

Vigneault, C., Thompson, J., & Wu, S. (2009). Designing container for handling fresh horticultural produce. *Postharvest Technologies for Horticultural Crops, 2*, 25–47.

Wade, N.L. (1984). Estimation of the refrigeration capacity required to cool horticultural produce. *International Journal of Refrigeration, 7*(6), 358–366.

Wang, Y., & Long, L.E. (2015). Physiological and biochemical changes relating to postharvest splitting of sweet cherries affected by calcium application in hydrocooling water. *Food Chemistry, 181*, 241–247.

Watkins, J.B. (1990). Forced-Air-cooling, 2nd Edition, Queensland Department of Primary Industries. Brisbane, 56 pp.

Yahia, E.K., & Jennifer Smolak (2014). Developing the Cold Chain for Agriculture in the Near East and North Africa (NENA). FAO Regional Office for the Near East and North Africa, http://neareast.fao.org

Yang, Z., Ma, Z., Zhao, C., & Chen, Y. (2007). Study on forced-air pre-cooling of Longan. American Society of Agricultural and Biological Engineers, Paper No. 076267. St. Joseph.

Yu, X., Wu, P., Han, W., & Zhang, Z. (2013). A remote SCADA system for keeping fruits and vegetables fresh with ozone. *Journal of Food, Agriculture and Environment, 11*(2), 187–192.

Yunus A. Cengel & Afshin J. Ghajar (2013). Refrigeration and Freezing of Foods. Chapter 17. In: Heat and Mass Transfer: Fundamentals and Applications. McGraw-Hill.

Yunus, C., & Michael, B. (2014). Thermodynamics: An Engineering Approach. McGraw-Hill.

Zhang, Z., Zhang, Y., Su, T., Zhang, W., Zhao, L., & Li, X. (2014). Heat and Mass Transfer of Vacuum Cooling for Porous Foods-Parameter Sensitivity Analysis. *International Journal of Agricultural and Biological Engineering*, 1–8.

CHAPTER 2

POSTHARVEST HANDLING AND STORAGE OF ROOT AND TUBERS

MUNIR ABBA DANDAGO

Department of Food Science and Technology, Faculty of Agriculture and Agricultural Technology, Kano University of Science and Technology, Wudil, Kano State, Nigeria

CONTENTS

2.1 Introduction 70
    2.1.1 Nutritional Importance of Root and Tuber Crops 71
    2.1.2 Production Statistics of Root and Tuber Crops 71
    2.1.3 Physiology of Root and Tuber Crops 72
    2.1.4 Dormancy and Spouting in Root and Tuber Crops 73
    2.1.5 Curing in Root and Tuber Crops 74
    2.1.6 Description of Major Root and Tuber Crops..................... 75
        2.1.6.1 Cassava... 75
        2.1.6.2 Yams 76
        2.1.6.3 Sweet Potato 77
        2.1.6.4 Potato 78
        2.1.6.5 Cocoyam 78
    2.1.7 Postharvest Losses in Root and Tuber Crops..................... 79
    2.1.8 Major Reasons for Postharvest
           Loss in Root and Tuber Crops... 79
    2.1.9 Harvesting, Handling, Transportation and
           Marketing of Root and Tuber Crops 81

2.1.9.1 Harvesting of Root and Tubers 81
2.1.9.2 Handling of Root and Tubers 83
2.1.9.3 Transportation of Root and Tubers 83
2.1.9.4 Marketing of Root and Tubers 83
2.2 Storage Method for Root and Tuber Crops 84
2.2.1 Traditional Storage Methods of Root
and Tuber Crops 84
2.2.2 Improved/Modern Storage Methods of Root and
Tuber Crops 86
2.2.3 Common Handling Practices and
Conditions Affecting Postharvest Life and
Quality of Root and Tuber Crops 86
2.2.4 Recommended Storage Conditions of
Root and Tuber Crops 88
Keywords 88
References 89

2.1 INTRODUCTION

Root and Tubers belong to the class of foods that provide energy in the human diet in form of carbohydrates. According to FAO (1990), root and tubers refer to any grown plant that stores edible material in the roots, corms, and tubers. They rank next to cereal crops in importance because they provide a major part of the daily calorie needs of the people in the tropics (Ihekoronye and Ngoddy, 1989).

The principles root and tuber crops of the tropics are cassava (*Manihot esculenta* Crantz), Yam (*Dioscorea* Spp), sweet potato (*Ipomea batatas* L.), Irish potato (*Solanum tubaeroson*) and edible aroids (*Colocasia* Spp and *Xanthosomonas sagattifolium*). They are widely grown and consumed as subsistence staples in many parts of Africa, Latin America, Pacific Islands and Asia (Dandago, 2009).

Root and Tuber crops also serve as a source of fermentable sugar required in the production of alcoholic beverages as well as source of raw materials for various industrial fermentations in pharmaceuticals, industrial enzymes, organic solvents and cosmetics. The edible green leaves of sweet potatoes, cocoyam and cassava are good sources of proteins, vitamins, and minerals are offer used to augment diets of local people as well as livestock feed (Eka, 1998).

2.1.1 NUTRITIONAL IMPORTANCE OF ROOT AND TUBER CROPS

Root and Tubers form a major staple food group for a large number of persons in most developing countries of Africa, Asia and Latin America. Nutritionally they are principal sources of calorie to the diets and many contribute a nominal quantity of protein. Cassava for example is the chief source of energy in Southern Nigeria, and D. R Congo. Yam, which is another popular staple tuber crop has less than 5% true protein while cocoyam has <7% true protein but a fair source of calcium. Sweet potatoes are poor source of protein but good source of β-vitamins, ascorbic acid and rich source of provitamin A. Root and tubers contribute about 21–46% of total calorie and about 6.6% protein to the diet of people sub-Saharan Africa (Okaka, 2009).

According to Sanni et al. (2009), root and tuber crops provides a staple carbohydrate source to an estimated population of over 500 million people and also contribute to energy and nutritional requirement of more than 2 billion people (Table 2.1).

2.1.2 PRODUCTION STATISTICS OF ROOT AND TUBER CROPS

It is estimated that about 300 million tons of root and tuber crops are produced in 1993 by developing countries of the world. Cassava and potatoes put up about 83% of the total production. The most important root and tubers in terms of production are cassava (48.8%) and sweet potato (39.8%); while yam (9.4%) and cocoyam (2%) are less important. Most cassava is produced in Africa, Asia, and South America while sweet potatoes production is heavily concentrated in Asia. Africa dominates the production of yams and Taro (Sanni et al., 2009).

TABLE 2.1 Nutritional Composition of R & T Crops on Fresh Weight Basis

Nutrient	Yam D. alata	Cassava M. esculuta	Taro cocoyam	Sweet potato	Irish potato	Tania cocoyam
Moisture (%)	77.3	62.8	69.2	71.1	78.0	67.1
Energy (KJ/100 g)	347.0	580.0	480.0	438.0	300.0	521.0
Protein (%)	2.2	0.5	1.1	1.4	2.1	1.6
Starch (%)	16.7	31.0	24.5	20.1	17.0	27.6
Sugar (%)	1.0	0.8	1.0	2.4	1.4	0.4
Dietery Fiber (%)	1.9	1.5	1.5	1.6	1.4	1.0
Fat (%)	0.1	0.2	0.1	0.2	0.1	0.1
Ash (%)	0.8	0.8	0.9	0.7	1.2	1.1

Source: Eka (1998).

2.1.3 PHYSIOLOGY OF ROOT AND TUBER CROPS

Root and Tuber crops are still living even after they have been harvested and as such they continue to respire by taking in oxygen (O_2) and passing out carbon dioxide (CO_2). The respiration process results in the oxidation of starch, which is contained in cells of the root/tuber into water, carbon dioxide and release of some energy in form of heat. The amount of starch (which is the dry matter in the tuber) is reduced during respiration. The factors affecting the rate of respiration in root and tuber crops include the following:

  i. Physiological age of the root or tuber crop.
  ii. Storage conditions of the root/tuber (mainly temperature).

Generally, when a root or tuber crop is harvested, the rate of respiration is normally high and this is then followed by a decrease (especially during storage) then followed by another increase once spouting begins.

Temperature is the single most important factor affecting the rate of respiration in root and tuber crops. The rate of respiration is almost doubled for every 10°C increase in temperature over the range of 25°C. This is mathematically expressed as $Q_w = 2$.

Postharvest Handling and Storage of Root and Tubers

Another important physiological factor is transpiration or water loss from the surface of the root or tuber crop. Root and Tubers are characterized by high water content even at ambient conditions; and they continue to lose water to the surrounding atmosphere. This loss of water even though may not affect the original food value but will affect the quality of the produce in many ways. It may reduce the market value and culinary property as well as increase peeling loss.

2.1.4 DORMANCY AND SPOUTING IN ROOT AND TUBER CROPS

During physiological development of root and tuber crops, they pass through different stages of growth, harvesting, storage, and subsequent planting as seeds. At all these stages of development, the physiology of the tuber varies and as such the respiration rate also varies.

Root and Tuber crops are normally propagated vegetatively and in their attempt to counter an unfavorable condition at the end of their growth period, they go into dormant phase (with exception of cassava). The beginning of dormancy is considered the point of physiological maturity in root and tuber crops.

Dormancy in root and Tuber crops is defined as the period of reduced endogenous metabolic activity during which the tuber shows no intrinsic or bud growth although it retains the potential for future growth. Dormancy is both a space and a varietal characteristic and it is affected by factors, such as:

- temperature;
- O_2 and CO_2 content of storage environment;
- extent of wounding or otherwise.

Cassava root is a plant of perennation and not propagation; therefore it has no dormancy (it senescence after harvesting).

Root and tuber crops can be satisfactorily stored for a period of time when in dormant phase provided they are not injured. When root crops are harvested, they go into dormancy as such they can be satisfactorily stored but as soon as the dormancy is broken, spouting begins.

Spouting is a period when the tuber dormancy is broken and tiny sprouts begin to appear from the eyes of the tuber crop. During sprouting the dry matter content of the tuber decreases because the formation of the spouts requires energy, which is normally drawn from the tubers' carbohydrate reserve. The rate of water loss from the tuber surface also increases causing the tuber to shrivel and exposed to attack from pathogens.

2.1.5 CURING IN ROOT AND TUBER CROPS

During harvesting, bruising and cutting of root and tubers are likely to occur. These fresh wounds are ideal entry point for disease causing organisms (Dandago, 2009) if not properly healed. The term curing refers to the operation of self-healing of wounds, cuts, abrasions and bruises in root and tuber crops.

According to Kitinoja and Kader (2003) curing root and tuber crops such as sweet potatoes, potatoes, cassava and yams is an important practice if these crops are to be stored for any length of time. Curing can be accomplished by holding the produce at high temperature and high relative humidity for several days while harvesting wounds heals and new protective layers of cells form. The best conditions for curing vary among root and tuber crops as shown in Table 2.2.

Curing can be done in specially heated storage house, which must be well ventilated to prevent accumulation of carbon dioxide. In the tropics, curing of root and tuber crops can be accomplished by piling the crops on the ground (under shade) and covering it with dark sheet of polyethylene for few days depending on the type of crop. The two steps involved in curing process of root and tuber crops are:

TABLE 2.2 Best Conditions for Curing Root and Tuber Crops

Commodity	Temperature °C	°F	Relative Humidity %	Days
Potato	15–20	59–68	90–95	5–10
Sweet potatoes	30–32	86–90	85–90	4–7
Yams	32–40	90–104	90–100	1–4
cassava	30–40	86–104	90–95	2–5

Source: Kitinoja and Kader (2003).

Postharvest Handling and Storage of Root and Tubers

i. **Cell suberisation**: This is a stage where a chemical substance called *suberin* is synthesized and deposited in the cell walls.
ii. **The formation of cork cambium**: The second stage of curing involved the production of cork tissue in the bruised area, which seals the cut or bruised area and prevents the entrance of decay causing organisms and reduce water loss. The factors that affect the wound healing ability during curing in root and tubers are:
   a. temperature of the commodity;
   b. oxygen and carbon dioxide concentration within the commodity;
   c. humidity within the commodity;
   d. use of sprout inhibitors.

Curing of root and tuber crops has several advantages such as increase sweetness and palatability (Wang et al., 1998); facilities synthesis of enzymes and improves flavor during cooking (Kays and Wang, 2000) and facilitates toughening of the skin (Ray and Balagopalan, 1997).

2.1.6 DESCRIPTION OF MAJOR ROOT AND TUBER CROPS

2.1.6.1 Cassava

Cassava (*Manihot esculenta* Crantz) is believed to have originated from eastern Brazil where it is grown as a major staple food. Cassava was introduced to Africa in the year 1600 and into Nigeria about 300 years ago. The Portuguese merchants were said to be responsible for the spread, cultivation and consumption of cassava (Oyebunji, 2004).

Cassava is almost entirely produced and consumed in developing countries and according to Oyebanli (2004) the major producers of cassava in the world are Nigeria, Ghana, Madagascar, Mozambique, Tanzania, Uganda, Zaire, China, India, Indonesia, Philippines, Thailand, Brazil, Columbia and Paraguay. Nigeria is the leading producer of cassava in the world since 1989 and coincidently the largest consumer of cassava as food.

Cassava is highly productive and tolerant crop. It is also a relatively disease free crop compared to other crops and can do fairly well on poor soil.

Cassava provides a major source of energy to over 500 million people. The energy content of cassava in the diets of people in tropical areas of Africa, America and Asia has been established as 37%, 12% and 7%, respectively. Cassava is the chief source of energy in southern Nigeria and DR Congo. It is consumed in many forms as Garri, Fufu, Lafun, etc.

Basically there are two important varieties of cassava based on the hydrogen cyanide content of the cassava root. The hydrogen cyanide content of cassava is not normally stable but fluctuating and several factors such as varietal and environmental differences are responsible for this. In Nigeria bitter varieties of cassava are found in the southern part of the country and can contain up to 250 mg/kg hydrogen cyanide. For the cassava to be safe for human consumption the HCN has to be removed through detoxification process.

While on the other hand, the sweet cassava variety is found generally in the Northern part of Nigeria and it contains around 50 mg/kg of Hydrogen cyanide. Because of the low Cyanide content of sweet cassava, it can be eaten fresh (to some limited extent) but it also undergoes some detoxification process in the course of processing.

2.1.6.2 Yams

Yams (*Dioscorea* Spp) are widespread in the humid tropics throughout the world. The genus *Dioscorea* contains about 600 species spread throughout the world. The edible and economically important species of yams include the following:

a. White yam (*Dioscorea rotundata* poir): This yam species is believed to have originated from Africa and it is the most widely grown and preferred yam specie. It is called white yam because of the flesh color. A number of cultivars of white yam exist.

b. Yellow yam (*D. cayenensis* Lam): This variety of yam also derives its name from the yellowish color of the flesh, which is believed to be due to the presence of carotenoids. Yellow yam is also believed to have originated from West Africa.

c. Water yam (*D. alata* L.): Unlike white yam the water yam is believed to have originated from South East Asia. Water yam is widely spread across the world and seeing only to white yam in Africa in popularity.

d. Bitter yam (*D. dumetorum* Pax): This specie of yam is characterized by bitterness of the flesh. It is believed to originate from West Africa and its cultivation is limited to Western Africa. Some species of Bitter yam are highly poisonous.
e. Chinese yam (*D. esculenta*): Chinese yam is believed to originate from South East Asia. Chinese yams are small and characteristically borne in clusters unlike other yams, which produce one or two large tubers per plant. The flesh of Chinese yam is also white but less fibrous then other yams. It is also sweeter and its starch grain is finer.

According to FAO (1998) the total world production of yam stand at 28.1 million tons in 1993, and 96% of this figure came from West Africa with Nigeria producing 71% of world production Cote d'ivoire 8.1%, Benin 4.3% and Ghana 3.5%. Nigeria, the leading yam producer is also the leading consumer.

2.1.6.3 Sweet Potato

Sweet potato (*Ipomea batatas*) is a root crop which is believed to have originated in central America and was probably introduced to Africa about the end of 19th century (Dandago, 2009).

It is the world's seventh most important food crop after wheat, rice, maize, potato, barley and cassava. The world sweet potato production has been established to be 140,903 x 10^3 m of which 92% is produced in Asia and the pacific Islands (FAO, 2002). China is the worlds' largest producer of sweet potato accounting for 86% of global production. (Ray and Ravi, 2005). The crop is new widely produced as an important staple food in a number of African countries, which includes Burundi, Rwanda, Uganda and Nigeria (Dandago, 2009).

The crop is usually planted in less fertile, marginal soils with limited water supply but despite this fact, sweet potato produces more calories/hectare/day than any other major food crop (Ray and Kari, 2005). In Africa, both the fleshy storage roots and tuber leaves of sweet potato are used as food animal feed and to some extent as raw materials for the production of starch and beverages.

There are many cultivars of sweet potato according to FAO (1998) each with its own characteristic size, shape, color, storage life, nutrition and suitability to processing. A single plant may produce 40–50 tubers weighing from 100 g–1 kg. The chemical composition of sweet potato varies according to genetic and environmental factors.

2.1.6.4 Potato

Potato (*Solanum tuberosom*) which is otherwise called Irish potato originated in tropical highlands of south America from where it was introduced into Europe towards the end 16th Century. Potato is the 14th most important crop (after wheat, rice and maize) in the world in term of production.

It was introduced to Nigeria (Jos–Plateau) by the missionaries. The production of Irish potato in Nigeria in commercial quantity is restricted to areas of high altitude (plateau and Mambilla in Taraba state). There are pockets of areas of production in Kaduna, Nassarawa, Kano, Gombe and Adamawa states. About 75% of the total production in Nigeria comes from plateau state, 15% from Taraba state and 10% from remaining states. Although Irish potato production is restricted to two states, its consumption is spread all over the country (Okunade, 2004).

The chemical composition of potatoes varies and is greatly influenced by environment, varietal type and also the farming practice in the production area. Irish potato is an important source of protein, Iron, riboflavin and ascorbic acid; and the starch content of it is in the range of 65–80% on dry weight basis.

2.1.6.5 Cocoyam

Cocoyams (*Colocasia esculenta* and *Xanthasomas sagattifolium*) are important staple foods in Nigeria particularly Southern and middle belt areas. *Colocasia esculenta* which is otherwise called *Taro* is by far more popular than *Xanthosomas Sagittifolium* called *Tannia* (Onuaha and Alfred, 2011). Cocoyam ranks third in importance after cassava and yam among root and tuber crops in Nigeria (NRCRI, 2012).

About 60% of the world production is grown in Africa while remaining 40% is produced in Asia and pacific Islands through Egypt (Eka, 1998).

Nigeria with an average annual production of 3.7 million metric tons is the leading cocoyam producer in the world (NRCRI, 2012).

The corms and cormels are prepared into food by simply boiling and roasting (Ofoeze, Ezeama and Awa, 2011). Young leaves of taro cocoyam are also eaten in Nigeria as in other West African countries as vegetables and food for ruminants particularly sheep and goats (Adetuyi and Ogundahunsi, 2009). Nutritionally cocoyam is more readily digested (NRCRI, 2012). Cocoyam is a good source of carbohydrate, energy and fair high source of crude and true proteins, rich in Ash, low in fiber and a fair source of lipids.

Cocoyam is high in P, Mg and Zn than any other root and tuber crops. The protein of taro cocoyam is well supplied with essential amino acids but sometimes low in only histidine and lysine (Adetuji and Ogundahunsi, 2009).

2.1.7 POSTHARVEST LOSSES IN ROOT AND TUBER CROPS

Root and Tuber crops incur high postharvest losses due to their perishable nature. They contain fairly high amount of moisture and the skins are delicate especially at harvest; and therefore any cut or bruise can speed up metabolic and physiological process and as such lead to postharvest loss. The cuts or bruise are also ideal entry points for spoilage micro-organisms.

Many researchers are of the opinion that postharvest losses in root and tuber crops can be high as 50% of the total output. Postharvest loss can simply be defined as any loss in quality or quantity of crop that occur between harvest and consumption.

2.1.8 MAJOR REASONS FOR POSTHARVEST LOSS IN ROOT AND TUBER CROPS

Root and tuber crops incur high loss and the reasons for the loss can be pre-or postharvest. Postharvest loss in root and tuber crops can occur due to the following factors:

I. Mechanical damage:
The skin of a mature root and tuber crop is an effective barrier against external factors such as disease causing organisms. When the root or tuber crop is intact, it is normally protected but once the skin is broken/

abraded or bruises micro-organisms gain entry into the root or tuber crop and deterioration sets in. An injury to the root or tuber crop also speeds up physiological process such as respiration within the material and the energy reserve of the tuber would be depleted and deterioration speed up. Moisture loss can also occur at the point of injury and this also is detrimental to the postharvest life of the tuber crop.

Root and tuber crops sustain injury at verities points within the production chain. The following illustrates the points:

- Injury can be sustained at the point of harvesting of root or tuber from harvesting instruments and/or rough handling.
- During transportation of the crop it can sustain injury on the skin due to rough vibrations from the vehicle and also due to rough roads.
- Loading and unloading of root and tuber crops can is also an important point where the crop can sustain injury or even broke.
- Crops can also sustain injury during storage. The injury can be as a result of attack by rodents or insects.

II. Physiological factors:

Physiological factors such as respiration and transportation are important factors in the postharvest life of root and tuber crops. Careless handling and poor management can speed up the physiological processes within the root and tuber and therefore lead to postharvest loss in the crop.

Root and tubers are living organs and therefore they respire by taking in oxygen and passing out water, carbon dioxide and heat energy. The rate of respiration in root and tuber crops is usually high at harvest time, which is followed by decrease during storage and another increase when spouting begins. Normally root and tuber at harvest through storage follows this natural pattern in respiration but once this pattern is disrupted, the rate of respiration include:

- mechanical injury to the root/tuber;
- over heating of storage environment;
- poor ventilation, etc.

Transpiration is another physiological factor that can affect the postharvest life of a root or tuber crop. The term transpiration is used to describe natural evaporation of moisture from the surface of the crop. Excessive transpiration due to exposure to high temperature during harvesting, handling, storage and distribution of root and tuber crops can lead to many farms of deterioration such as shriveling, loss of texture and subsequent loss of market value. All these contribute to deterioration of the tuber and postharvest loss.

Other physiological factors such as sun scorch, greening and sprouting also leads to various farms of deteriorations which all contribute to the postharvest loss in root and tuber crops.

III. Pathological factors:
Root and tuber crops are living organs and as such are subject to invasion by various microorganisms, which include bacteria, fungi and viruses. These organisms cause of direct postharvest loss in tropical root crops.

Attack of the root or tuber crop by the micro-organism can occur at pre or postharvest storage. For example, the microorganisms which are normally found in the soil, water or air can attack the crop through the natural pores when the crop's still underground and this can extend to the postharvest life of the root crop.

Cuts, abrasion and bruises on the root or tuber crop are ideal entry points for microorganisms. Injuries or cuts on the root or tuber can be sustained at various points of harvesting, handling, transportation, and storage. Once the organism gain entry into the crop, they establish themselves and subsequently cause deteriorations by breaking down of lost tissues thereby causing a loss in quantity and quality.

The microorganisms responsible for postharvest losses in root and tubers vary from one crop to the other. Many workers have isolated the different microorganisms responsible for postharvest losses or different root and tuber crops.

2.1.9 HARVESTING, HANDLING, TRANSPORTATION AND MARKETING OF ROOT AND TUBER CROPS

2.1.9.1 Harvesting of Root and Tubers

There are principally two methods of harvesting root and tuber crops. These are hand harvesting and machine harvesting. In developing countries, root and tuber crops are almost entirely harvested by hand. Hand harvesting when properly done is the best because the crops suffer less damages than machine harvesting.

There are different types of farm tools and implements that are used for harvesting root and tuber crops and they range from cutlasses, hoes, sticks and machetes. Root and tubers are likely to suffer some degree of mechanical injury at harvesting because of the nature of the tools used but harvesting is made easier where the crops are grown on heaps, mounds or ridges as is practiced in yam growing areas of Nigeria.

It is important to note that the quality of the root/tuber crop once harvested cannot be improved but only maintained. Therefore careful harvesting, harvesting at peak quality, removal of field heat, proper handling prompt transportation and curing are all important to successful postharvest life of root and tuber crops.

Harvesting should be done during cool hours of the day and produce kept shaded. Crops meant for storage shall be free from cuts, bruises, and abrasions and should be cured immediately (Siddiqui, 2015, 2016). Curing is so important in root and tubers because it helps to head the wounds sustained during harvesting and also assist in toughening of the delicate skin. Curing in root crops generally can be accomplished by under tropical conditions by subjecting the root to temperature of 27–29°C and relative humidity 85–95% for 4–5 days.

2.1.9.2 Handling of Root and Tubers

Handling refers to the way and manner the crop is treated between harvesting and marketing. Proper handling is a panacea to good postharvest life of the crop. Even if the crop is properly harvested but mishandled the storage or postharvest life will be short.

In developing countries like Nigeria most root and tuber crops are harvested badly, subjected to direct sunlight and transported in open and badly managed vehicles. In many instances a number of people sit on top of the produce. These are of course what lead to the short postharvest and storage life of root and tubers.

2.1.9.3 Transportation of Root and Tubers

Transportation of root and tuber crops from the farm to storage area or from tuber storage area to marketing or retail is also very important. This is so because the tubers in the course of transportation can sustain some injuries and as such get infected with a disease-causing organism.

The tubers during transportation can be exposed to temperature extreme, for example, direct sunlight. This in turn can speed up the rate of respiration in the crop and therefore shortens the postharvest life of the root or tuber.

Generally there is no organized means of transportation of agricultural produce from the farm to warehouse or market. All sorts of vehicles/means of transportation are used. For example donkeys, cars, vans, lorries, trucks, motorcycles and even bicycles are normally used. Most of the vehicles also are not in good condition and therefore it is a common sight to see a van or lorry loaded with a full load of an agricultural produce broken down and under harsh environmental condition.

2.1.9.4 Marketing of Root and Tubers

In most developing countries, the market is an open place where transactions are carried out. Root and tuber crops are not an exception, therefore marketing is normally carried out in an open place and exposed to all sorts of environmental factors such as wind, sunlight, rain etc. (Ahmad and Siddiqui, 2015) Marketing is not organized, and therefore their exposure increases the processes within the root or tuber and also exposed it to all sorts of accidents or damage from pests.

2.2 STORAGE METHOD FOR ROOT AND TUBER CROPS

Farmers through experience have learnt that root and tuber crops deteriorate rapidly once harvested and have develop methods to contract this problem using several techniques to ensure that the qualities of the root or tuber are preserved during storage. The basic requirements for the storage of root or tuber crops differ from one crop to the other and also from one region to another depending on several factors. Even for the same crop, storage methods can vary depending on the reason for storage and ultimate use (Ihekeronye and Ngoddy, 1985). Storage methods for root and tuber crops can briefly be classified into:

i. traditional storage methods for root and tubers;
ii. imported/modern storage methods for root and tuber crops.

2.2.1 TRADITIONAL STORAGE METHODS OF ROOT AND TUBER CROPS

The most common traditional method of root and tuber crops in the tropics after they are harvested are pit storage, storing in house, storing on platforms in the open, leaving the crop underground until needed, clamp storage and barns for yams particularly (Ihekeronye and Ngoddy, 1985).

a. Pit storage of root and tuber crops

Pit of various dimensions are usually dug on the ground for storage of root and tuber crops. The major advantage of pit storage method especially for root and tuber crops is that the tuber can be stored for fairly long period of time without serious alteration in the quality of the tuber. While the disadvantage of it is that it is not possible to inspect the crop and an infection on one tuber can migrate to other.

b. Barn storage

The barns are standing platforms elevated above the grand, and usually made of wood and covered on top with grass or thatch. The structure is specifically used for storage of yams. The advantages of yam barn include:

- product can be inspected while on storage;
- there is fairly enough air in circulation;
- spoilt is products can easily be spotted and isolated from the rest.

c. Underground storage of root and tuber crops

This is the method of leaving the root or tuber crop underground until needed and this method is commonly practiced in developing countries where there is general lack of storage facilities.

The problem of storing fresh cassava has led to the traditional practice of root underground until needed; and once harvested they are consumed or processed immediately. The major advantage of this method is that the crop can stay for a long period of time. The disadvantage of it is that large area of land is occupied, and the longer the storage period the more fibrous and woody the cassava is.

d. Clamp storage of root and tuber crops
The storage of root crops in field clamps for a period up to eight weeks is possible without serious deteriorations in the crop. The clamp consist of a layer of straw laid on a dry floor covered with a heap of 300–500 kg of roots followed by a day layer of straw and finally a layer soil. The clamp method is cheap because the materials of construction are readily available within the farmers reach. The disadvantage of the method is that it is difficult to manage in areas where there is a seasonal variation in climate.

e. Platform storage method of root and tuber crops
Root and Tuber crops can also be stored on raised platforms with various designs: A development by an FAO postharvest project in Benin comprised of a wooden shed with an elevated floor fitted with rat guards and covered with a hatch roof.

The Nigerian stored products research institute recommended a structure similar to the traditional yam barn where the tubers are placed on single layer on a shelf instead of being tied to a frame (FAO, 1998).

2.2.2 IMPROVED/MODERN STORAGE METHODS OF ROOT AND TUBER CROPS

More efficient methods of storage are essential if large quantities of root and Tuber crops are to be stored at regional center. The modern methods are somehow more sophisticated but cost effective and therefore cannot be practicable used by peasant farmers. The methods include:

a. Refrigeration method
Refrigerated storage of tuber crops like yams specifically at about 15°C in combination with the use of fungicides has been successful. The wide spread use of refrigeration for root and tuber crops is not yet feasible because of the high capital output, technical support and steady electricity (FAO, 1998).

b. Irradiation method

The used of gamma irradiation to inhibit sprouting has also been successfully used to reduced the losses and prevent sprouting of tubers for a period of up to eight months. The method has the advantage of keeping the tuber intact without spoilage for a long period of time but highly sophisticated and capital intensive. The Nigerian government has an irradiation facility at SHEDA near Abuja, but the facility is yet to be commercialized.

2.2.3 COMMON HANDLING PRACTICES AND CONDITIONS AFFECTING POSTHARVEST LIFE AND QUALITY OF ROOT AND TUBER CROPS

The common handling practices and conditions affecting postharvest life and quality of root and tuber crops can broadly be divided into pre-harvest and postharvest factors. The postharvest factors are further subdivided into different stages of operations. The practices are:

i. Pre-harvest
- Production of high yielding cultivars with short postharvest life or susceptible to pest and diseases.
- Poor field sanitation leading to infections and insect damage.
- Lack of pest management.

ii. Harvest
- Harvesting at improper stage (immature);
- Use of rough and/or unsanitary field containers;
- Harvesting at hot hours of the day and leaving harvested produce under direct sunlight;
- Rough handling, dropping or throwing produce;
- Over packing of field and marketing containers.

iii. Curing
- Lack of curing of root and tubers;
- Improper curing.

iv. Packing house operations
- Lack of proper sorting;
- Rough handling;
- Sitting on top of produce during handling and transportation.

v. **Packing and packaging materials**
   - Use of poorly ventilated packaging;
   - Loading of containers or use of too large containers;
   - Complete absence of packaging materials in some instances.

vi. **Cooling and humidity control**
   - General lack of any method of cooling produce during packing, transport, storage and marketing of root and tuber crops.

vii. **Storage**
   - Lack of storage facilities on farm, at wholesale and retail;
   - Poor sanitation and inadequate temperature management;
   - Over stacking of produce beyond permitted level.

viii. **Transportation**
   - Over loading of transportation vehicles;
   - Lack of adequate ventilation during transportation;
   - Rough handling during loading and unloading;
   - Carrying human passengers on top of produce.

ix. **Marketing**
   - Lack of packaging during marketing;
   - Lack of protection from direct sunlight during marketing;
   - Poor sanitation in marketing environment;
   - Produce is heaped on bare ground during marketing.

(Adapted from Kitinoja (2006). Postharvest CD).

2.2.4 RECOMMENDED STORAGE CONDITIONS OF ROOT AND TUBER CROPS

Root and tubers are perishable and therefore for longer storage life, they need to be stored at conditions optimum for their storage. Tropical root and tubers must be stored at temperatures that will protect the crops from chilling, since chilling injury can cause internal browning, surface pitting and increased susceptibility to decay (Kitinoja and Kader, 2003). The recommended storage conditions for root and tuber crops are listed in Table 2.3.

TABLE 2.3 Recommended Storage Conditions for Root and Tuber

S/N	Produce	Temperature °c	°F	RH(%)	Potential storage duration
1.	Potatoes	4–7	39–45	95–98	10 months
2.	Cassava	5–8	41–46	80–90	2–4 weeks
		0–5	32–41	85–95	6 months
3.	Sweet potatoes	12–14	54–57	85–90	6 months
4.	Yam	13–15	55–59	Near 100	6 months
		27–30	80–86	60–70	3–5 weeks
5.	Taro	13–15	55–59	85–90	4 months

Source: Cauntwell and Kasmire (2002).

KEYWORDS

- Curing operation
- Dormancy and Spouting
- Harvesting indices
- Postharvest harvest handling
- Postharvest quality
- Root and tuber crops

REFERENCES

Adetuyi, F.O., & Ogundahunsi, G.A. (2009). Nutrient composition of cocoyam (*colocacia esculenta*) based food. In: Nkama, I. (Ed.). *Proceedings of 33rd Annual Conference of Nigerian Institute of Food Science and Technology* held in Yola 12–16 October, 2009.

Ahmad, M.S., & Siddiqui, M.W. (2015). *Postharvest Quality Assurance of Fruits: Practical Approaches for Developing Countries*. Springer, New York. pp. 265.

Contwell, M.I., & Kasmire, R.F. (2009). Postharvest handling systems: underground vegetables. In: Kader, A.A. (Ed). *Postharvest Technology of Horticultural Crops*. University of California. ANR Publication 3311. 435–443.

Dandago, M.A. (2009). Effects of various storage methods on the quality and Nutritional composition of sweet potatoes (*Ipomea batatas* L.) Unpublished MTech Postharvest Technology Degree Thesis, Federal University of Technology Yola, Nigeria.

Eka, O.U. (1998). Roots and Tubers. In: Osagie, A.U., & Eka, O.U. (Eds.). *Nutritional Quality of Plant Foods*. Postharvest Research Unit University of Benin Nigeria, pp. 1–31.

FAO (1990). *Roots, Tubers, Plantains and Banana in Human Nutrition*. Food and Agricultural Organization of United Nations, Rome.

FAO (1998). *Storage and Processing of Roots and Tubers in the Tropics*. Available at www.fao.org/docrep/x5415e/x541500.htm 03/02/12.

FAO (2002). *Production Year Book*. Volume 54. Food and Agricultural organization Rome Statistics Section.

Ihekoronye, A.I., & Ngoddy, P.O. (1985). *Integrated Food Science and Technology for the Tropics*. Macmillan Education Ltd., London, pp. 266–270.

Kays, S.J., & Wang, Y. (2001). *Thermally Introduced Flavor Compounds*. HortScience 35, 1002.

Kitinoja, L., & Kader, A.A. (2003). *Small Scale Postharvest Handling Practices: A Manual for Horticultural Crops*. Fourth Edition, p. 257.

NRCRI (2012). *Cocoyam Program*. Retrieved from www.ncrri.gov.ng/pages/cocoyam.htm on 3rd March 2012.

Ofoeze, M.O., Ezeama, C.F., & Awa, E. (2011). Effect of fermentation on the Nutritional and anti-nutritional properties of the flours from two varieties of cocoyam. In: Abu, J.O. (Ed). *Proceedings of Annual Conference of Nigeria Institute of Food Science and Technology* held in Makurdi Nigeria, 10–14th October, 2011.

Okaka, J.C. (2009). *Handling, Storage and Processing of Plant Foods*. Second Edition, OCJ Academic Publishers, Enugu Nigeria, pp. 250–262.

Okunade, S.O. (2004). Indigenous knowledge in Irish potato preservation, processing and utilization. In: Olokesusi, F. (Ed). *Indigenous Knowledge in Root and Tuber Crops*. Postharvest Handling. Seminar Proceedings of Nigerian stored products Research Institute held in Ilorin, Nigeria, 23–25th August, 2004.

Onuoha, O.G., & Alfred, A.H. (2011). Effects of High-pressure – high temperature cooking on the acceptability of Achicha: a cocoyam based product. In: Abu, J.O. (Ed). *Conference Proceedings of Nigerian Institute of Food Science and Technology* held in Makurdi, Nigeria, 10–14th October, 2011.

Oyebanji, A.O. (2004). Indigenous knowledge in cassava processing, utilization and preservation. In: Olokesusi, F. (Ed.). *Indigenous Knowledge in Root and Tuber Crops*. Postharvest products Research institute held in Ilorin, Nigeria 23–25th August 2004.

Ray, R.C., & Balagopalan, C. (1997). *Postharvest Spoilage of Sweet Potato*. Technical Bulletin 23. Central Tuber Crops Research Institute Bulletin, India.

Ray, R.C., & Ravi, V. (2005). Postharvest spoilage of sweet potatoes in Tropics and control measures. *Critical Reviews in Food Science and Nutrition* 45, 623–644.

Sanni, L.O., Adebowale, A.A., Idowu, M.A., Sawi, M.K., Kamara, N.R., Olayiwola, L.O., Egunleti, M., Dipeolu, A., Aiye Laagbe, I.O.O., & Fomba, S. (2009). *West African Foods from Root and Tuber Crops: A Brief Review*. University of Agriculture Abeokuta Nigeria. AAU/MRCI/08/F07/P033 project.

Siddiqui, M.W. (2015). *Postharvest Biology and Technology of Horticultural Crops: Principles and Practices for Quality Maintenance.* CRC Press, Boca Raton, Florida, USA. pp. 550.

Siddiqui, M.W. (2016). *Eco-Friendly Technology for Postharvest Produce Quality.* Academic Press, Elsevier Science, USA. pp. 324.

Wang, Y., Horvat, R.J., White, R.A., & Kays, S.J. (1998). Influence of postharvest curing treatment on the synthesis of the volatile flavor components in sweet potatoes. *Acta Hort. 64,* 207.

CHAPTER 3

POSTHARVEST MANAGEMENT OF COMMERCIAL FLOWERS

SUNIL KUMAR,[1] KALYAN BARMAN,[2] and SWATI SHARMA[3]

[1]*Department of Horticulture, North Eastern Hill University, Tura Campus, West Garo Hills District, Tura – 794002, Meghalaya, India, E-mail: sunu159@yahoo.co.in*

[2]*Department of Horticulture (Fruit and Fruit Technology), Bihar Agricultural University, Sabour, Bhagalpur – 813210, Bihar, India*

[3]*ICAR-National Research Centre on Litchi, Mushahari Farm, Mushahari, Muzaffarpur – 842002, Bihar, India*

CONTENTS

3.1 Introduction 93
3.2 Reasons for Decline in Vase Life of Cut Flowers 94
3.3 Factors Affecting Postharvest Life of Commercial Flowers 95
    3.3.1 Genotype 96
    3.3.2 Pre-Harvest Factors 96
    3.3.3 Temperature 97
    3.3.4 Controlled and Modified Atmospheres 97
    3.3.5 Chilling Injury 98
    3.3.6 Water Relations 98
    3.3.7 Cut Flowers 99
    3.3.8 Desiccation 99

3.3.9 Ethylene and Other Hormones 100
    3.3.9.1 Ethylene 100
    3.3.9.2 Abscisic Acid.. 102
    3.3.9.3 Cytokinins 103
    3.3.9.4 Other Hormones and Regulators..................... 105
3.3.10 Disease 105
3.3.11 Growth and Tropic Responses.................................... 106
3.3.12 Carbohydrate Supply.. 107
3.4 Causes for Decline in Vase Life of Cut Flowers........................ 108
  3.4.1 Cultural Influences.. 108
  3.4.2 Insufficient Water Uptake... 109
  3.4.3 Ethylene..110
  3.4.4 Harvest 112
    3.4.4.1 Stages of Harvest 113
    3.4.4.2 Bud Harvesting 113
    3.4.4.3 Handling 116
  3.4.5 Conditioning..117
  3.4.6 Quality and Grading..118
    3.4.6.1 Quality of Flowers and Ornamental Plants......119
    3.4.6.2 Roses 119
    3.4.6.3 Orchid... 120
    3.4.6.4 Chrysanthemum 120
    3.4.6.5 For Standard Chrysanthemum 120
    3.4.6.6 Gladiolus 120
    3.4.6.7 Carnation 121
  3.4.7 Quality Standards 121
  3.4.8 Pre-Cooling 123
  3.4.9 Pulsing.. 123
    3.4.9.1 Preservatives for Extending Vase Life 125
  3.4.10 Vase-Solution 127
  3.4.11 Carnation 128
  3.4.12 Alstroemeria 128
  3.4.13 Packaging 129

3.4.14	Packing Boxes		129
	3.4.14.1	Wet Packing of Flowers	130
	3.4.14.2	Polyethylene Foil as Protective Cover	131
	3.4.14.3	Special Care for Exotic Flowers	131
	3.4.14.4	Protection Against Geotropic Bending	131
	3.4.14.5	Care for Ethylene Sensitive Flowers	131
3.4.15	Packaging of Different Flowers		132
	3.4.15.1	Rose	132
	3.4.15.2	Chrysanthemum	132
	3.4.15.3	Carnation	133
	3.4.15.4	Gladiolus	133
	3.4.15.5	Anthurium	133
	3.4.15.6	Gerbera	133
	3.4.15.7	Orchid	134
3.4.16	Storage of Cut-Flowers		134
	3.4.16.1	Reason of Storage	134
	3.4.16.2	Advantages of Storage	135
	3.4.16.3	Factors Affecting the Post-Storage Life of Flowers	135
3.4.17	Storage Methods		135
	3.4.17.1	Refrigeration with Wet or Dry Storage	135
	3.4.17.2	Refrigerated Storage	136
	3.4.17.3	Dry Storage	137
	3.4.17.4	Controlled Atmosphere (CA) Storage	141
	3.4.17.5	Modified Atmosphere (MA) Storage	142
	3.4.17.6	Low Pressure Storage (LPS)	142
Keywords			144
References			144

3.1 INTRODUCTION

A fresh flower is a living specimen even though it has been cut from the plant. Its maximum potential vase life is very short. Whether we grow fresh flowers for the local farmers' market and retail florist or have a large operation that sells truckloads to the national wholesale market, growers need to move product from the field to the consumers in a manner that ensures a high quality product. Many impinging forces can interact to reduce vase life of cut-flowers and requires successful postharvest management for preserving the potential life of fresh flowers (Siddiqui, 2015). There are various reasons and factors, which influence the postharvest life of cut flowers.

3.2 REASONS FOR DECLINE IN VASE LIFE OF CUT FLOWERS

- Food depletion;
- Attack by bacteria and fungi;
- Normal maturation and aging;
- Wilting due to water stress and xylem blockage;
- Bruising and crushing;
- Fluctuating temperatures during storage and transit;
- Color change-bluing;
- Accumulation of ethylene;
- Poor water quality;
- Suboptimal cultural practices or conditions.

As a grower, we must be aware of these problems and how to solve them with good postharvest care. Cold storage and proper attention to maintaining optimum cold storage temperatures will slow normal maturation and aging, bacterial and fungal growth and bluing of flowers, besides solving any improper temperature control problems. Consistent use of floral preservatives, careful handling and good sanitation practices will solve food depletion, poor water quality, bruising and crushing, wilting, and bacterial and fungal attack problems. Ethylene accumulation can be reduced by using silver thiosulfate (STS), good sanitation practices and good ventilation. Sub-optimal cultural practices and conditions can only produce substandard flowers and it is difficult to improve the quality of flowers after harvest. Thus, we need proper postharvest management practices for cut flowers that include both harvest and handling. Harvest includes the decisions like when, how and where to cut and the actual act of cutting the flowers. Handling is everything else involved in preparing the flowers for market. Exactly how these steps are done depends on the crop, the market and the operation size. Since flowers are delicate, graceful and highly perishable, it needs some important post-harvest treatments like conditioning or hardening, pulsing or loading and pre-cooling etc. These treatments keep the flowers fresh while under packaging and lengthen the postharvest life of cut flowers. Also, postharvest qualities of cut flowers are evaluated on the basis of following criterion:

- Final size and shape of flower;
- Development of florets in spike and of lateral florets or spike;
- Changes in fresh weight of flowers;
- Turgidity and freshness of flowers at the consumer end;
- Objective measurements of changes in petal color;
- Stability of the stem or pedicel;
- Yellowing or browning of foliage or stem;
- Water uptake on day after harvest;
- Finally, "Vase Life" represents the potential useful longevity of the flowers to the consumer.

Therefore, for proper post-harvest handling and prolonging vase life of cut flowers, we need several steps to make the commercial floriculture venture as a profitable trade.

3.3 FACTORS AFFECTING POSTHARVEST LIFE OF COMMERCIAL FLOWERS

The majority of the commercial flowers have relatively short lives. The delicate petals of flowers are easily damaged and are often highly susceptible to diseases. Even under optimum conditions, their biology leads to early wilting, abscission or both. The post-harvest life of commercial flowers is affected by physical, environmental and biological factors. Choice of plant material and pre-harvest factors plays an important role. After harvest, temperature become dominant and affects plant water relations, growth of disease, response to physical stresses, carbohydrate status and the relationship among endogenous and exogenous growth regulators. The role of these factors and the response of commercial flowers to them have now been established. Some of the research findings have led to the technologies that can greatly improve marketing and postharvest quality of commercial flowers.

3.3.1 GENOTYPE

The postharvest life of cut flowers varies enormously from the ephemeral flowers of the daylily to the extremely long-lived flowers of some orchid genera. Marked variations are also observed within genera and species that provide a great opportunity for breeders to develop flowers, which remain fresh for longer duration. Color, form, productivity and disease resistance continue to be the targets of breeding programs. This can be seen by comparing the postharvest life of different cultivars from the same breeder. It has been observed that Alstroemeria showed variation of more than 100% in lines in respect of time of petal fall and leaf yellowing. Elibox and Umaharan (2008) reported vase lives of anthurium cultivars ranging from 14 to 49 days. A simple model, based on abaxial stomatal density and flower color accurately predicted the relative vase life ranking of different cultivars that provides an excellent tool for future breeding. Variations in other important postharvest characteristics have also been reported for ethylene sensitivity in carnations. Five out of the thirty eight cultivars of carnation tested were insensitive to ethylene indicating the breeding opportunities not only for extending vase life but also eliminating the problem of ethylene-induced senescence and abscission (Woltering and Van Doorn, 1988; Wu et al., 1991; Reid and Wu, 1992) and in roses a difference in vase life of modern rose cultivars from 5 to 19 days were noticed (Evans and Reid, 1988; Macnish et al., 2010c). Mokhtari and Reid (1995) analyzed the difference in vase life between two rose cultivars and noted several morphological and anatomical characteristics that correlated with improved water uptake and longer vase life. Whereas, Clements and Atkins (2001) characterized a single-gene recessive mutant (Abs) of *Lupinus angustifolius* L. 'Danja' in which no organs abscise in response to continuous exposure to high concentrations of ethylene. A long-lived delphinium mutant also showed no ethylene-induced sepal abscission (Tanase et al., 2009). These mutants indicated the opportunity for a genetic approach to prevent flower abscission and petal abscission that is a common postharvest problem in cut flowers.

3.3.2 PRE-HARVEST FACTORS

It seems axiomatic that pre-harvest factors would strongly affect the post-harvest performance of cut flowers. Marissen and Benninga (2001) studied a range of pre- and post-harvest factors using multivariate analysis and regression techniques and demonstrated that mean relative humidity in the greenhouse was the most important variable determining the differences in vase life of roses. The number of branches per square meter at harvest time also influenced the vase life of rose, since it represents alternative sinks. However, In et al. (2009) accurately predicted the vase life of greenhouse grown cut roses using a neural network approach with 29 environmental, morphological and physiological parameters as the input layer.

3.3.3 TEMPERATURE

The marked effects of temperature were first quantified on the vase life of carnation cut flowers in 1973 (Maxie et al., 1973). Respiration of cut flowers has a very high Q_{10} value. The Q_{10} value observed for narcissus is more than 7 between 0 and 10°C (Cevallos and Reid, 2001). The close link between respiration, growth and senescence in these poikilotherms means that a narcissus flower held at 10°C may lose as much vase life in 1 day as does a similar flower held for one week at 0°C. For critical measurement of the effect of temperature on flower respiration, a dynamic system has been adopted in which the effect of a chosen temperature was measured on replicate single flowers (Cevallos and Reid, 2000, 2001; Celikel and Reid, 2002, 2005). The cut flower industry is aware of the importance of cool temperatures in improving long distance marketing of flowers by adoption of forced air pre-cooling and the use of cool rooms.

3.3.4 CONTROLLED AND MODIFIED ATMOSPHERES

The close association between flower respiration during storage and vase life after storage suggests the potential usefulness of controlled (CA) or modified (MA) atmospheres, in which the O_2 content of the storage atmosphere is reduced, sometimes with an increase in the CO_2 content. The beneficial effects are attributed to reduced respiration (resulting from low internal oxygen concentrations) and reduced ethylene sensitivity (attributed largely to elevated CO_2 levels) (Kader et al., 1989; Kader, 2003). Use of an inducible silencing system to block glycolysis synthesis or some other rate limiting process can be important tool for postharvest management of commercial flowers. Joyce and Reid (1985) demonstrated that high levels of CO_2 and low levels of O_2 could be used in reducing diseases or killing quarantine insects in cut flower shipments. Hammer et al. (1990) studied the potential use of high CO_2 atmospheres and found a significant reduction in *Botrytis* incidence both in naturally inoculated and in artificially inoculated flowers.

3.3.5 CHILLING INJURY

It is a general recommendation that most of the commercial flowers should be stored close to the freezing point (0°C) except anthurium, heliconia, poinsettia, African violet etc. These tropical species must be transported and handled at temperatures above 10°C. Also, summer flowers zinnia, celosia, and cosmos perform better when stored at temperatures above 0°C (Dole et al., 2009). Chilling injury causes wilting, necrosis, browning of colored bracts and petals in the cut flowers during storage and transportation. Harvest time can have a significant impact on the severity of chilling symptoms. Presence of intracellular calcium in cytoplasm provides valuable information about chilling sensitive and insensitive species. Woods et al. (1984a, b) studied cytoplasmic streaming and structure in hair cells from flowers and other organs of chilling sensitive and insensitive species to know about the role of intracellular calcium in the response to chilling stress. They observed that the immediate cessation of streaming and loss of cytoplasmic structure resulting from exposure of sensitive cells to chilling temperatures was accompanied by a change in cytosolic calcium and could be evoked by perturbing cytosolic calcium with a calcium ionophore. The dramatic effects of chilling temperatures on structure and cytoplasmic movement were suggested to be due to depolymerization of F-actin, all events that would certainly upset metabolic homeostasis and lead to the accumulation of toxic metabolites into visible damage to chilled tissues. Such research efforts open a new arena for the development of chilling resistant commercial flowers.

3.3.6 WATER RELATIONS

Adequate water relation in cut flowers is an important aspect of their postharvest management. Water balance is determined by the difference between water supply and water loss and optimal postharvest handling includes managing both sides of this relationship. The primary tool in reducing water loss is temperature control. The water content of saturated air rises in an exponential fashion (doubling for every 11°C). Therefore, depending on the humidity, water loss can also rise with temperature in a similar fashion. Sealed bags or perforated polyethylene wraps can maintain higher humidity and thus reduce water loss after harvest. But at higher temperature, proliferation of disease is greatly accentuated in such packages due to condensation of moisture.

3.3.7 CUT FLOWERS

Intuitively, providing adequate water to a cut flower is an easy practice to maintain adequate turgidity into cut flower for their utmost freshness. Since, the vase solution has direct access to the xylem of cut flower, water uptake is frequently impeded by the desiccation due to extended dry handling of the flowers, air emboli that form when the water column in the xylem is broken, and very commonly by microbial occlusion and/or the formation of physiological plugs, tyloses, and gels (Van Doorn and Reid, 1995). Brodersen et al. (2010) observed differences among species and varieties of the same species in the structure of the xylem, size of emboli and cavitation, embolism repair ability and colonization of the stem by microbes.

3.3.8 DESICCATION

The floral tissues, devoid of functional stomata's, relatively lose little water. However, water loss can occur rapidly through the stomata of stems and leaves during postharvest handling. It has been observed that the opening and vase life of flowers in roses (Macnish et al., 2009a) and gypsophila (Rot and Friedman, 2010) is not affected unless desiccation is in excess of 15% of the fresh weight. The apoplastic fluorescent dye

8-hydroxypyrene-1,3,6-trisulfonic acid (HPTS) were used to measure water uptake by florets and whole stems. These dye showed positive effects of anionic detergents (such as Triton X-100) in improving water uptake in dehydrated flowers (Jones et al., 1993).

3.3.9 ETHYLENE AND OTHER HORMONES

It has long been known that plant hormones and plant growth regulators can have dramatic effects on floral longevity. The dramatic effects of pollination on orchid flowers (anthocyanin accumulation, wilting) have long been explained in terms of a response to plant hormones and the interplay among them (Arditti, 1975). Ethylene is certainly foremost among the hormones affecting flower longevity, but other hormones can affect sensitivity to ethylene and a large group of flowers is insensitive to ethylene. The nature of the senescence signal in ethylene-insensitive flowers remains to be established, but there is evidence that ABA and GA may respectively play accelerating and retarding roles.

3.3.9.1 Ethylene

Sleepiness of carnations, premature wilting of petals before the flowers even open was known to be the result of gas leaks in greenhouses long before the active principle was shown to be ethylene (Crocker, 1913) and the effects of ethylene on the senescence of flowers and abscission of flowers and flower parts was well documented in the first half of the 20th century by researchers at the Boyce Thompson institute and others. The role of endogenous ethylene in triggering senescence has been well documented with reports on dynamics of ethylene production, changes in activity of the biosynthetic enzymes (Bufler, 1984, 1986) and up-regulation of the genes encoding these enzymes (Woodson et al., 1992). The key role of ethylene has been corroborated by studies with long-lived carnation cultivars (Wu et al., 1991) and with transgenic or VIGS constructs silencing the biosynthetic pathway (Savin et al., 1995; Bovy et al., 1999; Chen et al., 2004). The discovery that the action of ethylene could be inhibited by Agþ (Beyer, 1976) and the subsequent development of the stable, non-toxic silver thiosulfate complex (Veen and Vande Geijn, 1978) has provided an important commercial tool as silver thiosulfate, still in wide spread use for preventing ethylene-mediated senescence and abscission in cut flowers and potted plants. Other inhibitors of ethylene synthesis (amino ethoxyvinyl glycine, aminooxyacetic acid) and action (2,5-norbornadiene) were also effective to varying degrees, but none is presently being used commercially. Sisler et al. (1984) synthesized diazo-cyclopentadiene (DACP), a cyclic diolefin with an attached reactive diazo group, and found that it was very effective in inhibiting ethylene action when dissociated with UV light after being applied to the tissue. Curiously, the activity only required exposure to fluorescent light, not the expected shorter wavelength UV (Sisler and Blankenship, 1993), and DACP treated with fluorescent light was just as active as DACP itself (Blankenship and Sisler, 1993; Sisler and Lallu, 1994). Examination of the mixture of breakdown products in the irradiated DACP revealed the presence of 1-MCP which these researchers found to be a potent inhibitor of ethylene action (Sisler and Blankenship, 1996). This material has now become a standard treatment for ethylene-sensitive flowers and potted plants (Serek et al., 1994b, 1995a, b) applied either as a gas in an enclosed space, or through the use of sachets or nano-sponges (Seglie et al., 2011) that are placed in boxes prior to transportation. However, the volatile nature of 1-MCP restricts its application to an airtight environment. A non-volatile 1-MCP formulation N, N-dipropyl (1-cyclopropenylmethyl) amine (DPCA) has recently been successfully tested for improvement of postharvest quality of ornamental crops (Seglie et al., 2010). Spray application of this new formulation could provide a major advantage for handling ornamental crops, since they could be treated prior to harvest in the field or greenhouse. The response of ethylene-sensitive ornamentals to treatment with 1-MCP varies widely due to the reason that the inhibitory effects are quickly lost at room temperature and wears off quite quickly.

Cameron and Reid (2001) studied the ethylene-induced petal abscission in Pelargonium and measured the response to ethylene by determining percentage petal abscission from detached flowers after a 2 h ethylene exposure. The half-life of 1-MCP activity was determined to be 2, 3, and 6 days after 1-MCP treatment at 25, 20, and 12°C, respectively, and there was no evidence for a residual effect after 4 or 5 days at the two warmer temperatures. The effects of temperature and perhaps differences among species in the persistence of inhibition may reflect differences in the rate of turnover of the ethylene-binding site. In studies using carnation (Dianthus caryophyllus L. 'White Sim') petals to determine the optimal conditions for commercial treatment, Reid and Selikel (2008) noted some aspects of the inhibition response that were not consistent with the competitive inhibition model of 1-MCP action. They suggested an alternative model in which 1-MCP binds to a site that is exposed during the allosteric changes that accompany the enzymatic activities of the binding site in the absence of ethylene. Using their response to exogenous ethylene, pollination, and 1-MCP, flowers have been broadly classified into two groups ethylene-sensitive and ethylene-insensitive. However, this classification is undoubtedly too simplistic, since some flowers show an intermediate behavior. In daffodil for example, pollinated flowers, or flowers exposed to ethylene senesce rapidly, indicating an ethylene-sensitive senescence pattern (Hunter et al., 2004a). However inhibitors of ethylene action have minimal effect on the senescence of daffodil flowers held in ethylene-free air indicating that natural senescence is initiated by regulators other than ethylene. There is still considerable need for research to identify the role of other hormones in flower senescence.

3.3.9.2 Abscisic Acid

There is substantial published evidence implicating ABA in the regulation of perianth senescence. Several researchers shown a close association between petal senescence and increased petal ABA concentrations (Nowak and Veen, 1982; Hanley and Bramlage, 1989; Onoue et al., 2000) but exogenously applied ABA has also been shown to accelerate the senescence of a number of flowers (Arditti, 1971; Arditti et al., 1971; Mayak and Halevy, 1972; Mayak and Dilley, 1976; Panavas et al., 1998b). Such application results in many of the same physiological, biochemical, and molecular events that occur during normal senescence (Panavas et al., 1998b). In ethylene-sensitive flowers such as carnation and roses, ABA-accelerated senescence appears to be mediated through induction of ethylene synthesis, since it is not seen in flowers that are pretreated with ethylene (Mayak and Dilley, 1976; Ronen and Mayak, 1981; Muller et al., 1999). This is consistent with the pattern of endogenous ABA content in rose petals, where the increase in ABA concentration occurs 2 days after the surge in ethylene production (Mayak and Halevy, 1972). Also, Daylilies are ethylene-insensitive (Lay-Yee et al., 1992), ABA presumably induces senescence independently of ethylene (Panavas et al., 1998b). The fact that ABA accumulates in daylily tepals before any increase in activities of hydrolytic enzymes and even before the flowers has opened was considered evidence that the hormone may coordinate early events in the transduction of the senescence signal (Panavas et al., 1998b). Application of ABA to pre-senescent daylily tepals resulted in a loss of differential membrane permeability, an increase in lipid peroxidation, increase in the activities of proteases and nucleases, and the accumulation of senescence-associated mRNAs (Panavas et al., 1998b). However, during senescence of daffodil flowers, although ABA accumulated in the tepals as they senesced, it did not appear to play a signaling role in natural senescence (Hunter et al., 2002). The increase in ABA concentrations in the tepals occurred after the induction of senescence-associated genes. They concluded that the increase in ABA content is therefore most likely a consequence of the cellular stresses that occur during senescence and suggested that the hormone does not trigger senescence but may help drive the process to completion.

3.3.9.3 Cytokinins

The striking effects of cytokinins in delaying senescence of leaves were known (from the effects of benzyl adenine) long before the first isolation of zeatin. Given the homology between leaves and petals, it is perhaps not surprising that cytokinins were also found to delay petal senescence (Mayak and Kofranek, 1976; Eisinger, 1977), an effect that was shown to be associated both with reducing the sensitivity of the corolla to ethylene (Mayak and Kofranek, 1976) and with delaying the onset of ethylene biosynthesis (Mor et al., 1984). Endogenous cytokinins content shows a pattern consistent with its putative role in delaying senescence buds and young flowers contain high cytokinins levels, which fall as the flower ages and commences senescence (Mayak and Halevy, 1970; Van Staden and Dimalla, 1980; Van Staden et al., 1990). The interplay between cytokinins content and senescence in ethylene-sensitive flowers was elegantly demonstrated by Chang et al. (2003), who transformed petunia with a SAG12-IPT construct designed to increase cytokinins synthesis at the onset of senescence in leaves (Gan and Amasino, 1995). Cytokinins content of corollas in the transformed plants increased after pollination, ethylene synthesis was delayed and flower senescence was delayed 6–10 days. As in flowers treated with exogenous cytokinins, the flowers from the IPT-transformed plants were less sensitive to exogenous ethylene and required longer treatment times to induce endogenous ethylene production and the symptoms of flower senescence.

Leaf senescence is also an important component of loss of quality in floricultural crops, particularly members of the Liliaceae. Commercial pretreatments containing cytokinins and/or gibberellins are recommended as a prophylaxis in sensitive genera such as Alstroemeria and Lilium. The non-metabolized cytokinins like thidiazuron (TDZ) have proven very useful as an amendment in tissue culture and transformation/regeneration media. Ferrante et al. (2001) reasoned that TDZ might be a useful tool for preventing leaf yellowing in cut flowers. Pulse treatment of cut Alstroemeria stems with as little as 5 mM TDZ essentially prevented leaf yellowing in flowers of the cultivar 'Diamond,' where yellowing normally starts after 4–5 days (Ferrante et al., 2001). The flowers of Alstroemeria are ethylene-insensitive; the TDZ treatment had only a minor effect on Alstroemeria flower life, although cytokinins have been shown to increase the life of iris, whose natural senescence is ethylene-independent (Wang and Baker, 1979; Mutui et al., 2003). In Iris, TDZ treatment at considerably higher concentrations (200–500 mM) significantly improved flower opening (including the opening of axillary flowers, if present) and flower life (Macnish et al., 2010b). The treatment was of particular value in that it reduced the loss of vase life that results from cool storage. While control iris that were held in cool storage for two weeks had only a very short display life, those pretreated with TDZ had the same vase life as freshly harvested controls. Most experiments with TDZ have been conducted with flowers that are insensitive to ethylene, but in lupins and phlox, TDZ has been shown to improve flower opening and reduce ethylene-mediated flower abscission and senescence (Sankhla et al., 2003, 2005) indicating that TDZ acts like other cytokinins in decreasing ethylene sensitivity and that this regulator should be tested on a broader range of ornamentals.

TDZ has also proved to have remarkable effects in improving the postharvest life of potted flowering plants. Leaf yellowing is a common postharvest problem with potted flowering crops and low concentrations of TDZ are very effective in preventing this symptom in a wide range of crops. Researchers showed that the TDZ treatment appears to maintain the photosynthetic ability of the plants, since fresh and dry weights of TDZ-treated plants are much higher than those of the controls. After 2 months, potted cyclamen plants treated with 5 mM TDZ maintained full display value, while control plants had almost ceased flowering and were showing obvious etiolation in response to the low light of the display environment.

3.3.9.4 Other Hormones and Regulators

Gibberellins, auxins and other plant hormones and regulators have positive and negative effects on floral longevity. Auxin is considered an important component for the rapid senescence response in orchids and other flowers to pollination (Arditti, 1975). Several workers observed that plant growth regulators shows positive response in longevity of the cut flowers. Saks and Staden (1993) showed an increase in longevity of carnation flowers treated with 0.1 mM gibberellic acid (GA). Commercially, GA (sometimes in combination with BA) is used in solutions to prevent leaf yellowing in cut bulb flowers and potted flowering plants. GA treatments may have the undesirable side effect of increased stem or scape length. But, floral longevity are largely absent by the use of jasmonic acid, brassinosteroids, and salicylic acid on plant growth, development and responses to biotic and abiotic stress (Ashraf et al., 2010). Similar to auxins, brassinosteroids stimulate ethylene biosynthesis and their effects on ethylene-sensitive flowers would be expected to be negative. Jasmonic acid reduced life of petunias and Dendrobium through stimulation of ethylene production (Porat et al., 1993). However, the salicylic acid signaling pathway has shown response for up-regulation of genes in leaf senescence (Morris et al., 2000), but the effects of down-regulating this pathway on flower senescence have not been studied.

3.3.10 DISEASE

Improper temperature management in the absence of proper pre-cooling techniques, results in condensation and accelerated growth of pathogens on delicate petals and other floral parts, particularly when the flowers are packed under conditions that limit air movement. B. cinerea, a relatively weak pathogen, is the major pathogen of cut flowers and a range of chemicals have been used for postharvest protection. The push for organic or sustainable production and the loss of established chemicals has led to an effort to identify alternative strategies for controlling disease. High CO2 levels provide effective control for species whose leaves (or petals) are not damaged by the gas. Also, SO2 has shown good control of this pathogen, but damages the host (Hammer et al., 1990). Recently, Macnish et al. (2010d) reported the efficacy of a simple dip in a solution of NaHClO4, which has excellent performance as well as considered for the commercial fungicides. Methyl jasmonate (MJ), a natural plant growth regulator has been tested for postharvest control of *B. cinerea* in cut flowers of a range of rose cultivars (Meir et al., 1998). Pulse applications of 200–400 mM MJ following either natural or artificial infection seemed to provide systemic protection. MJ applications significantly reduced lesion size and appearance of the infection apparently due to inhibition of *B. cinerea* spore germination and germ-tube elongation. Effective concentrations of MJ caused no loss of flower quality or longevity.

3.3.11 GROWTH AND TROPIC RESPONSES

Elongation of many cut flowers occurs in response to environmental cues, particularly gravity and there has been considerable research effort devoted to understanding the mechanisms for these responses and to devise strategies to prevent them. Researchers agreed that the primary driver for gravitropic responses is the redistribution of auxin in response to its polar transport, and differential growth in response to that redistribution (McClure and Guilfoyle, 1989; Vanneste and Friml, 2009). Vanneste and Friml (2009) also observed the rate-limiting step in changed auxin distribution is the activity of the auxin efflux carriers (called PIN, based on a mutant phenotype of the gene). Some research has suggested a role for ethylene and/or calcium in the response (Philosoph-Hadas et al., 1996; Friedman et al., 1998). It has been observed that the gravitropic response of *Antirrhinum majus* could be avoided by a pretreatment with silver thiosulfate. The importance of auxin redistribution in the gravitropic response is well demonstrated by the impressive effects of pretreatment with naphthyl phthalamic acid, an auxin transport inhibitor (Teasetal, 1959), but has not been developed as a commercial pre-treatment for flowers such as antirrhinum, gladiolus that have pronounced gravitropic responses. Unwanted stem elongation can be a problem even for flowers that are held vertical to prevent gravitropic responses. In some (ethylene insensitive) flowers, this problem can be overcome by the treatment with ethylene or ethephon. Van Doorn et al. (2011) noticed that the negative effects of ethylene could be overcome by simultaneous treatment with GA (to overcome inhibition of opening and stimulation of leaf yellowing) in tulip, BA (to prevent ethylene-stimulated tepal abscission), and Caþþ (to prevent BA-induced stem browning). This has become a standard treatment for flowers transported from the Netherlands to the United States in refrigerated marine containers. Clearly a genetic approach that would select for tulips with minimal scape elongation after floral maturation would be a preferable long-term strategy.

3.3.12 CARBOHYDRATE SUPPLY

The high respiration of flowers and the energy required for flower growth, bud opening and floral display requires substantial energy reserves in harvested cut flowers. The primary component in floral preservatives is a simple sugar like fructose, glucose or sucrose that reflects the profound effects of added carbohydrates on flower development, opening and display life. Responses to sugar in the vase solution include improved floral opening (Doi and Reid, 1995), improved pigmentation and size of the opening flowers (Choetal, 2001), improved water relations (Acock and Nichols, 1979) and even reduced sensitivity to ethylene (Nichols, 1973). In gladiolus, senescing flowers appear to supply carbohydrate to those still developing. Therefore, removal of senescing florets on gladiolus spikes, significantly reduced opening and size of florets (Serek et al., 1994a). The most striking effects of carbohydrate stress in harvested cut flowers is the blackening of leaves of cut flower proteas (Reid et al., 1989). These bird-pollinated flowers produce copious nectar; in the postharvest environment there is insufficient photosynthate to meet the demands of the flower, resulting in necrotic death of the leaves. Girdling the stem just below the flower (Newman et al., 1989), holding the flowers in high light conditions (Bieleski et al., 1992) or providing supplementary carbohydrate (Newman et al., 1989) prevents the blackening symptoms. This study highlighted the importance of the leaves in supplying carbohydrate to the flower. It appears that sugar in the flower preservative is transported in the xylem to the leaves, where it enters the symplast and is transported to the flowers via the phloem (Halevy and Mayak, 1979). The potential benefits of trehalose has also been reported to mitigate the damaging effects of ionizing radiation and to extend the postharvest life of gladiolus flowers (Otsubo and Iwaya-Inoue, 2000). Trehalose delayed the symptoms of senescence by protective effect on membranes and associated programmed cell death events including nuclear fragmentation. One of the remarkable technologies that have been successful in improving the opening and vase life of cut flowers is the provision of additional carbohydrate in high concentration or "pulse" pretreatments (Halevy and Mayak, 1979). Pulsing has shown successful treatment in gladiolus (Mayak et al., 1973), the opening of Strelitzia flower (Halevy et al., 1978), Eustoma (Halevy and Kofranek, 1984; Cho et al., 2001) and tuberose (Naidu and Reid, 1989; Waithaka et al., 2001). In lisianthus, the pretreatment greatly improves the color of the newly opened blooms and in tuberose, it ensures satisfactory bud opening which normally is inhibited by even brief periods of cool storage.

3.4 CAUSES FOR DECLINE IN VASE LIFE OF CUT FLOWERS

3.4.1 CULTURAL INFLUENCES

Basically, those forces which improve crop quality before and after harvest usually improve vase life. Light intensity is very important. A crop grown under low light, such that light is a limiting factor for photosynthesis, will be low in carbohydrate content. Respiration continues after the flower is harvested, but little photosynthesis occurs, because light is limited in the packing house, florist shop, and consumer's home. When carbohydrates are low, respiration is very low and flower senescence (deterioration) occurs. Optimum light intensity during growth of the crop is very important to vase life. Temperature also influences photosynthesis and respiration, which in turn influence carbohydrate accumulation. During hot periods of the year, crops sensitive to high temperatures, have shorter vase life because flowers contain low carbohydrate levels. When the temperature is raised to an adversely high level to force earlier flowering, the same problem occurs. Nutrition of the crop likewise has an effect on flower longevity. Shortages or toxicities of nutrients that retard photosynthesis will reduce vase life. Deficiency in a number of nutrients, including nitrogen, calcium, magnesium, iron, and manganese, result in a reduction in the chlorophyll content, which in turn reduces photosynthesis. The net result is a low carbohydrate supply for the flower. High levels of nitrogen at flowering time can have an adverse effect on keeping quality. Diseases and insects reduce the vigor of the plant, directly reducing vase life. Diseases also reduce vase life indirectly: injured tissue releases large quantities of ethylene gas, which hastens senescence or deterioration of the flower.

In fact, fresh flowers deteriorate for one or more reasons. Five of the most common reasons for early senescence are:

1. Inability of stems to absorb water due to blockage;
2. Excessive water loss from the cut flower;
3. A short supply of carbohydrate to support respiration;
4. Diseases;
5. Ethylene gas.

3.4.2 INSUFFICIENT WATER UPTAKE

Inability to absorb water is a very common reason for premature wilting. The water-conducting tubes in the stem (xylem) become plugged. Bacteria, yeast, and/or fungi living in the water or on the flower or foliage proliferate in the containers holding the flowers. These microorganisms and their chemical products plug the stem ends, restricting water absorption. They continue to multiply inside and eventually block the xylem tubes. Chemical blockage also can occur. Chemicals present in some stems, upon cutting, change into a gum like material, which blocks the end of the stem.

Excessive water loss from flowers can lead to wilting and reduction in quality and vase life. After harvest, flowers should be removed from the field or greenhouse and refrigerated as soon as possible. Leaving the flowers out of water, in warm air or in warm drifts such as from a heater, causes considerable damage. Flowers should be in water and under cool temperatures as much as possible from the time they are cut until they reach the final customer.

Low carbohydrates are another reason for flower deterioration. A low carbohydrate supply can occur as a result of improper storage temperature and handling. Respiration continues to be governed by temperature after harvest. Low temperatures reduce respiration and conserve carbohydrates, thereby prolonging quality and vase life. Each of the many stages in the marketing channel must be watched. Flowers should be placed in cold storage as soon after harvesting as possible. They should be refrigerated during surface transport and during holding periods at the wholesaler and retailer. Serious damage occurs when flowers are left on a heated loading dock at the motor or air-freight terminal or when they are left sitting in a hot warehouse for a day or so.

The harmful effect of ethylene with fruits, especially apples, gives off large quantities of ethylene gas, making it inadvisable to store containing flowers in coolers. Ethylene is evolved from plant tissue, particularly injured and old plant tissue. The cooler should be kept clean of plant debris such as cut stems and leaves that might accumulate on the floor. Old unsalable flowers should be discarded. Ethylene gas has many deleterious effects. Generally it causes premature deterioration of flowers. Ethylene can cause flower wilting and is generally not reversible.

3.4.3 ETHYLENE

Ethylene, called the ripening, senescence and wound hormone, is a naturally occurring plant hormone. It is important in the reproductive cycle of plants. It triggers the ripening and senescence of flowers and fruit and is also produced when plants are wounded. Many decay and disease organisms also produce ethylene. Ethylene damages some cut flower species by causing flowers to drop prematurely, flower buds to not open, and flower petals to close. There are three strategies to prevent ethylene damage:

1. Keep ethylene from flowers by preventing ethylene pollution;
2. Remove ethylene from the atmosphere;
3. Inhibit the effect of ethylene on flowers.

Some specific measures to prevent ethylene damage on flowers are:

1. Make sure CO_2 generators in greenhouses and oil or gas heaters in greenhouses and handling areas are working properly and well vented.
2. Protect plants against pest and diseases.
3. Prevent pollination of flowers.
4. Harvest flowers at optimum stage.
5. Avoid physical injury to flowers during handling.
6. Cool flowers as soon as possible after harvest.
7. Keep storage and handling facilities clean, and remove diseased and dying plant material.
8. Do not use internal combustion engines in any handling work or production areas.
9. Have good air circulation and ventilation in handling and storage areas.
10. No smoking in handling and storage areas.
11. Do not store flowers with ethylene producing fruits and vegetables.
12. Do not store newly harvested flowers in bud stages with fully open flowers.
13. Use ethylene scrubbers in cold storage area.
14. Use STS treatment on sensitive species.
15. Use other chemical treatments in floral preservatives.

There are various chemicals that can inhibit the effect of ethylene. The most common is the metal ion silver. It usually is applied to flowers in the form silver thiosulfate, STS. It acts on both ethylene receptors and production sites in the flower. This protects the flowers from ethylene in the environment and it stops the flower from producing ethylene itself. Other chemicals are 1-MCP (1-methylcyclopropene) a gas which acts only on receptors, but is not available commercially, and EVB (Pokon and

Chrystal) and Vita Flora which act on the flower's ethylene production sites. To treat or pulse flowers with STS, stems are placed in STS solution for 20 min at 18°C. Floralife sells a two-part solution that makes STS.

Sanitation is of utmost importance in handling fresh cut flowers. The handling area and cold storage should be cleaned and sanitized after each use. Equipment, cutting utensils, containers and handling surfaces should be cleaned and disinfected with a 1:10 bleach solution. Unmarketable flowers should be disposed of after each harvest. Dirty harvest and holding containers and cutting utensils spread disease. Dying plant material is a reservoir for plant disease organisms and produces ethylene. All shorten the vase life of flowers.

3.4.4 HARVEST

The most important factors for harvest are when, how and where, "when" the plant material will reach the optimum stage of development and "when" during the day to harvest. Each plant material has its own best harvest stage and this can vary depending on the use of, and market for, the plant material. Materials for preserving usually are harvested more mature than those for fresh, wholesale markets.

The other "when" is, when the best time of day for harvesting flowers is? The best time is the coolest part of the day and when there is no surface water from dew or rain on the plants. Also, harvesters need enough light to see what they are harvesting. This usually is in the cool of the morning after the dew has dried. Late afternoon or evening also has possibilities because the plants have stored carbohydrates from the day, which will provide a food reserve for the plant material. "How" and "where" go together? Besides knowing at what stage of development to harvest, where and how to cut the flower on the plant also is important. This is most important on plants that produce multiple flowers/crops per season. The growers want to harvest the longest stem possible without sacrificing future production. For this, leave at least two- to five-nodes (growing points) below cut to ensure new growth. Very vigorous plants can be cut back to fewer nodes, while less vigorous plants should have more nodes left. Most stems should be at least 15 to 18 inches long. Longer lengths usually are better. It does not matter if the cut is slanted or squared, but it does matter that one should use sharp, clean cutting knife/secateur. Sharp cutting knife/secateur will not crush the xylem and block the flow of water up the stem. Clean knife/secateur will not introduce harmful microbes to the cut stems. Some shears are designed to hold the flower after it is cut. Inexperienced harvesters may find shears less dangerous than knives. Cutting knife/secateur should be cleaned daily with disinfectant, for example, 1:10 solution of chlorine bleach in water. Flowers with sticky sap require special treatment immediately after harvest. To prevent the flow of the sticky sap, which can block the xylem, dip the cut ends in boiling water for 10 sec or sear with a flame, immediately after harvest. For example, poppies, mignonette and poinsettia.

3.4.4.1 Stages of Harvest

For better vase life and quality, the flowers should be harvested at optimum developmental stage. Flowers harvested at immature stage do not develop properly in water or vase solution after harvest. If, however, harvested late, flowers last only for a short duration. Although, there is no general rule to decide the optimum stage of harvest of flowers, the stage for cutting flowers for direct sale varies with the types of flowers and their varieties. The flowers of gladiolus, lilium, hippeastrum, iris and tulip should be harvested when their flower buds show color, while the flowers of cymbidium, cattleya, dendrobium, phalaenopsis, gaillardia, helianthus, cyclamen, dahlia, calendula and rudbeckia when they are fully opened.

3.4.4.2 Bud Harvesting

Bud harvesting is a procedure that is used infrequently but is fairly well proven and has a tremendous potential. Carnations and chrysanthemums can be harvested and shipped in the bud stage, which cuts down greatly on their volume and hence lowers the cost of shipping. The wholesaler may then store the buds or open them immediately for resale. Once open, the

TABLE 3.1 Optimal Stage of Development for Harvest of Fresh Cut Flowers

Common Name	Species	Stage of Development
African Marigold	*Tagetes erecta*	Fully open flowers
Annual Gaillardia	*Gaillardia pulchella*	Fully open flowers
Astilbe	*Astilbe hybrids*	One-half florets open
Bachelor's Button	*Centaurea spp.*	Flowers beginning to open
Pot Marigold	*Calendula officinalis*	Fully open flowers
Canterbury Bells	*Campanula spp.*	One-half florets open
China Aster	*Callistephus chinensis*	Fully open flowers
Clarkia	*Clarkia unquiculata*	One-half florets open
Glory Lily	*Gloriosa superba*	Almost fully open flowers
Cockscomb	*Celosia argentea var. cristata*	One-half florets open
Stock	*Matthiola incana*	One-half florets open
Tulip	*Tulipa cvs.*	Half-colored buds
Foxglove	*Digitalis purpurea*	One-half florets open
Sunflower	*Helianthus annuus*	Fully open flowers
Coreopsis	*Coreopsis grandiflora*	Fully open flowers
Daffodils	*Narcissus cvs.*	Goose neck stage
Dahlia	*Dahlia cvs.*	Fully open flowers
Daylily	*Hemerocallis cvs.*	Half-open flowers
Delphinium	*Delphinium spp.*	One-half florets open
Dutch Iris	*Iris x hollandica*	Colored buds
English Daisy	*Bellis perennis*	Fully open flowers
Freesia	*Freesia hybrids*	First bud beginning to open
Gladiolus	*Gladiolus cultivars*	1 to 5 buds showing color
Golden rod	*Solidago spp.*	One-half florets open
Amaranth	*Amaranthus tricolor*	One-half florets open
Larkspur	*Consolida ambigua*	2 to 5 florets open
Lily-of-the-Valley	*Convallaria majalis*	One-half florets open
Lisianthus	*Eustoma grandiflorum*	5 to 6 open flowers
Lupine	*Lupinus cvs. Russell*	One-half florets open
Nasturtium	*Tropaeolum majus*	Fully open flowers
Pansy	*Viola x wittrockiana*	Almost open flowers
Peony	*Peonia cvs.*	Colored buds
Perennial Gaillardia	*Gaillardia x grandiflora*	Fully open flowers

TABLE 3.1 Continued

Common Name	Species	Stage of Development
Baby's Breath	*Gypsophila spp.*	Flowers open but not overly mature
Snapdragon	*Antirrhinum majus*	One-third florets open
Statice or Sea-lavendar	*Limonium spp.*	Almost fully open flowers
Garden Phlox	*Phlox paniculata*	One-half florets open
Sweet Pea	*Lathyrus odoratus*	One-half florets open
Sweet William	*Dianthus barbatus*	One-half florets open
Florists Violet	*Viola odorata*	Almost open flowers
Lilium	*Lilium spp.*	Colored buds
Torch-Lily	*Kniphofia uvaria*	Almost all florets are showing color
Tuberose	*Polianthes tuberosa*	Majority of florets open
Yarrow	*Achillea filipendulina*	Fully open flowers
Zinnia	*Zinnia elegans*	Fully open flowers

flower has at the least the same vase life potential as a flower cut mature (Table 3.1).

Bud harvesting enables a grower to produce more crops per year in the greenhouse space. There is a significant increase in net return to the grower. There are other advantages to this system. Buds are more immune to handling injuries and ethylene toxicity, making a higher-quality final product possible. As in the case of mature harvested flowers, buds will dry store very well, enabling one to build up the inventory for higher-priced market dates. Bud harvesting is not a new concept for all crops, since roses, gladiolus, iris, tulips, peony, etc., have always been cut in the bud stage.

When needed, buds are removed from the storage box, one-half inch of stem is cut off, and they are placed in a floral preservative solution. The buckets of buds are held in an opening room at 21–24°C until the buds are fully open. A low light intensity is provided in the opening room. The open flowers may be held under refrigeration in the preservative solution or they may be sold directly. The quality and longevity of these flowers has been reported to be superior to those harvested at maturity.

Bud harvesting is becoming important. Growers who ship flowers great distances recognize its value and find it necessary to use this system. Greater cooperation among growers, wholesalers, and retailers will foster it even more.

Advantages of harvesting cut flowers in bud stage:

1. Reduction in sensitivity of flowers to drastic condition and C_2H_4 during handling and transit.
2. Saving space during shipment and storage.
3. Extending the vase life of flowers.
4. Reducing the time, the crop remain in the greenhouse enabling a 'once over' harvesting of a crop.
5. Improving the size, opening, color and longevity of cut flower especially those grown under poor light or high temperature condition.
6. Minimizing the hazard of damage to field grown flower by adverse external condition.

3.4.4.3 Handling

Once harvested, there are a series of steps or tasks done to prepare the flowers for market. These are collectively called handling. These handling steps include:

1. Grading
2. Leaf Removal
3. Bunching
4. Recutting
5. Hydration
6. Special Treatments
7. Packing
8. Precooling
9. Cold Storage
10. Delivery to Market

Not all of these are done to all flowers and whether they are used or not depends on the market the flowers are going to be sold to. Where and how the steps are done depends on the market and the facilities of the operation. Flowers can have all the handling steps performed in the field, only some done in the field with the rest in the packaging shed or have all handling steps done in the packaging shed. Field handling usually is limited to leaf removal, grading, bunching, hydrating, and packing with immediate transport to market or cold storage for brief holding. Flowers for local retail markets often are packed this way since they are marketed immediately after harvest. Flowers also can have these steps performed in the field and then be transported to a packing shed where re-cutting, special treatments, pre-cooling and dry packing can be performed. All the handling steps can be done in a packaging shed. It often makes for a better flow of activities if they are all done in the same place. Some of the steps can only be feasibly done in the packaging shed, such as special treatments, pre-cooling, cold storage and re-cutting. These extra steps usually are done for flowers going to wholesale markets. The packaging shed may be an ultra modern air-conditioned building or an open air covered porch. The handling space should:

- Be shaded or covered to keep temperatures lower and prevent direct sunlight on the flowers.
- Be well lit so you can see well when grading the flowers.
- Have a clean water source for preparing harvest, treatment and holding solutions, and for use in cleaning the area.
- Have ample space so all handling activities can be performed smoothly, such that workers are not crossing over each other.
- Have a cold storage or at least a cooler, shaded place to store the flowers until they are ready for market.
- Have a place to prepare for harvest activities.

Although, the first step after cutting the stem, whether to be handle them in the field or in the packaging shed, should be to place in water or a harvest solution. This solution may be acidified (pH 3.5) either with citric acid or a floral preservative. The harvest containers should be clean and disinfected after each use. Flowers should never be laid on the bare ground. After the harvest container is full of flowers, place them in a cool place until they can be handled or taken to market. The cool place can be a shady area in the field or a refrigerated cold storage. Do not over fill the containers. This will bruise flowers and cause some to tangle with each other. Leaves should be stripped off from the stem. If the flowers are being field handled this can be done before they are placed in the harvest containers or before they are bunched into marketable bouquets. Usually, leaves are stripped off from the bottom one- third of the stem or at least the ones that would be in any holding solution.

3.4.5 CONDITIONING

While handling in the field or in grading room, flowers suffer from water stress. Conditioning is normally done by saturating the cut flowers with water, initially with warm water at room temperature and then overnight in the cool room. When flowers are wilted, turgidity can be restored by immersing the entire flower in water for 1 h before conditioning. Hydration is considerably promoted when water is de-aerated or acidified or when wetting agent is added.

Conditioning of the flowers is required to rehydrate and to overcome slight wilting. The purpose is to load the flower with water to ensure maximum turgidity at the time of sale and utilization. Steps for proper conditioning:

1. Remove about 5–8 cm or 2–3 inches or more of the stem ends if the stems have been out of water for a long time.
2. Soak the cut stem ends in warm water (40–43°C, pH 3.5), preferably in a cool room, until flowers are fully rehydrated.

3.4.6 QUALITY AND GRADING

Grading starts with deciding which flowers to harvest. Only marketable flowers should be harvested. Marketable flowers are free of blemishes, including both leaves and petals. The flowers can be grouped or graded by stem length if there are differences and also by developmental stage. More mature ones should be sold as soon as possible, while others can be held in cold storage for later sales. How the flowers are bunched and packaged depends on the market, where it is being used. In a local retail market, lot of flexibility has been seen, but customers has to decide what sells are the best. Mixed bunches and single type bunches are both popular. Larger flowers such as lilies, gladiolus and sunflowers often are sold as single stems. Sleeving or wrapping the bunch helps prevent the different bunches and flowers from becoming tangled. Columbine, larkspur, delphinium, baby primrose, forget-me-nots and buddleia are flowers that should be wrapped or sleeved prior to marketing to prevent tangling. Wholesale markets have a set of guidelines for the methods of bunching and packaging flowers. Most are bunched by 10's or 5's. Some, like roses and carnations, are bunched by 25's. Lilies-of-the-Valley are bunched in 25's and Sweet Violets are bunched in 100's with a collar of leaves underneath the flowers. Large, expensive to grow flowers can be sold by single stems. Most are boxed and shipped dry. Proper pre-shipping handling is important in order to get flowers to the market in good shape. The flowers should be well hydrated but not wet when packed. The flowers like snapdragons and gladiolus need to be packed upright to prevent the tips from curving. Special boxes or hampers are made for these types of flowers. Once bunched, flowers should be hydrated, placed in water for a while before they are packed dry. The hydrating step should include a step where, after the flowers are bunched, the stems are recut under water to eliminate any air bubbles in the xylem that can block the uptake of water. These air bubbles can occur when the flowers were harvested. Once recut, the flower can be placed in a general holding solution used to hydrate the flowers or receive a special treatment such as silver thiosulfate. Flowers usually are not packed dry into boxes in the field but are in the packing shed for distant wholesale markets. When flowers are packed into boxes, the bunch are sleeved or wrapped and then packed tightly so the bunches do not move or vibrate in transit (causes bruising). The standard flower box is 12 × 12 × 48 inches. There are smaller sizes, too, called half or quarter boxes that are 6 × 12 × 48 inches and 6 × 6 × 48 inches, respectively.

1. Sort the flowers according to the following: cultivar, stage of maturity, extent of damage due to pests and diseases, malformed floral parts and color defects.
2. Grade according to stem length or size.
3. Bunch flowers according to number, cost, susceptibility to injury, and display quality of individual flower heads.
4. Tie bunches below the flower head, and about two inches from the cut stem ends. Tying should not be too loose or too tight. Rubber bands are best, because they can hold the bunch securely. They are easier to use and cheaper than tape or wire.

3.4.6.1 Quality of Flowers and Ornamental Plants

a. **Subjective evaluation of flower quality (Qualitative):** It includes color, fragrance, cleanliness and form.
b. **Objective evaluation of flower quality (Quantitative):** It includes flower diameter, leaf size, stem length, bulb circumference, plant height and bunch weight.

Normally, flowers should be free from brushing, injury dirt or foreign materials, nutritional, chemical or mechanical abnormality, free from diseases and pests or petal discoloration. Sizes of flowers should be the representative of the required cultivar. Bright, clean and firm flowers and leaves of uniform stem length must be selected. Cut stem should be straight and strong, capable of holding the flowers in upright position. Flowers are graded according to the standards of the society of American florist (SAF) and EEC standards in Europe. Stem length is considered the most important point for grading:

3.4.6.2 Roses

Stem length is considered the most important point for grading:

Code	Difference between shortest and longest stem (cm)
1. For stem length 0–15<20 cm	2.5 cm
2. For stem length 20–60 cm	5 cm
3. For stem length 60 cm and above	10 cm

3.4.6.3 Orchid

Grading is done mainly on length of the flower spike, flower no. size and arrangement of flowers on spike.

3.4.6.4 Chrysanthemum

Metric grade specification for spray mum:

Grade	Stems per sleeve	Specifications
Gold	10	6 flowers or more out and some to come
Silver	15	4 or 5 flowers out and some to come
Bronze	20	3 flowers out and some to come
Make-up	(—)	All stems not covered above, filling sleeves to some extent as other grades

3.4.6.5 For Standard Chrysanthemum

Stem length should not be less than 66 cm and those less than 51 cm should be marked as short. Society of American Florist (SAF) specifications:

Quality parameters	Blue	Red	Green	Yellow grade
Min. stem length (cm)	75	75	60	60
Min. flower diameter (cm)	15	12.5	10.0	–
Stem strength	Strong	Strong	Strong	Strong

Pompons are graded into 250–340 g bunches having several stems. For pot-mums, there is no standard grade. Plants should be bushy with good growth, 2.0–2.5 times as tall as their pots, having a minimum of 15 flowers and free from pest and diseases. A plant having 20–25 good quality flowers would be desirable.

3.4.6.6 Gladiolus

Spikes are graded based on overall quality, length of spikes and number of florets in each spike.

Grade	Spike length (cm)	Minimum number of florets
Fancy	>107 cm	16
Special	>96≤ 107 cm	15
Standard	>81 to ≤ 96 cm	12
Utility	≤ 81	10

3.4.6.7 Carnation

White and pink standard carnations are in great demand followed by Red, Yellow, Sky blue and Bicolor. Society of American Florist (SAF) suggested the following inspection standards for cut carnation. Standard specifications are bright, clean, firm flowers and leaves, fairly tight petals near the center of the unopened flowers, Symmetrical flower shape and size characteristics of the cultivar, no split or mended calyx, no lateral bud or sucker, no decay or damage, straight stem and normal growth.

For standard carnation	Grade			
Quality parameters	Blue	Red	Green	White
Min. flower diameter (cm)	7.0	5.7	–	–
Stem length (cm)	55	42.5–55	25–42.5	any
Stem strength	Strong	Strong	Unrestricted	–

3.4.7 QUALITY STANDARDS

Extra Class
Produce which qualifies for Class I without the aid of any tolerance. This excludes American Carnations with split calyx.

Class-I
Flowers must be of good quality, characteristics of the species variety. They must be whole, fresh, unburied, free of animal or vegetables parasites and resultant damage, free of residues of pesticides and other extraneous matter affecting appearance and free of development defects. Tolerance permitted up to 7 per cent.

Class-II
Flowers which do not meet all requirements of Class-I but are whole, fresh, free of animal parasites. Slight defects such as malformation, brushing, damage, small marks, weaker and less rigid stems may be present provided they do not impair appearance. Tolerance permitted upto 10%.

EC's standards are generally applicable to all cut flowers – no separate standard for different species of cut flowers have been established with the exception of mimosa. Whereas, the United Nation's Economic Commission for Europe has recommended general as well as specified standards for a number of flowers.

The U.S.A. has no official standards for cut flowers. The society of America Florists has recommended standards for certain cut flowers which included carnations. The grades are known as Blue, Red and Green and are based on flower diameter and length of stem.

The ECE of UN or EEC standards ignore length of stem and flower diameter in making class selections. Thus, Extra Class may contain classified flowers with both long and short stems. Nonetheless flowers must be sorted out according to stem length as given below:

Description Code	Minimum and Maximum Stem Length (in cm)
0	Less than 5 cm or flowers marked without stems
5	5–10
10	10–15
15	15–20

Description Code	Minimum and Maximum Stem Length (in cm)
20	20–30
30	30–40
40	40–50
50	50–60
80	80–100
100	100–120
120	120

In any unit of presentation (e.g., bunch, bouquet or box, etc.) the maximum permitted difference between shortest and largest stem lengths is as follows:

Description Code	Stem length (cm)
For stem lengths 0–15 < 20 cm	2.5
For stem lengths 20–60	5
For stem lengths 60 and over > 60 cm	10

3.4.8 PRE-COOLING

Pre-cooling is a step that rapidly brings the temperature of the flowers down from the field temperature to a proper storage temperature. A low temperature slows the respiration rate of the flowers which in turn helps them last longer. Forced-air-cooling is the best method for flowers; cool air is actively forced with fans through the bunched flower. This can be done when the flowers are in a bucket or when they are packed dry into boxes. The pre-cooling of flowers is a very important step for individuals selling to a large wholesale market, distant markets and if their crop is to be stored for a long time such as peonies. Individuals who sell at a local retail market usually do not need to worry about this step since their flowers will be in the customer's home the day they are picked.

All flowers should be pre-cooled after harvest as quickly as possible. Flowers may be pre-cooled by placing them in a cold storage without packing or in open boxes, until they reach the desired temperature. Pre-cooling slows down respiration, checks breakdown of nutritional and other stored materials in the stem, leaves and petals, slows bud opening and inhibit

TABLE 3.2 Pre-Cooling Temperature for Various Species and Cultivar

Sl. No.	Flowers	Pre-cooling temperature
1.	Alstroemeria	4°C
2.	Anthurium	13°C
3.	Cattleya	7–10°C
4.	Chrysanthemum	–0.5 to 4°C
5.	Cymbidium	–0.5 to 4°C
6.	Paphiopedilum	–0.5 to 4°C
7.	Dendrobium	5–7°C
8.	Carnation	1°C
9.	Gerbera	4°C
10.	Gladiolus	4–5°C
11.	Rose	1–3°C
12.	Bird of Paradise	7–8°C

flower senescence. It also prevents rapid water loss and decreases flower sensitivity to ethylene. The pre-cooled flower should be packed in cold room to prevent rapid rise of their temperature (Table 3.2).

3.4.9 PULSING

It improves the quality of cut flowers. This is a short-term treatment given to the cut flowers before packing. Its effect normally lasts for the entire shelf life of the flower even when the flowers are held in water. The cut flower may be pulsed with flower preservatives containing sugar, anti-microbial substances and anti-ethylene substances. The main ingredient of pulsing solution is however, sugar. Sucrose replaces the depleted endogenous carbohydrate utilized during the post-harvest life of flowers. It helps in continuation of normal metabolic activities after harvest and inhibition of the process associated with senescence. Excessive concentration of sugar in floral preservatives may be harmful, whereas, too low concentration may not produce an optimal response (Table 3.3).

Sugar should be added to the water of flowers at the bud stage or if flowers are to be shipped to distant markets or stored for an extended period. The stalks should be put in a 10–20% sugar concentration for between

TABLE 3.3 Percentage of Sucrose for Different Cut Flowers

Sl. No.	Flowers	Percentage of Sucrose
1.	Rose	2–6%
2.	Chrysanthemum	2–6%
3.	Bird of Paradise	10%
4.	GYPSOPHILA	10%
5.	CARNATION	10%
6.	GLADIOLUS	20%
7.	GERBERA	20%

12 and 24 h. Too much sugar may result in leaf yellowing, although there may be no noticeable injury to the flowers. To acidify the solution, it is suggested to add a small amount of citric acid as follows:

For soft water, use 0.1 g/L
For medium hard water, use 0.3 g/L
For hard water, use 0.45 g/L

3.4.9.1 Preservatives for Extending Vase Life

Fresh flower preservatives are chemicals added to water to make flowers last longer. They contain a germicide, a food source, a pH adjuster, water, and sometimes surfactants and hormones. Germicides are used to control bacteria, yeasts and molds. These microorganisms harm flowers by producing ethylene, blocking the xylem, producing toxins and increasing sensitivity to low temperatures. Bacterial counts of 10 to 100 million per 1 mL impairs uptake, while counts of 3 billion per 1 mL causes wilting. Some common germicides are listed on the following page. 8-HQC is the most common one used in commercial floral preservatives. Sucrose is the most common food source used in floral preservatives. It provides energy to sustain flowers longer and to open flowers in the bud stage. One-to-two percent sucrose is the standard amount in preservatives. Never use sucrose without a germicide, as it is the primary food source for microorganisms, too. Acids or acid salts are added to adjust the pH of the water to 3.5–5.0. At this pH, less microbes can grow and water is taken up by the flowers more easily. Surfactants and wetting agents like Tween 20 and Triton reduces water tension. Water is the most important component of floral preservatives. pH, temperature, soluble salts, alkalinity and hardness are to be ensured in normal range before adding floral preservatives. Acidic water with pH 3.5 to 5.0 is considered best. Water with a low pH is taken up by the flowers quicker and more easily. The lower pH inhibits the growth of xylem blocking microbes. Citric acid, an organic acid, usually is used to acidify water. Warm water has less dissolved gases that enhance the postharvest life of commercial flowers than cool water possessing more dissolved gas bubbles that can cause blockages in the xylem like microbes.

Fluoride is one specific ion that causes many problems in postharvest life of cut flowers. It is commonly added to municipal water supplies to prevent tooth decay in humans. Flowers in the lily family and other monocots are more sensitive to fluoride than others. Fluoride toxicity is more of a problem at a lower pH, which is best for holding flowers.

Floral preservatives perform three functions:

1. Provide sugar (carbohydrate);
2. Supply a bactericide to prevent microbial growth and blockage of the water-conductive cells in the stem;
3. Acidify the solution.

The most popular preservatives today contain 8-hydroxyquinoline citrate (8-HQC) and sucrose (common table sugar). The 8-HQC is a bactericide and an acidifying agent. Besides suppressing bacterial development and lowering the pH, 8-HQC also prevents chemical blockage, thus aiding in the absorption of water. Sucrose taken up by the stem maintains quality and turgidity and extends vase life by supplementing the carbohydrate supply.

There are a number of commercial preservatives in the market as Floralife®, Petalife®, Oasis®, Rogard®, and Everbloom®. These work well. One can also purchase 8-HQC under the name oxine citrate from florist companies and add sucrose to make the preservatives.

The bactericide 8-HQC is not totally effective in preventing the build up of bacteria in floral solutions. Chlorine is a very effective bactericide but dissipates quickly from solution unless provided in a slow-release form. Two slow-release forms sold extensively in products including bleaches, deodorizers, detergents, dishwashing compounds, and swimming-pool additives are DICA (sodium dichloroisocyanurate) and DDMH (1,3-dichloro-5,5-dimethyhydantoin). Both are highly effective bactericides for floral preservation. Each is used at a concentration of 300 ppm in the place of 8-HQC. DICA or DDMH is used with sucrose at a concentration of 2%. These chlorine compounds will bleach stems and leaves immersed in the preservative solution. They may also injure outer petals. These disadvantages are outweighed by the superior bactericidal effects of these materials.

Floral preservatives are very effective in maintaining quality and extending longevity. On the average, they can double the vase life of cut flowers when compared to water. Snapdragons with a life expectancy of 5- to 6-days last up to 12 days in preservatives.

Besides the standard floral preservative solutions for holding flowers there are some specific solutions and treatments that serve different needs.

- A harvesting solution often will simply be water acidified with citric acid to a pH of 3.5–5.0.
- A conditioning, hardening, or hydrating solution is used to restore the turgor of wilted flowers and dry packed flowers. It usually is warm water with a germicide, acidified to pH 3.5–5.0 with citric acid and a wetting agent, for example, Tween 20 at 0.01–0.1%.
- Impregnation is a treatment that protects stems against the blockage of water vessels by microbes. Stems are dipped in 1000 ppm silver nitrate solution for 10 minutes. The stems should not be recut after treatment.
- Pulsing or loading is a type of treatment used to extend the vase life of flowers held in water, stored wet or dry for long periods or shipped long distances. It is called a pulse because it is only done for a short period of time or called loading because the flowers are loaded up with food for a long storage period. Stems are placed in solutions with germicide and a higher concentration of sugar for specific treatment periods depending on the species. Because the higher concentration of sugar can act like a soluble salt causing petal and leaf injury, the treatment is only a few hours or a day. The temperature should be 18 to 19°C and light intensity should be 2000 lux.
- Bud opening solutions are used to open flowers harvested in a tight bud stage. Flowers harvested in the tight bud stage will keep longer in storage and will ship better. Stems are placed in solutions

containing higher concentrations of sugar, plus a germicide and hormonal compounds that facilitate bud growth and development. High light and humidity, and room temperatures are used. The high sugar content can injure the flowers and leaves.

3.4.10 VASE-SOLUTION

Several methods have been developed for bud-opening in rose, mum, carnation and gladiolus. It is necessary to select optimal bud stage for each flower types. Bud smaller than optimum do not open to full size and do not produce best quality flowers. The bud-opening solutions contain sucrose, germicide and hormonal compound. These floral preservative solutions have maintained manifold actions. It maintains turgidity in cut flowers, provides substrate for continued respiration, prevents vascular blockage in the stem, stimulates normal flower opening, prevents bacterial growth, and prevents undesirable changes in petal color. The place for bud-opening should be equipped with artificial light, humidity, temperature control and proper ventilation.

For bud cut carnation 10% sucrose + 200 ppm 8-HQC + 25 ppm $AgNO_3$ + 75 ppm citric acid is advised. For tight bud rose, vase solution may contain 2% sucrose + 300 ppm 8-HQC. For chrysanthemum, 2% sucrose + 200 ppm 8-HQC + 150 ppm citric acid. Treatment with 200 ppm 8-HQC + 75 ppm citric acid + 25 ppm $AgNO_3$. Flowers of bird-of-paradise cut at tight bud stage can be opened successfully in the vase solution of 10% sucrose +250 ppm 8-HQC + 150 ppm citric acid. Treatment with 200 ppm 8-HQC + 5% sucrose improve the flower quality and vase life of dendrobium, vanda, aranthera, oncidium and aranchnis. The vase life of cymbidium, phalaenopsis is prolonged by using chrysal VB and sucrose. For gladiolus, a solution containing 600 ppm 8-HQC or $Al_2(SO_4)_3$ (0.1%) + 4% sucrose is effective for extending postharvest life of flowers. Sucrose at 3–6% in preservative solution of pH3.0 extended the vase life of gladiolus flower. For Dahlia, vase solution contains 10% sucrose + 0.2 mM $AgNO_3$ or 10% sucrose + 0.2 mM $AgNO_3$ + 200 ppm 8-HQS. For Gerbera, solution contains 7% sucrose + 25 ppm $AgNO_3$ + 200 ppm 8-HQC. A pre-shipping dip of anthurium flower stems in solution of 2.25% 7 UP (Carbonated beverage) + 500 ppm benzoic acid/7.3 ppm Sodium hypochlorite remarkably extended the vase life of anthurium flower.

3.4.11 CARNATION

Pulsing solution: Sucrose (10%) + STS (2 mM) for 8 hours.
GA_3 (10 ppm) + Kinetin (2 ppm) + Alar (900 ppm) + AOA (450 ppm+ Triton–x (1000 ppm) for 6 hours.
Holding solution: 2% sucrose + 50 ppm STS + 50 ppm 8-HQC.

3.4.12 ALSTROEMERIA

Postharvest studies in cvs. of Serena and Aladdin reveals that pulsing (sucrose 20% + STS 1.0 mM) followed by keeping flowers in holding solution sucrose 2% + STS 1.0 mM + GA_3 50 ppm increasing the vase of flowers as well other postharvest attributes.

3.4.13 PACKAGING

Cut flowers are packed in rectangular bamboo baskets in India. The sides and tops of baskets are lined with hessian cloth. Loose flowers are packed in gunny bags or cloth bags and also in bamboo baskets tapering towards bottom and lined with newspaper. In such packages flowers are compressed due to improper size and stacking of the packages cannot absorb the load weight, fresh flowers are damaged considerably. Therefore, it is necessary to adopt modern methods of packaging, which can ensure protection of flowers against physical damages. Water can ensure protection of flowers against physical damages. Water loss and external condition are detrimental to the transported flowers. The main principles of packaging towards long storage life and keeping quality are to lower the rate of transpiration, resp. and cell division during transportation and storage. Hence, the ideal packing should be air tight, water proof and strong enough to withstand handling. For long distance transportation and storage, flowers should be harvested 1–3 days earlier than those harvested for the direct sale.

1. Place the bunched flowers in sleeves to prevent them from becoming entangled with each other. The sleeves may be made of plastic sheets, absorbent paper, wax paper or cellophane. Do not close the top of the sleeve.

2. Place the flowers in a fiberboard or styrofoam box. Arrange in layers according to type. Alternate layers of flowers should have the heads at the same end. Use rolled newspapers placed below the flower heads to minimize mechanical injury.

3.4.14 PACKING BOXES

The most suitable package material for cut blooms is corrugated fiber board (CFB) boxes since they have isothermic properties. These boxes are light in lit and can be made water resistant by using suitable adhesive or coating with wax or plastic film. The boxes must be strong enough to support the weight of at least 8 full boxes placed atop one another under high humidity. The minimum length of the boxes should be about twice the width and its width about twice the height. The dimension of the boxes however, depends upon the length of cut flowers stems, type of flowers to be packed and also on the space available inside standard refrigerated trucks (Reefer van).

Size of the package should be just sufficient to hold the flowers or flower holders tightly. If the package is oversized for the contents, the empty space results in higher transpiration rate of the flowers and too much space also allows the contents enough room for movement during transportation, which may damage the flowers.

The boxes must be able to withstand low temperature and high humidity and should preferably be wax coated from inside. Generally, the tests conducted after conditioning the boxes at 0°C and 100% RH are Drop test, compression test and vibration test. Small sized boxes are generally preferred on account of its higher compression strength/unit area. Use of telescope style boxes made of CFB is ideal.

The boxes /container should have an ISI marking also. For forced air-cooling of flowers, packing boxes with vents on either side is used. In such cases, the total vent area should equal to 4–5% of the area of the end walls of the packing boxes. Excessive number of holes reduces the strength of the package and increase transpiration loss. Ventilation holes near the corners are to be avoided. If the cut flowers are wrapped with a paper or foil, care must be taken that wrapping material do not impede air-flow through the box.

3.4.14.1 Wet Packing of Flowers

Since cut flowers like orchid and anthurium are susceptible to damage by chilling, they should be transported in water. For a prolonged transport, it is recommended that each individual stalk should be placed in a plastic vial or rubber balloon filled with water and tied to the flower stem with twine. The end of flower stem can also be placed in absorbent cotton saturated with water and enclosed in wax paper or polyethylene foil securing with twine.

3.4.14.2 Polyethylene Foil as Protective Cover

Polyethylene foil is commonly used to protect flowers from water loss. Tight packing in foil maintains relative humidity (RH) owing to its gas proof nature. It helps to maintain a high level of CO_2 and low level of O_2. Such condition keeps the resp. rate low and increase the longevity of the cut blooms. High conc. of CO_2 may also cause injuries in some flowers. Hence, very thin polyethylene foil, 0.04–0.06 mm thick permits partial gas exchange.

3.4.14.3 Special Care for Exotic Flowers

While packing, special care should be taken for delicate flower buds and flower heads. These buds or flower heads are protected by wrapping them in soft paper or plastic mesh or by placing them in specially molded forms made of plastic or cardboard.

In anthurium, damaged caused by the contact of the petals with the squamon pistil is avoided by covering each blossom with cellulose wood or cellulose acetate (used as plastic sleeve). The flowers heads of Gerbera are specially protected by transparent PVC covers. In cattleya, shredded wax papers should be placed around the flowers to prevent physical damage. Also, foil sacks filled with air or N_2 are also used for packing delicate exotic flowers.

3.4.14.4 Protection Against Geotropic Bending

Cut flowers of gladiolus, larkspur, lupin, snapdragon are sensitive to geotropic bending. These flower spikes must be transported in upright position. Since, they bend when transported horizontally. Special type of packing boxes that hold the flowers vertically should be used for transporting these flowers.

3.4.14.5 Care for Ethylene Sensitive Flowers

Cut flowers of carnation, orchid, alstroemeria, narcissus, lily, snapdragon and delphinium are highly sensitive to ethylene. Ethylene scrubber containing $KMnO_4$ may be added to package having such flowers. Direct contact with $KMnO_4$ causes flower injury, so material impregnated with $KMnO_4$ may be packed separately from flowers.

3.4.15 PACKAGING OF DIFFERENT FLOWERS

3.4.15.1 Rose

Graded flowers are grouped together in bunches of 10, 12, 20 or 25. A bundle of 20 stems is usually preferred. Each bundle is tied up sufficiently and the upper half is wrapped with tissue or parchment paper or cellophane sleever or grease proof paper. The wrapped bunches are then placed in appropriate boxes of dimension 100 cm x 6.50 cm (accommodates 80 roses having 65–70 long stem). Size of the carton used depends upon the length of the rose stem. To avoid mechanical injury, paper pillows should be placed at the bottom of the flower neck. To prevent the effect of desiccation or fluctuation in temperature, shaved or crushed ice should be placed on the stem. Normally, 4 bundles of roses containing 80 blooms are accommodated in a standard sized carton. The bundles are placed in two parallel rows, two in one row and other two in second row. The direction of buds in two rows should be opposite to each other. Paper shavings are also used in the interspaces as a cushion and during winter, adequate insulation should be provided in the container to prevent freezing and chilling injury to the flowers. Care should be taken to ensure that the packing materials do not absorb water from the flowers. Finally the box lid should be closed under slight pressure and the box is sealed.

3.4.15.2 Chrysanthemum

It requires special attention after harvest since mum start wilting rapidly. Blooms are packed in bundles of 10–20 flowers, loosely tied with string or a rubber band. In a standard sized carton, 400–600 flowers can be accommodated. Dimension of packing box for standard mum should be 91 cm × 43 cm × 15 cm and for spray mum should be 80 cm × 50 cm × 23 cm.

Flowers of standard cultivar are individually packed by covering them with protective sleeves or sealing them in plastic bags. Different types of wrapping materials are used to cover blooms in boxes like polysterol, polypropylene and perforated polypropylene sheet. To prevent crushing or shattering of florets or breaking of flowers stems, paper pillows are placed below the flower heads.

3.4.15.3 Carnation

Usually, 600–800 flowers of carnation can be accommodated in Std. carton box (122 cm × 50 cm × 30 cm). A bundle of 20–25 carnation can be made. Treating flower with 1200 ppm $AgNO_3$ for 10 min. before packing is beneficial. The pre-cooled flowers are wrapped in paper and then placed in foil bags.

3.4.15.4 Gladiolus

Bundles of 12–20 spikes are made. About 60–100 spikes of gladiolus are packed in a standard box (120 cm × 60 cm × 30 cm). Absorbent cotton is placed near the swollen florets to avoid their flattening. Bunches should be wrapped in paper and packed in gas tight polyethylene bags to protect flowers from sudden fluctuation in temperature, bruising and moisture loss. To avoid negative geotropic curvature of cut stems flowers are shipped in upright position in boxes.

3.4.15.5 Anthurium

Since the flowers are transported wet, hence the cut ends of each flower stem should be wrapped with cotton pad soaked with water and covered with wax paper and security tied or stem-ends are fitted in plastic vial containing water. In one method, spadix is dipped in melted paraffin to reduced moisture loss and packed in polythene bags and thereafter placed in carton. In another method, flowers are packed in moist boxes placing the soft protective material between the spathe and spadix. They are sometimes fixed to the bottom of the box with tape or separation paper between the layers. Foam or plastic supports are provided in the box. To maintain high humidity polythene lining is provided in the box. Usually, 120 anthurium cut spikes are accommodated in Std. sized packing box (21.6 cm × 50.8 cm × 91.4 cm or 27.9 cm × 43.2 cm × 101.6 cm).

3.4.15.6 Gerbera

Flowers are packed in flat boxes containing protective inserts with holes for individual flower stem. The stems are taped to the protective inserts while placing in insulted flat box to protect them from cold or freezing temperature. The flower heads of gerbera are specially protected by transparent PVC covers, cardboard cups or plastic mini sleeves or plastic coated metal grids measuring 70 × 50 cm with a mesh size of 2 × 2 cm.

3.4.15.7 Orchid

These flowers are packed either as intact sprays or as individual flower. Each flower stalk is put into a plastic vial or tube containing moistened cotton wool. For large delicate orchid flowers a few shreds of the wax paper are woven between the sepals and petals and around the lip or labellum. They are packed in polythene bags. The air containing these bags provides a protective cushion. The flower stalk along with tube of water is then laid on a bed of shredded wax paper in the packing box.

When individual flowers of Cymbidium are harvested from spikes, their peduncles should be immediately placed in a tube of water with a rubber or plastic cover. Flowers in tubes are packed to groups of 6, 8 or 12. In cattleya, peduncles are dipped in plastic vials containing water. The vials are taped to the bottom of the box. Flower spikes of aranda are packed individually and wrapped in tissue paper. Aranthera and oncidium are bundled together and then enclosed in polyethene bags.

Cut ends of the individual spikes of phalaenopsis, vanda and dendrobium are also dipped in the water of the plastic vial or tied with moistened cotton in small plastic bags. They are then kept in polythene bags and packed, taking sufficient care that no flower or flower part is damaged.

Normally, 50 flower spikes of these types can be packed in a box 75 cm × 25 cm × 17.5 cm size. The boxes are lined for better insulation during winter by several layers of newspaper or other insulating materials. High RH should be maintained inside the package by using waxed carton or polythene lining inside the box.

3.4.16 STORAGE OF CUT-FLOWERS

3.4.16.1 Reason of Storage

Some flower shows irregular cycle in the blooming season and to extending the season and delaying marketing in times of one production.

3.4.16.2 Advantages of Storage

Regulating market flow; reducing loss from demand decline; anticipating holidays; improving production efficiency; eliminating greenhouse production in winter; saving energy; making possible long term shipment.

3.4.16.3 Factors Affecting the Post-Storage Life of Flowers

High flower quality and proper maturity at the time of harvest, lower respiration rate, decreased water loss, inhibited ethylene production and action, retarded bacterial and fungal infections are the factors affecting the postharvest life of cut flowers.

3.4.17 STORAGE METHODS

3.4.17.1 Refrigeration with Wet or Dry Storage

Cold storage is recommended for all flowers that will not be in the market immediately and any flowers sold wholesale. Low temperatures slow the respiration rate of the flowers and prolong the vase life of the flowers. In general, temperatures should be 0–4°C and have a relative humidity of

85–90%, for most flowers. Flowers should never be stored with fruits and vegetables. Some fruits and vegetables produce ethylene that can dramatically shorten the life of the flowers. Once flowers are bunched into marketable units they should be placed in cold storage.

If flowers have to be stored before marketing, a cool place (preferably a refrigerated cold storage, especially for flowers) should be used. There are many flowers that are not commonly found in the wholesale market because they do not store well, ship well or last long. These should only be used for local markets. These include foxglove, garden phlox, lupine, clarkia, stevia, common stocks, candytuft, cornflower, feverfew, blue lace flower, English daisy, calendula, pot marigold, sweet violets and gaillardia.

It is advisable that store the high-quality, disease free and sorted cut flowers at the lowest temperature as possible. Avoid chilling injury. For wet storage, keep the cut stem ends in water or flower preservative. For dry-pack storage, seal flowers inside plastic bags or airtight cylinders made of either plastic or metal. Gently press the sides of plastic containers to remove as much air as possible before sealing the bags. Wet storage is ineffective in inhibiting the biological activity of flowers. However, iris, gerbera, lily, snapdragon responds well in wet storage. Plastic bags or boxes are used for dry storage. Flowers that show geotropic bending like gladiolus, snapdragon, larkspur, lupin must be stored at vertical position.

3.4.17.2 Refrigerated Storage

The most common system for handling harvested flowers is refrigerated storage, which involves the following steps:
1. Flower stems should be cut with a sharp knife or shears to prevent crushing of stem and water conduction cells.
2. The cut flowers should be placed in a preservative solution as soon as possible to prevent wilting. The flowers should not be allowed to be out of water while they are waiting to be transferred to the storage or grading rooms. If cut in the field, buckets containing solution can be brought out on trailers to hold the harvested flowers. Flowers cut in the greenhouse should not be left in the sun or out of water for more than a few minutes. One person should

be assigned to carry these flowers to the grading room or storage cooler immediately.
3. As soon as flowers arrive at the storage room they should be placed in preservative solution inside the refrigerated storage room. If wilted, they should be placed in a warm preservative solution at room temperature until turgid. They should then be placed in the cooler.
4. The temperature of the refrigerated room should be 0.5–4.4°C. The lower the temperature, the better, because the respiration rate falls off with diminishing temperature. Low respiration rates have an effect similar to that resulting from adding sucrose to the preservative solution in that they conserve carbohydrates within the flower. A temperature range of 0.5–4.5°C is usually encountered in flower coolers.
5. Air should be gently circulated inside the cooler only to the extent necessary to insure uniform temperatures in all areas. Unprotected flowers placed in a direct air stream will be desiccated. Flowers immediately adjacent to a cooling coil may freeze even though the air temperature is above freezing. Since the coil itself is below the freezing point, radiant heat is lost from the flower to the coil, and the flower can be colder than the surrounding air.
6. Potential sources of ethylene gas should be avoided by keeping fruit and vegetables out of the cooler. Discard old flowers. Wash the inside of the cooler periodically.
7. Replace the preservative solution at two to seven days' intervals. The preservative should be checked periodically for bacterial growth, which is apparent when the solution becomes cloudy. In spite of the bactericides in preservatives, microorganisms will develop and need to be eliminated periodically. To accomplish this, wash the buckets with a disinfectant such as bleach.
8. The wholesaler and retailer should hold the flowers under refrigeration. Whenever possible, flowers should be transported under refrigeration. Encourage the wholesaler and retailer to cut one-half inch from the base of the stems whenever it has been necessary to leave the flowers out of water for a period of time and then to place them in warm water at a cool air temperature to avoid the ends of the stems drying out and restricting water movement. Use a preservative solution throughout the entire marketing channel.

3.4.17.3 Dry Storage

Flowers can be held in refrigerated storage for one to three weeks, depending on the species. Refrigerated storage is more generally used as an aid for maintaining quality as flowers pass through the market channel. Dry storage is used when flowers must be held for periods longer than one to five days.

Only the best-quality flowers should be dry stored. Those of poor quality will have a short vase life when they are removed from storage. Flowers should be cut and packaged for storage immediately without being placed in water. Standard cardboard flower boxes are suitable, but a lining of polyethylene film should be placed in them to cover the flowers and seal in moisture. Desiccation can be a problem in long-term storage, especially when an absorbent container such as cardboard is used.

A common problem of dry storage is the presence of free water on the flowers, which encourages the development of diseases. While flowers freeze only at temperatures below −1.5°C, the free water will freeze at 0°C. Resulting ice crystals on the petals can be injurious. Boxes and flowers packed at warm temperatures develop condensation (free water) as the plants and air inside are cooled. Because of the polyethylene barrier, the water cannot escape. Disease, enhanced by this moisture, is a common cause of failure in dry storage. Boxes of flowers should be cooled open in a 3.3–4.4°C cooler, then sealed and placed in a 0.5°C cooler.

Most flowers freeze at −2.8 to −1.7°C, so it is essential that the temperature stay above this point. Flower life expectancy is lessened at 0.5°C and drops rapidly at temperatures above that point. Many of the failures of this system have been due to high temperatures or fluctuating temperatures. Since the dry storage cooler should not be open too often, another cooler is needed for regular refrigerated storage. The −0.5°C cooler is often built inside the 1.6–4.4°C cooler to provide for a more uniform temperature.

Space should be left between boxes of flowers when they are placed in storage initially. Respiration is occurring, and this produces heat. A large stack of boxes can generate enough heat and provide sufficient insulation to prevent thorough cooling of the inner flowers. Leave space between each stack of boxes and between every other box in a stack to permit the absorption of heat by circulating cool air. Flowers removed from dry storage need to be hardened. Cut one-half inch from the bottom of each stem. Place the flower in a preservative solution inside a 3.3–4.4°C cooler. Allow the flowers to become fully turgid before marketing them; this will take 12–24 hours. When properly handled, dry stored flowers should have reasonable quality and the same longevity as fresh flowers. Poor temperature control or disease will decrease quality and longevity.

Dry storage is used only to a limited degree by the industry and works best with chrysanthemums. Chrysanthemums, carnations, and roses are the crops to which it is primarily applied. Much more potential exists here than is being realized. The main reason for its low level of acceptance has probably been failures due to poor handling of the system (Tables 3.4 and 3.5).

TABLE 3.4 Method, Storage Temperature and Period for Cut Flowers

Flower	Method	Storage Temperature	Period
Anthurium	Dry	13°C	4 weeks
Bird of paradise	Dry	8°C	4 weeks
Carnation	Wet	4°C	4 weeks
Carnation	Dry[a]	0–1°C	4–6 months
Carnation	Dry	0.5–0°C	8 weeks
Chrysanthemum	Dry	0.5–1°C	3–5 weeks
Cyclamen	Dry	0–1°C	3 weeks
Cyclamen	Wet	1°C	1 weeks
Daffodils	Dry	1°C	14 days
Gerbera	Wet	4°C	3–4 weeks
Gladiolus	Dry[a]	2–4°C	4 weeks
Lilium	Wet	1°C	4 weeks
Lilium	Dry[a]	0–1°C	6 weeks
Daisy	Dry	2°C	2 weeks
Peony	Dry	(0–1°C) 0.5 3°C	4 weeks
Rose	Dry	0–1°C	3 weeks
Rose	Wet	2–5°C	5–7 days
Snapdragon	Wet	1–4°C	3–8 weeks
Orchid	Dry	7–10°C	10–14 days

NOTE: Tight pack creating a modified atmosphere condition.

TABLE 3.5 General Storage Temperature Recommendations and Approximate Storage Life for Fresh Cut Flowers

Flower	Scientific Name	Storage Temperature recommendation	Storage life
China Aster	*Callistephus chinensis*	0 to 4°C	7 to 21 days
Calendula	*Calendula officinalis*	4°C	3 to 6 days
Candytuft	*Iberis umbillata*	4°C	3 days
Clarkia	*Clarkia unquiculata*	4°C	3 days
Coreopsis	*Coreopsis grandiflora*	4°C	3 to 4 days
Cornflower	*Centaurea cyanus*	4°C	3 days
Cosmos	*Cosmos bipinnatus*	4°C	3 to 4 days
Dahlia	*Dahlia variabilis*	4°C	3 to 5 days
English Daisy	*Bellis perennis*	4°C	3 days
Delphinium	*Delphinium majus*	4°C	1 to 2 days
Foxglove	*Digitalis purpurea*	4°C	1 to 2 days
Freesia	*Freesia hybrids*	0 to 0.5°C	10 to 14 days
Gaillardia	*Gaillardia pulchella*	4°C	3 days
Gladiolus	*Gladilous grandiflorus*	2 to 5°C	5 to 8 days
Godetia	*Godetia sp.*	10°C	7 days
Gypsophila	*Gypsophylla paniculata*	4°C	7 to 21 days
Iris	*Iris sp.*	–0.5 to 0°C	7 to 14 days
Lilium	*Lilium sp.*	0 to 1°C	14 to 21 days
Lily-of-the-valley	*Convallaria majalis*	–0.5 to 0°C	14 to 21 days
Lupin	*Lupinus sp.*	4°C	3 days
Marigold	*Tagetes erecta*	4°C	7 to 14 days
Daffodils	*Narcissus sp.*	0 to 0.5°C	7 to 21 days
California Poppy	*Eschscholzia californica*	4°C	3 to 5 days
Peony	*Peonia sp.*	0 to 1°C	14 to 42 days
Phlox	*Phlox paniculata*	4°C	1 to 3 days
Snapdragon	*Antirrhinum majus*	4°C	7 to 14 days
Statice	*Limonium spp*	2 to 4°C	21 to 28 days
Stock	*Matthiola incana*	4°C	3 to 5 days
Strawflower	*Helichrysum bracteatum*	2 to 4°C	21 to 28 days
Sweet pea	*Lathyrus odoratus*	–0.5 to 0°C	14 days

TABLE 3.5 Continued

Flower	Scientific Name	Storage Temperature recommendation	Storage life
Sweet William	*Dianthus barbatus*	7°C	3 to 4 days
Tulip	*Tulipa cvs.*	–0.5 to 0°C	14 to 21 days
English Violet	*Viola odorata*	1 to 5°C	3 to 7 days
Zinnia	*Zinnia elegans*	4°C	5 to 7 days

3.4.17.4 Controlled Atmosphere (CA) Storage

It is based on storage at lower concentration of O_2 and higher concentration of CO_2 with precise control of atmospheric gasses. It inhibits C_2H_4 production and action by increasing CO_2 conc., slows/retards resp. rate and conservation of respirable substrates, for example, delays softening's; inhibit conversion of ACC (1-amino cyclopropane-1-carboxylic acid) to C_2H_4 as a result of decreased O_2 level. CA refers to increased CO_2, decreased O_2 and high N_2 in general as compared to normal atmosphere. All CA condition involves low temperature to reduce the velocity of enzymatic reaction ant to retard, resp. (Table 3.6).

Disadvantages of CA storage: High cost of application and inconvenience for handling various flowers in such a chamber.

TABLE 3.6 Optimal CA Storage Condition for Certain Flowers

Flower	CO_2%	O_2%	Temperature (°C)	Period
Freesia	10	21	1–2	3 weeks
Carnation	5	1–3	0–1	4 weeks
Gladiolus	5	1–3	1.5	3 weeks
Lily	10–20	21	1	3 weeks
Mimosa	0	7–8	6–8	10 days
Rose	5–10	1–3	0	20–30 days
Tulip	5	21	1	10 days

3.4.17.5 Modified Atmosphere (MA) Storage

MA condition is a less precise type of CA storage created by sealing plant material air tight with moisture proof packaging material *viz.* cellophane, polyethylene, and other films. It also requires increase in CO_2 and N_2 and decrease in O_2 but differ only in degree of control and methods of maintaining that control. Cellulose acetate is the most promising film for wrapping of cut flowers. Proper pre cooling prior to storage may reduce the risk of flower damage.

In MA condition RH is not essential. High air humidity is created by flower transpiration into tightly wrapped moisture retentive foil packaged. Moisture absorbent paper is used to prevent flowers from having contact with condensed water on the foil.

3.4.17.6 Low Pressure Storage (LPS)

It was described first by Burg and Burg (1966). This method consists of maintaining a product under constant sub atmospheric pressure (40–60 mm Hg) combined with low temperature and ventilation with fresh humid air; reduction in O_2, continuous removal of C_2H_4 by creating C_2H_4 free environment; acceleration of the outward diffusion of the different gases from within the tissue. This system is beneficial for cuttings, pet plants, cut flowers viz. carnation, mum, gladiolus, daffodils, rose, orchid and snapdragon. Maintenance of low air pressure of 40–60 mm Hg in the storage cut flowers; prolongs storage period; improves post storage life; larger bloom; reduced C_2H_4 production; reduced blaring and bent neck in rose, retard growth of botrytis. PS provides a unique system for creating air atmosphere nearly free from C_2H_4. Longest LPS period of 9 weeks was noted for carnation (highly susceptible to C_2H_4) (Tables 3.7 and 3.8).

Disadvantages of LPS:
It incurred high cost of installation, so best for laboratory.

TABLE 3.7 Flowers Sensitive to Ethylene

Achillea	*Aconitum*	*Agapanthus*
Allium	*Alstroemeria*	*Anemone*
Antirrhinum	*Aquilegia*	*Asclepias*
Astilbe	*Bouvardia*	*Campanula*
Carnation	*Celosia*	*Centaurea*
Chelone	*Consolida*	*Delphinium*
Dianthus	*Dicentra*	*Digitalis*
Eremurus	*Eustoma*	*Freesia*
Godetia	*Gypsophila*	*Iris*
Kniphofia	*Lathyrus* (Sweet Pea)	*Lavatera*
Lilium	*Limonium*	*Lupinus*
Lysimachia	*Matthiola* (Stock)	*Phlox*
Penstemon	*Physostegia*	*Ranunculus*
Rosa	*Rudbeckia*	*Salvia*
Saponaria	*Scabiosa*	*Sedum*
Silene	*Solidago*	*Thalictrum*
Trachelium	*Tricyrtis*	*Triteleia*
Trollius	*Veronica*	*Veronicastrum*
Snapdragons	Delphinium	Sweet Peas
Freesia	Larkspur	Alstroemeria

TABLE 3.8 Fluoride Sensitivity for Select Species

Species	Sensitivity Concentration
Freesias	1 ppm
Gladiolus	1 ppm
Gerberas	1 ppm
Chrysanthemum	5 ppm
Snapdragon	5 ppm
Roses	5 ppm

KEYWORDS

- Cut flower microbiology
- Cut flowers
- Flower packaging
- Flower storage
- Senescence regulation
- Vase life
- Vase solution

REFERENCES

Acock, B., & Nichols, R. (1979). Effects of sucrose on water relations of cut senescing carnation flowers. *Am. Bot. 44*, 221–230.
Arditti, J. (1975). Orchids, pollen poison, pollen hormone and plant hormones. *Orchid Rev. 83*, 127–129.
Arditti, J., Flick, B., & Jeffrey, D. (1971). Post-pollination phenomena in orchid flowers. *New Phytol. 70*, 333–341.
Armitage, A.M. (1991). "The Georgia Report—Stage of Flower Development at Harvest." *The Cut Flower Quarterly 3*(1), 13.
Armitage, A.M. (1993). *Specialty Cut Flowers*. Timber Press, Portland OR, USA.
Ashraf, M., Akram, N.A., Arteca, R.N., & Foolad, M.R. (2010). The physiological, biochemical and molecular roles of brassinosteroids and salicylic acid in plant processes and salt tolerance. *Crit. Rev. Plant Sci. 29*, 162–190.
Beyer, E. (1976). A potent inhibitor of ethylene action in plants. *Plant Physiol. 58*, 268–271.
Bieleski, R.L., Ripperda, J., Newman, J.P., & Reid, M.S. (1992). Carbohydrate changes and leaf blackening in cut flower stems of *Protea eximia. J. Am. Soc. Hort. Sci. 117*, 124–127.
Bovy, A.G., Angenent, G.C., Dons, H.J.M., & van Altvorst, A.C. (1999). Heterologous expression of the Arabidopsis etr1-1; allele inhibits the senescence of carnation flowers. *Molec. Breed. 5*, 301–308.
Brodersen, C.R., McElrone, A.J., Choat, B., Matthews, M.A., & Shackel, K.A. (2010). The dynamics of embolism repair in xylem: In vivo visualizations using high-resolution computed tomography. *Plant Physiol. 154*, 1088–1095.
Bufler, G. (1984). Ethylene-enhanced 1-aminocyclopropane-1-carboxylic acid synthase activity in ripening apples. *Plant Physiol. 75*, 192–195.
Bufler, G. (1986). Ethylene-promoted conversion of 1-aminocyclopropane-1-carboxylic acid to ethylene in peel of apple at various stages of fruit development. *Plant Physiol. 80*, 539–543.

Cameron, A.C., & Reid, M.S. (2001). 1-MCP blocks ethylene-induced petal abscission of Pelargonium peltatum but the effect is transient. *Postharvest Biol. Technol. 22*, 169–177.
Celikel, F.G., & Reid, M.S. (2002). Storage temperature affects the quality of cut flowers from the Asteraceae. *Hort Science 37*, 148–150.
Celikel, F.G., & Reid, M.S. (2005). Temperature and postharvest performance of rose (*Rosa hybrida* L. 'First Red') and gypsophila (*Gypsophila paniculata* L. 'Bristol Fairy') flowers. *Acta Hort. 682*, 1789–1794.
Cevallos, J.C., & Reid, M.S. (2000). Effects of temperature on the respiration and vase life of Narcissus flowers. *Acta Hort. 517*, 335–342.
Cevallos, J.C., & Reid, M.S. (2001). Effect of dry and wet storage at different temperatures on the vase life of cut flowers. *Hort. Technol. 11*, 199–202.
Chang, H., Jones, M.L., Banowetz, G.M., & Clark, D.G. (2003). Over production of cytokinins in petunia flowers transformed with PSAG12-IPT delays corolla senescence and decreases sensitivity to ethylene. *Plant Physiol. 132*, 2174–2183.
Chen, J.C., Jiang, C.Z., Gookin, T.E., Hunter, D.A., Clark, D.G., & Reid, M.S. (2004). Chalcone synthase as a reporter in virus-induced gene silencing studies of flower senescence. *Plant Mol. Biol. 55*, 521–530.
Cho, M.C., Celikel, F.G., Dodge, L., & Reid, M.S. (2001). Sucrose enhances the postharvest quality of cut flowers of *Eustoma grandiflorum* (Raf.) Shinn. *Acta Hort. 543*, 305–315.
Clements, J., & Atkins, C. (2001). Characterization of a non-abscission mutant in *Lupinus angustifolius*. I. Genetic and structural aspects. *Am. J. Bot. 88*, 31–42.
Crocker, W. (1913). The effects of advancing civilization upon plants. *School Sci. Math. 13*, 277–289.
Doi, M., & Reid, M.S. (1995). Sucrose improves the postharvest life of cut flowers of a hybrid Limonium. *HortScience 30*, 1058–1060.
Dole, J.M., Viloria, Z., Fanelli, F.L., & Fonteno, W. (2009). Postharvest evaluation of cut dahlia, linaria, lupine, poppy, rudbeckia, trachelium, and zinnia. *Hort. Technol. 19*, 593–600.
Eisinger, W. (1977). Role of cytokinins in carnation flower senescence. *Plant Physiol. 59*, 707–709.
Elibox, W., & Umaharan, P. (2008). Morphophysiological characteristics associated with vase life of cut flowers of anthurium. *HortScience 43*, 825–831.
Evans, R.Y., & Reid, M.S. (1988). Changes in carbohydrates and osmotic potential during rhythmic expansion of rose petals. *J. Am. Soc. Hort. Sci. 113*, 884–888.
Ferrante, A., Hunter, D., Hackett, W., & Reid, M. (2001). TDZ: A novel tool for preventing leaf yellowing in Alstroemeria flowers. *HortScience 36*, 599.
Friedman, H., Meir, S., Rosenberger, I., Halevy, A.H., Kaufman, P.B., & Philosoph Hadas, S. (1998). Inhibition of the gravitropic response of snap dragon spikes by the calcium channel blocker lanthanum chloride. *Plant Physiol. 118*, 483–492.
Gan, S., & Amasino, M.R. (1995). Inhibition of leaf senescence by autoregulated production of cytokinin. *Science 270*, 1986–1988.
Halevy, A., & Mayak, S. (1979). Senescence and postharvest physiology of cut flowers— Part 1. *Hort. Rev. 1*, 204–236.
Halevy, A.H., & Kofranek, A.M. (1984). Evaluation of lisianthus as a new flower crop. *HortScience 19*, 845–847.

Halevy, A.H., Kofranek, A.M., & Besemer, S.T. (1978). Postharvest handling methods for bird of paradise flowers (Sterlitzia reginae Ait.). *J. Am. Soc. Hort. Sci. 103*, 165–169.
Hammer, P.E., Yang, S.F., Reid, M.S., & Marois, J.J. (1990). Postharvest control of Botrytis cinerea infections on cut roses using fungi static storage atmospheres. *J. Am. Soc. Hort. Sci. 115*, 102–107.
Hanley, K., & Bramlage, W. (1989). Endogenous levels of abscisic acid in aging carnation flower parts. *J. Plant Growth Regul. 8*, 225–236.
Hardenburg, R.E., Watada, A.E., & Wang, C.Y. (1986). USDA–ARS Agricultural Handbook Number 66, The Commercial Storage of Fruits, Vegetables, and Florist and Nursery Stocks.
Hunter, D., Lange, N.E., & Reid, M.S. (2004a). Physiology of flower senescence. In: Nooden, L.D. (ed.), *Plant Cell Death Processes*. Elsevier. 307–318.
Hunter, D., Steele, B.C., & Reid, M.S. (2002). Identification of genes associated with perianth senescence in Daffodil (*Narcissus pseudonacissus* L. 'Dutch Master'). *Plant Sci. 163,* 13–21.
In, B.-C., Inamoto, K., & Doi, M. (2009). A neural network technique to develop a vase life prediction model of cut roses. *Postharvest Biol. Technol. 52*, 273–278.
Jones, R.B., Serek, M., & Reid, M.S. (1993). Pulsing with Triton X-100 Improves hydration and vase life of cut sunflowers (*Helianthus annuus* L.). *HortScience 28*, 1178–1179.
Joyce, D.C., & Reid, M.S. (1985). Effect of pathogen-suppressing modified atmospheres on stored cut flowers. p. 185–198. In: S. Blankenship (ed.), Controlled atmospheres for storage and transport of perishable agricultural commodities. North Carolina State Univ., Raleigh.
Kader, A.A. (2003). A perspective on postharvest horticulture (1978–2003). *HortScience 38*, 1004–1008.
Kader, A.A., Zagory, D., & Kerbel, E.L. (1989). Modified atmosphere packaging of fruits and vegetables. *Crit. Rev. Food Sci. Nutr. 28*, 1–30.
Kalkman, E.Ch. (1986). Post-harvest treatment of Astilbe Hybr. *Acta Hort. 181,* 389–392.
Lay-Yee, M., Stead, A.D., & Reid, M.S. (1992). Flower senescence in daylily (Hemerocallis). *Physiologia Plantarum 86*, 308–314.
Macnish, A., deTheije, A., Reid, M.S., & Jiang, C.-Z. (2009a). An alternative postharvest handling strategy for cut flowers-dry handling after harvest. *Acta Hort. 847,* 215–221.
Macnish, A., Jiang, C.-Z., & Reid, M.S. (2010b). Treatment with thidiazuron improves opening and vase life of iris flowers. *Postharvest Biol. Technol. 56*, 77–84.
Macnish, A., Leonard, R.T., Borda, A.M., & Nell, T.A. (2010c). Genotypic variation in the postharvest performance and ethylene sensitivity of cut rose flowers. *HortScience 45*, 790–796.
Macnish, A., Morris, K.L., de Theije, A., Mensink, M.G.J., Boerrigter, H.A.M., Reid, M.S., Jiang, C.-Z., & Woltering, E.J. (2010d). Sodium hypochlorite: A promising agent for reducing Botrytis cinerea infection on rose flowers. *Postharvest Biol. Technol. 58*, 262–267.
Marissen, N., & Benninga, J. (2001). Nursery comparison on the vase life of the rose 'First Red': Effects of growth circumstances. *Acta Hort. 543*, 285–291.

Maxie, E., Farnham, D., Mitchell, F., Sommer, N., Parsons, R., Snyder, R., & Rae, H. (1973). Temperature and ethylene effects on cut flowers of carnation (*Dianthus caryophyllus*). *J. Am. Soc. Hort. Sci. 98*, 568–572.

Mayak, S., & Dilley, D. (1976). Regulation of senescence in carnation (*Dianthus caryophyllus*), effect of abscisic acid and carbon dioxide on ethylene production. *Plant Physiol. 58*, 663–665.

Mayak, S., & Halevy, A. (1972). Interrelationships of ethylene and abscisic acid in the control of rose petal senescence. *Plant Physiol. 50*, 341–346.

Mayak, S., & Halevy, A.H. (1970). Cytokinin activity in rose petals and its relation to senescence. *Plant Physiol. 46*, 497–499.

Mayak, S., & Kofranek, A.M. (1976). Altering the sensitivity of carnation flowers (*Dianthus caryophyllus* L.) to ethylene. *J. Am. Soc. Hort. Sci. 101*, 503–506.

Mayak, S., Bravdo, B., Gvilli, A., & Halevy, A.H. (1973). Improvement of opening of cut gladioli flowers by pretreatment with high sugar concentrations. *Scientia Hort. 1*, 357–365.

McClure, B.A., & Guilfoyle, T. (1989). Rapid redistribution of auxin-regulated RNAs during gravitropism. *Science 243*, 91–93.

Meir, S., Droby, S., Davidson, H., Alsevia, S., Cohen, L., Horev, B., & Philosoph-Hadas, S. (1998). Suppression of Botrytis rot in cut rose flowers by postharvest application of methyl jasmonate. *Postharvest Biol. Technol. 13*, 235–243.

Mokhtari, M., & Reid, M.S. (1995). Effects of postharvest desiccation on hydric status of cut roses. 489–495. In: Ait-Oubahou, A., & El-Otmani, M. (Eds.), Postharvest physiology, pathology and technologies for horticultural commodities: Recent advances. Institut Agronomique et Veterinaire Hassan II, Agadir, Morocco.

Mor, Y., Halevy, A., Kofranek, A., & Reid, M. (1984). Postharvest handling of lily of the Nile flowers. *J. Am. Soc. Hort. Sci. 109*, 494–497.

Morris, K., Mackerness, S.A.H., Page, T., John, C.F., Murphy, A.M., Carr, J.P., & Buchanan-Wollaston, V. (2000). Salicylic acid has a role in regulating gene expression during leaf senescence. *Plant J. 23*, 677–685.

Muller, R., Stummann, B.M., Andersen, A.S., & Serek, M. (1999). Involvement of ABA in postharvest life of miniature potted roses. *Plant Growth Regul. 29*, 143–150.

Mutui, M., Emongor, V.N., & Hutchinson, M.J. (2003). Effect of benzyl adenine on the vase life and keeping quality of Alstroemeria cut flowers. *J. Agric. Sci. Technol 5*, 91–105.

Naidu, S.N., & Reid, M.S. (1989). Postharvest handling of tuberose (*Polianthestuberosa* L.). *Acta Hort. 261*, 313–318.

Newman, J.P., van Doorn, W., & Reid, M.S. (1989). Carbohydrate stress causes leaf blackening. *J. Intl. Protea Assoc. 18*, 44–46.

Nichols, R. (1973). Senescence and sugar status of the cut flower. *Acta Hort. 41*, 21–27.

Nowak, J., & Rudnicki, R.M. (1990). Postharvest Handling and Storage of Cut Flowers, Florist Greens and Potted Plants. Timber Press, Portland OR, USA.

Nowak, J., & Rudnicki, R.M. (1990). Postharvest handling and storage of cut flowers, florist greens, and potted plants. Timber Press, Portland, Oregon.

Nowak, J., & Veen, H. (1982). Effects of silver thiosulfate on abscisic acid content in cut carnations as related to flower senescence. *J. Plant Growth Regul. 1*, 153–159.

Onoue, T., Mikami, M., Yoshioka, T., Hashiba, T., & Satoh, S. (2000). Characteristics of the inhibitory action of 1,1-dimethyl-4-(phenylsulfonyl) semicarbazide (DPSS) on ethylene production in carnation (*Dianthus caryophyllus* L.) flowers. *Plant Growth Reg. 30*, 201–207.

Otsubo, M., & Iwaya-Inoue, M. (2000). Trehalose delays senescence in cut gladiolus spikes. *HortScience 35*, 1107–1110.

Panavas, T., Walker, E.L., & Rubinstein, B. (1998b). Possible involvement of abscisic acid in senescence of daylily petals. *J. Exp. Bot. 49*, 1987–1997.

Philosoph-Hadas, S., Meir, S., Rosenberger, I., & Halevy, A.H. (1996). Regulation of the gravitropic response and ethylene biosynthesizing gravity stimulated snapdragon spikes by calcium chelators and ethylene inhibitors. *Plant Physiol. 110*, 301–310.

Porat, R., Borochov, A., & Halevy, A.H. (1993). Enhancement of petunia and dendrobium flower senescence by jasmonic acid methyl ester is via the promotion of ethylene production. *Plant Growth Regul. 13*, 297–301.

Reid, M.S., & Scelikel, F.G. (2008). Use of 1-Methylcyclopropene in ornamentals: Carnations as a model system for understanding mode of action. *HortScience 43*, 95–98.

Reid, M.S., & Wu, M.J. (1992). Ethylene and flower senescence. *Plant Growth Reg. 11*, 37–43.

Reid, M.S., van Doorn, W., & Newman, J.P. (1989). Leaf blackening in proteas. *Acta Hort. 261*, 81–84.

Ronen, M., & Mayak, S. (1981). Interrelationship between abscisic-acid and ethylene in the control of senescence processes in carnation flowers *Dianthus caryophyllus* cultivar White Sim. *J. Exp. Bot. 32*, 759–766.

Rot, I., & Friedman, H. (2010). Desiccation-induced reduction in water uptake of gypsophila florets and its amelioration. *Postharvest Biol. Technol. 57*, 189–195.

Sacalis, J.N. (1993). Cut Flower Prolonging Freshness-Postproduction Care and Handling 2nd ed., edited by Joseph L. Seals. Ball Publishing, Batavia, IL-USA.

Saks, Y., & Staden, J. (1993). Evidence for the involvement of gibberellins in developmental phenomena associated with carnation flower senescence. *Plant Growth Regul. 12*, 105–110.

Sankhla, N., Mackay, W.A., & Davis, T.D. (2003). Reduction of flower abscission and leaf senescence in cut phlox inflorescence by thidiazuron. *Acta Hort. 628*, 837–841.

Sankhla, N., Mackay, W.A., & Davis, T.D. (2005). Effect of thidiazuron on senescence of flowers in cut inflorescences of *Lupinus densiflorus* Benth. *Acta Hort. 669*, 239–243.

Savin, K.W., Baudinette, S.C., Graham, M.W., Michael, M.Z., Nugent, G.D., Lu, C.-Y., Chandler, S.F., & Cornish, E.D. (1995). Antisense ACC oxidase RNA delays carnation petal senescence. *HortScience 30*, 970–972.

Seglie, L., Martina, K., Devecchi, M., Roggero, C., Trotta, F., & Scariot, V. (2011). The effects of 1-MCP in cyclodextrin-based nanosponges to improve the vase life of *Dianthus caryophyllus* cut flowers. *Postharvest Biol. Technol. 59*, 200–205.

Seglie, L., Sisler, E.C., Mibus, H., & Serek, M. (2010). Use of a non-volatile 1-MCP formulation, N, N-dipropyl(1-cyclopropenylmethyl)amine, for improvement of postharvest quality of ornamental crops. *Postharvest Biol. Technol. 56*, 117–122.

Serek, M., Jones, R.B., & Reid, M.S. (1994). Role of ethylene in opening and senescence of Gladiolus sp. flowers. *J. Amer. Soc. Hort. Sci. 119*, 1014–1019.

Serek, M., Jones, R.B., & Reid, M.S. (1994a). Role of ethylene in opening and senescence of Gladiolus sp. flowers. *J. Am. Soc. Hort. Sci. 119*, 1014–1019.

Serek, M., Sisler, E., Tirosh, T., & Mayak, S. (1995a). 1-Methylcyclopropene prevents bud, flower and leaf abscission of Geraldton wax flower. *HortScience 30*, 1310.

Serek, M., Sisler, E.C., & Reid, M.S. (1994b). Novel gaseous ethylene binding inhibitor prevents ethylene effects in potted flowering plants. *J. Am. Soc. Hort. Sci. 119*, 1230–1233.

Serek, M., Sisler, E.C., & Reid, M.S. (1995b). Effects of 1-MCP on the vase life and ethylene response of cut flowers. *Plant Growth Reg. 16*, 93–97.

Siddiqui, M.W. (2015). Postharvest Biology and Technology of Horticultural Crops: Principles and Practices for Quality Maintenance. CRC Press, Boca Raton, Florida, USA. pp. 550.

Sisler, E.C., & Blankenship, S.M. (1993). Diazocyclopentadiene (DACP), a light sensitive reagent for the ethylene receptor in plants. *Plant Growth Reg. 12*, 125–132.

Sisler, E.C., & Blankenship, S.M. (1996). Method of counteracting an ethylene response in plants. USPTO.U.S. Patent 5, 518,988.

Sisler, E.C., & Lallu, N. (1994). Effect of diazocyclopentadiene (DACP) on tomato fruits harvested at different ripening stages. *Postharvest Biol. Technol. 4*, 245–254.

Sisler, E.C., Goren, R., & Huberman, M. (1984). Effect of 2,5-norborbadiene on citrus leaf explants. *Plant Physiol. Suppl. 75*, 127.

Stimart, D.P., & Brown, D.J. (1982). Regulation of postharvest flower senescence in Zinnia elegans Jacq. *Scientia Hort. 17*, 391–396.

Tanase, K., Tokuhiro, K., Amano, M., & Ichimura, K. (2009). Ethylene sensitivity and changes in ethylene production during senescence in long-lived Delphinium flowers without sepal abscission. *Postharvest Biol. Technol. 52*, 310–312.

Teas, H.J., Sheehan, T.J., & Holmsen, T.W. (1959). Control of gravitropic bending in snapdragon and gladiolus inflorescences. *Proc. Florida State Hort. Soc. 72*, 437–442.

van Doorn, W.G., & Reid, M.S. (1995). Vascular occlusion in stems of cut rose flowers exposed to air: Role of xylem anatomy and rates of transpiration. *Physiol. Plantarum 93*, 624–629.

Van Doorn, W.G., Perik, R.R.J., & Belde, P.J.M. (1993). Effects of surfactants on the longevity of dry-stored cut flowering stems of rose, Bouvardia, and Astilbe. *Postharvest Bio. Tech. 3*, 69–76.

van Doorn, W.G., Perik, R.R.J., Abadie, P., & Harkema, H. (2011). A treatment to improve the vase life of cut tulips: Effects on tepal senescence, tepal abscission, leaf yellowing and stem elongation. *Postharvest Biol. Technol. 61*, 56–63.

Van Staden, J., & Dimalla, G.G. (1980). Endogenous cytokinins in bougainvillea 'San Diego Red': I. Occurrence of cytokinin glucosides in the root sap. *Plant Physiol. 65*, 852–854.

Van Staden, J., Upfold, S.J., Bayley, A.D., & Drewes, F.E. (1990). Cytokinins in cut carnation flowers transport and metabolism of isopentenyl adenine and the effect of its derivatives on flower longevity. *Plant Growth Reg. 9*, 255–262.

Vanneste, S., & Friml, J. (2009). Auxin: A trigger for change in plant development. *Cell 136*, 1005–1016.

Veen, H., & van de Geijn, S.C. (1978). Mobility and ionic form of silver as related to longevity of cut carnations. *Planta 140*, 93–96.

Waithaka, K., Reid, M.S., & Dodge, L. (2001). Cold storage and flower keeping quality of cut tuberosa (*Polianthes tuberosa* L.). *J. Hort. Sci. Biotechnol. 76*, 271–275.

Wang, C.Y., & Baker, J.E. (1979). Vase life of cut flowers treated with rhizobitoxine analogs, sodium benzoate, and isopentenyl adenosine. *HortScience 14*, 59–60.

Woltering, E.J., & van Doorn, W.G. (1988). Role of ethylene in senescence of petals—morphological and taxonomical relationships. *J. Expt. Bot. 39*, 1605–1616.

Woods, C.M., Polito, V.S., & Reid, M.S. (1984a). Response to chilling stress in plant cells II. Redistribution of intracellular calcium. *Protoplasma 121*, 17–24.

Woods, C.M., Reid, M.S., & Patterson, B.D. (1984b). Response to chilling stress in plant cells I. Changes in cyclosis and cytoplasmic structure. *Protoplasma 121*, 8–16.

Woodson, W.R., Park, K.Y., Drory, A., Larsen, P.B., & Wang, H. (1992). Expression of ethylene biosynthetic pathway transcripts in senescing carnation flowers. *Plant Physiol. 99*, 526–532.

Wu, M.J., van Doorn, W.G., & Reid, M.S. (1991). Variation in the senescence of carnation (*Dianthus caryophyllus* L.) cultivars. I. Comparison of flower life, respiration and ethylene biosynthesis. *Scientia Hort. 48*, 99–107.

Wu, M.J., van Doorn, W.G., & Reid, M.S. (1991). Variation in the senescence of carnation (*Dianthus caryophyllus* L.) cultivars. I. Comparison of flower life, respiration and ethylene biosynthesis. *Scientia Hort. 48*, 99–107.

CHAPTER 4

POSTHARVEST MANAGEMENT AND PROCESSING TECHNOLOGY OF MUSHROOMS

M. K. YADAV,[1] SANTOSH KUMAR,[2] RAM CHANDRA,[1]
S. K. BISWAS,[3] P. K. DHAKAD,[1] and MOHAMMED WASIM SIDDIQUI[4]

[1]*Department of Mycology and Plant Pathology, Institute of Agricultural Sciences, Banaras Hindu University, Varanasi, 221005, Uttar Pradesh, India*

[2]*Department of Plant Pathology, Bihar Agricultural University, Sabour, Bhagalpur, 813210, Bihar, India, E-mail: santosh35433@gmail.com*

[3]*Department of Plant Pathology, C.S. Azad University of Agriculture and Technology, Kanpur, 208002, Uttar Pradesh, India*

[4]*Department of Food Science and Postharvest Technology, Bihar Agricultural University, Sabour, Bhagalpur, 813210, Bihar, India*

CONTENTS

4.1 Introduction 152
4.2 Nutritional Importance of Mushrooms 156
    4.2.1 Vitamins 156
    4.2.2 Proteins 157
    4.2.3 Minerals 158
4.3 Medicinal Mushroom 158

4.4 Mushroom Processing and Postharvest Technology 160
4.5 White Button Mushroom (*Agaricus bisporus*) 163
    4.5.1 Washing 163
    4.5.2 Packing and Packaging 164
        4.5.2.1 Modified Atmosphere Packaging 165
        4.5.2.2 Modified Humidity Packaging 166
    4.5.3 Storage 167
        4.5.3.1 Optimum Storage Conditions 167
        4.5.3.2 Controlled Atmospheric Storage 167
    4.5.4 Drying of Mushrooms 168
    4.5.5 Cooling 169
        4.5.5.1 Pre-Cooling or Refrigeration 169
        4.5.5.2 Vacuum Cooling 170
        4.5.5.3 Ice-Bank Cooling 170
        4.5.5.4 Steeping Preservation 170
        4.5.5.5 Radiation Preservation 171
    4.5.6 Canning 171
4.6 Oyster Mushroom (*Pleurotus* sp.) 172
4.7 Milky Mushroom (*Calocybe indica*) 174
4.8 Paddy Straw Mushroom (*Volvariella volvacea*) 174
4.9 Transportation 174
4.10 Diseases and Disorders 175
Keywords 175
References 175

4.1 INTRODUCTION

The global food and nutritional security of growing population is a great challenge, which looks for new crop as source of food and nutrition. In this context, mushrooms find a favor, which can be grown even by landless people, that too on waste material and could be a source for proteinous food. Use of mushrooms as food and nutraceutical have been known since time immemorial, as is evident from the description in old epics Vedas and Bible. Earlier civilizations had also valued mushrooms for delicacy and therapeutic value. In the present time, it is well recognized that mushroom is not only rich in protein, but also contains vitamins and minerals, whereas, it lacks cholesterol and has low calories. Furthermore, it also has high medicinal attributes like immune-modulating, antiviral, antitumor, antioxidants and hepatoprotective properties. With the growing awareness for nutritive and quality food by growing health conscious population, the demand for food including mushrooms is quickly rising and will continue to rise with increase in global population which will be 8.3 million by 2025 and flexible income (Singh, 2011). The mushroom cultivation has grown up in almost all the parts of the world and during last three decades, the world mushroom production achieved the growth rate of about 10%. Globally, China is the leading producer of mushrooms with more than 70% of the total global production, which is attributed to community, based farming as well as diversification of mushrooms. In India, owing to varied agro-climate and abundance of farm waste, different types of temperate, tropical and subtropical mushrooms are cultivated throughout the country.

Huge quantities of lingo-celluloid crop residues and other organic wastes are generated annually through the activities of agricultural, forest and food processing industries. It is estimated that India is generating 600 million tons of agricultural waste besides, fruit and vegetable residue, coir dust, husk, dried leaves, pruning, coffee husk, tea waste which has potential to be recycled as substrate for mushroom production leading to nutritious food as well as organic manure for crops. If even 1% of these crop residues are used to produce mushroom, India will become a major mushroom producing country in the world (Tewari and Pandey, 2002). Edible mushroom production represented an attractive method of improving the nutritional quality of lingo-celluloid wastes for use as an animal feed stock. Among the various physical, chemical and biological methods used for upgrading the digestibility and nutritive value of agricultural wastes, biodegradation by using white rot fungi including mushrooms have been found promising. The species of *Pleurotus* has ability to excrete hydrolyzing and oxidizing enzymes (Toyama and Ogawa, 1974; Daugulis and Bone,

1977), which enables them to grow and flourish over wide range of natural lingo-celluloid waste materials.

Around 800 million people living in 46 countries are malnourished, 40,000 die every day of hunger and hunger-related diseases (Siddiqui, 2015; Swaminathan, 1995). In this context, mushroom cultivation represent one of the economically viable processes for the bioconversion making it a potent weapon against malnutrition in developing countries like India which have lowest per capita consumption of protein in the world (Sohi, 1982; Wood, 1989; Chang and Miles, 1989). It is also consistent with the emerging view that an ecologically oriented society must use its wastes and resources rather than discarding them as useless materials.

Even in case that the world population would not increase any more, there is an enormous amount of waste from field, agro-industry, and wood industry. Only using 25% of the yearly burned cereal straw in the world could result in a mushroom yield of 317 million metric tons (317 milliard kg) of fresh mushroom per year (Chang and Miles, 1989). But on this moment, the yearly mushroom production is only 6 milliard persons or 1 kg per year or 3 gram per day (Courvoisier, 1999). In fact counting the early available world waste in agriculture (500 milliard kg) and forestry (100 milliard kg), we can easily grow 360-milliard kg of fresh mushroom on the total of 600 milliard kg dry wastes. This would bring us a yearly mushroom food of 6 kg per head per year containing 4% protein in fresh mushroom and we know that 30% of the world population is protein deficient. On the other hand, we all know the high risk of further population growth with again more need for food (field crops) and wood (forest) swelling up the already so high mountains of wastes.

More than 2000 mushroom species exist in nature, but only approximately 22 species are intensively cultivated (Manzi, et al., 2001). Around 20 genera of mushrooms are being cultivated throughout the world, only four types, *viz*., white button mushroom (*Agaricus bisporus*), oyster mushroom (*Pleurotus* spp.), paddy straw mushroom (*Volvariella volvacea*) and milky mushroom (*Calocybe indica*) are grown commercially in India with the white button mushroom contributing about 90% of total country's production as against its global share of about 40% (Mehta et al., 2011). Button mushroom (*Agaricus bisporus*) is the most popular variety, fetches high price, still dominating the Indian and International market. Mushrooms are now getting significant importance due to their nutritional and medicinal value and today their cultivation is being done in about 100 countries. Production and consumption of mushrooms have tremendously increased in India mainly due to increased in awareness of the commercial and nutritional significance of this commodity. At present production of mushrooms has crossed lakh tone with annual growth rate of above 15% (Sharma and Dhar, 2010). In India, white button mushroom still contributes more than 85% of the total mushroom production, though its share is below 40% in the global trade (Prakasam, 2012).

Mushrooms are the fruiting bodies of macro fungi. They include both edible/medicinal and poisonous species. However, originally, the word "mushroom" was used for the edible members of macro fungi and "toadstools" for poisonous ones of the "gill" macro fungi. Scientifically the term "toadstool" has no meaning at all and it has been proposed that the term is dropped altogether in order to avoid confusion and the terms edible, medicinal and poisonous mushrooms are used.

Edible mushrooms once called the "Food of the Gods" and still treated as a garnish or delicacy can be taken regularly as part of the human diet or be treated as healthy food or as functional food. The extractable products from medicinal mushrooms, designed to supplement the human diet not as regular food, but as the enhancement of health and fitness, can be classified into the category of dietary supplements/mushroom nutriceuticals (Chang and Buswell, 1996). Food with appropriate nutritional value will remain the essential and most important need for nutritional security of the mankind.

Dietary supplements are ingredients extracted from foods, herbs, mushrooms and other plants that are taken without further modification for their presumed health-enhancing benefits. There is an old Chinese saying, which states that "Medicines and Foods Have a Common Origin." Mushrooms are source of quality protein having essential amino acids and high digestibility. No cholesterol and low fat with ergosterol and polyunsaturated fatty acids: good for heart. Low calorific food with no starch, low sugars: delight of diabetics. High fiber, low sodium-high potassium diet: anti-hypertensive. Good source of vitamin B-complex and vitamin C; only vegetable source of vitamin D. Rich in minerals like copper (cardio-protective) and selenium (anti-cancer), anti-HIV, anti-viral, anti-histaminic, hypo-cholesterolemic, hepatic- and nephro-protective, anti-oxidant, stamina enhancer, etc. (Chu et al., 2002).

4.2 NUTRITIONAL IMPORTANCE OF MUSHROOMS

Mushrooms are usually eaten for their culinary properties, providing a flavoring and garnish for other foods. They are cultivated with special technique and usually consumed by the rich people because the price of mushroom is usually much higher than that of the most common vegetables. This may give one the impression that mushrooms constitute a luxury food and that their promotion would only benefit relatively rich people. Actually, mushrooms are rich in protein and contain several vitamins and mineral salts and should thus be considered as high protein vegetables to enrich all human diets (Table 4.1).

Mushrooms have been found effective against cancer, cholesterol reduction, stress, insomnia, asthma, allergies and diabetes (Bahl, 1983). Due to high amount of proteins, they can be used to bridge the protein malnutrition gap. Mushrooms as functional foods are used as nutrient supplements to enhance immunity in the form of tablets. Due to low starch content and low cholesterol, they suit diabetic and heart patients. One third of the iron in the mushrooms is in available form. Their polysaccharide content is used as anticancer drug. Even, they have been used to combat HIV effectively (Namba, 1993; King, 1993).

4.2.1 VITAMINS

Mushrooms are good source of vitamins (Table 4.2) especially Vitamin B such as vitamin 'B1' (Thiamine), vitamin 'B2' (Riboflavin), niacin, biotin

TABLE 4.1 Composition of Common Edible Mushrooms (g/100 fresh weight)

Mushroom	Moisture	Protein	Fat	Carbohydrate	Fiber	Ash	Calories
A. bisporus	90.1	2.9	0.3	5.0	0.9	0.8	36
V. volvacea	90.1	2.1	1.0	4.7	1.1	1.0	36
P. sajor-caju	90.2	2.5	0.2	5.2	1.3	0.6	35
Cabbage	91.9	1.8	0.1	4.6	1.0	0.6	27
Cauliflower	90.0	2.6	0.4	4.0	1.2	1.0	30
Potato	74.7	1.6	0.1	22.6	0.4	0.6	97

Source: Sohi (1988).

TABLE 4.2 Vitamin Content of Edible Mushrooms

Mushrooms	Content mg/100 g D. wt.			
	Thiamine	Riboflavin	Niacin	Ascorbic acid*
Agaricus bisporus	1.1	5.0	55.7	81.9
Lentinus edodes	7.8	4.9	54.9	0.0
Volvariella volvacea	0.32–0.35	1.63–2.97	64.8	20.2
Pleurotus spp.	1.16–4.8	4.7*	46.1	0.0

*Adapted from Eli V. Crisan and Anne Sands (Chang and Hayes, 1978)
(Source: Chadha and Sharma, 1995).

and vitamin 'C' (Ascorbic acid) (Chang and Buswell, 1996; Mattila et al., 2000). Vitamin content of edible mushrooms has been reported by Esselen and Fellers (1946), and Litchfield (1964).

4.2.2 PROTEINS

Mushroom contains 20–35% protein (dry weight), which is higher than those of vegetables and fruits and is of superior quality. The value of protein is determined by the kinds of amino acids that form protein. Mushrooms contain all the essential amino acids as well as the most commonly occurring non-essential amino acids and amides. Mushrooms are rich in essential amino acids lysine, which is deficient in cereals. The most nutritious mushrooms are almost equal in nutritional value to meats and milk. Protein is the main body building constituent of our food. Protein content of mushrooms depends on the composition of the substratum, size of pileus, harvest time and species of mushrooms (Bano and Rajarathnam, 1982).

Verma et al. (1987) reported that mushrooms are very useful for vegetarian because they contain some essential amino acids, which are found in animal proteins. The digestibility of *Pleurotus* mushrooms proteins is as that of plants (90%) whereas that of meat is 99% (Bano and Rajarathnam, 1988). Rai and Saxena (1989a) observed decrease in the protein content of mushroom on storage. The protein conversion efficiency of edible mushrooms per unit of land and per unit time is far more superior compared to animal sources of protein (Table 4.3) (Bano and Rajarathnam, 1988).

TABLE 4.3 Essential Amino Acid (% crude protein) in Edible Mushrooms

Amino acid	*Agaricus bisporus*	*Pleurotus sajor-caju*	*Volvariella volvacea*
Leucine	7.5	7.0	4.5
Isoleucine	4.5	4.4	3.4
Valine	2.5	5.3	5.4
Tryptophan	2.0	1.2	1.5
Lysine	9.1	5.7	7.1
Threonine	5.5	5.0	3.5
Phenyl alanine	4.	5.0	2.6
Methionine	0.9	1.8	1.1
Histidine	2.7	2.2	3.8

Sources: Bano and Rajarathnam (1982); Li and Chang (1982).

4.2.3 MINERALS

Mushrooms are also good source of minerals (Table 4.4) such as potassium, phosphorus, sodium, calcium, and contain low but available form of Iron. Sodium and potassium ratio is very high which idea is for patient of hypertension.

The fruiting bodies of mushrooms are characterized by a high level of well-assimilated mineral elements. Major mineral constituents in mushrooms are K, P, Na, Ca, Mg and elements like Cu, Zn, Fe, Mo, Cd form minor constituents (Bano and Rajarathanum, 1982; Bano et al., 1981; Chang, 1982). K, P, Na and Mg constitute about 56–70% of the total ash content of the mushrooms (Li and Chang, 1982) while potassium alone forms 45% of the total ash. Abou-Heilah et al. (1987) found that content of potassium and sodium in *A. bisporous* was 300 and 28.2 ppm respectively. *A. bisporus* ash analysis showed high amount of K, P, Cu and Fe (Anderson and Fellers, 1942). Varo et al. (1980) reported that *A. bisporus* contains Ca (0.04 g), Mg (0.16), P (0.75 g), Fe (7.8 g), Cu (9.4 mg), Mn (0.833 mg) and Zn (8.6 mg) per kilogram fresh weight.

4.3 MEDICINAL MUSHROOM

Medicinal mushrooms or extracts from mushrooms that are used as possible treatments for diseases (Table 4.5). Some mushroom compounds,

TABLE 4.4 Minerals Content in Button Mushroom

Species	K	P	Mg	Na	Ca	Fe	Cd	Zn	Cu	Pb	Hg
	(mg/100 g. Dry wt.)					(ppm)					
P. Sajor-caju	3260	760	221	60	20	12.4	0.3	29	12.2	3.2	0
P. eous	4570	1410	242	78	23	9.0	0.4	82.7	17.8	1.5	0
P. flabellatus	3760	1550	292	75	24	12.4	0.5	56.6	21.9	1.5	0
P. florida	4660	1850	192	62	24	18.4	0.5	11.5	15.8	1.5	0

Source: Rai (1995).

including polysaccharide, glycoprotein and proteoglycons modulate immune system responses and inhibit tumor growth. Some medicinal mushroom isolates that have been identified also show cardiovascular, antiviral, antibacterial, antiparasitic, anti-inflammatory, and antidiabetic properties. Currently, several extracts have widespread use in Japan, Korea and China, as adjuncts to radiation treatments and chemotherapy.

Historically, mushrooms have long had medicinal uses, especially in traditional Chinese medicine. Mushrooms have been a subject of modern medical research since the 1960s, where most modern medical studies concern the use of mushroom extracts, rather than whole mushrooms. Only a few specific mushroom extracts have been extensively tested for efficacy. Polysaccharide-K and lentinan are among the mushroom extracts with the firmest evidence. The available results for most other extracts are based on in vitro data, effects on isolated cells in a lab dish, animal models like mice, or underpowered clinical human trials. Studies show that glucan-containing mushroom extracts primarily change the function of the innate and adoptive immune system, functioning as bioresponse modulator, rather than by directly killing bacteria, viruses, or cancer cells as cytocidal agents. In some countries, extracts like, Polysaccharide-K schizophyllan, polysaccharide peptide and lentinan are government-registered adjuvant cancer therapies.

Pharmaceuticals worth $700 million are produced annually in Japan from *Lentinus*, *Coriolus*, *Schizophyllum* and *Ganoderma*. According to Mizuno et al. (1990) Mushroom extracts have a high amount of retene that

TABLE 4.5 Various Pharmaceutical Components Isolated From Several Mushrooms

Pharmacodynamics	Component	Species
Antibacterial effect	Hirsuitic acid	Many species
Antibiotic	E-B, Methoxyacrylate	O. radicata
Antiviral effect	Polysoceharid, Protein	L. edodes and Polyporaceal
Cardoic tonic	Volvatoxin, Flammutoxin	Volvariella, Flammulina
Decrease cholesterol	Eritadenine	Collybia velutipes
Decrease blood pressure	Triterpene	G. lucidum
Decrease level of blood	Peptide, glycogen, Ganoderan Glucon.	G. lucidum
Antifrombus	5' AMP, 5'GMP	Psalliota hartensis
Inhibition of PHA	r-GHP	P. hartensis, L. edodes
Antitumor	B-glucan RNA complex	Hysizygus marmoreus etc.
Increase secretion of bile	Armillarisia A	Armillaria tabescens
Analgestic and Sedetive effect	Marasmic acid	Marasmices androsaceus

Source: Chadha and Sharma (1995).

has an antagonistic effect on some form of tumor. Some mushroom extracts induce formation of interferon, a defense mechanism against viral infection and have hypo-cholesteroemic activity (lowering cholesterol levels). Further, compounds extracted from mushroom have antifungal and antibacterial properties. The low fat content and cholesterol free mushroom diets comfortable for patients of hyperlipemia, blood pressure and hypertension and for diabetes due to lack of sugar in mushroom. Mushroom diets are also effective against diarrhea patient due to presence of fiber and its more digestibility and tumors due to some medicinal properties.

4.4 MUSHROOM PROCESSING AND POSTHARVEST TECHNOLOGY

Fresh mushrooms have very short self-life and hence processing is recommended to increase their shelf-life. Initially, fresh mushrooms are washed

Postharvest Management and Processing Technology of Mushrooms 161

in cold water and then blanched in boiling water for about 3–4 min. Then they are dehydrated in a drier and packed. It is advisable to pre-treat fresh mushrooms in a solution containing brine to prevent discoloration. Packing is very critical as formation of moisture contaminates mushrooms very quickly. Yield of dried mushroom depends upon many factors like moisture content in fresh mushrooms, type of dryer, process employed, moisture content required in the finished product etc. Hence average yield is taken at 25%. Plain cans and a brine of 2% salt and 0.2% citric acid are used for packing. The cans are exhausted at 19°C for 7–8 min, sealed and processed under pressure for 20–25 min. The flow chart of processing is given in Figure 4.1.

Mushrooms have very short shelf life – these cannot be stored or transported for more than 24 h at the ambient conditions prevailing in most parts of the country. Due to presence of more than 90% moisture content, mushrooms are highly perishable and start deteriorating immediately after harvest. They develop brown color on the surface of the cap due the enzymatic action of phenol oxidase, this results in shorter shelf life. Loss of texture, development of off flavor and discoloration results in poor marketable quality and restricts trade of fresh mushrooms. Browning, veil-opening, weight-loss and microbial spoilage are the most common postharvest changes in the mushrooms which often result into huge economic losses. Almost all the mushrooms have very short shelf-life but the paddy straw

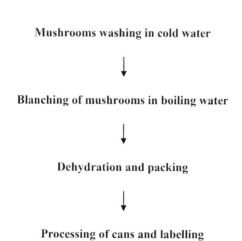

FIGURE 4.1 Flow chart of processing.

mushroom has the shortest (few hours at the ambient) and Milky has very good shelf-life (3–5 days) if microbial spoilage is taken care of. Most damaging postharvest changes in mushrooms vary with species—it is blackening in the button mushroom, cap-opening in the paddy straw mushroom and mucilage in the oyster mushroom, which affect their marketability significantly. Weight loss is very serious problem in all the mushrooms as these contain very high moisture (85–90%) and are not protected by the conventional cuticle. Due to very high moisture and rich nutritive value, microbial spoilage in mushrooms is also a problem.

Proper, sound and appropriate postharvest practices of storage and processing are needed to sustain the budding mushroom farming and industry in the country. In India, more than 90% of the total mushroom production is still contributed by the common button mushroom (*Agaricus bisporus*). Information about proper postharvest care and processing of such a perishable commodity is therefore of vital importance to keep the wheels of this industry moving at the right speed; with the adoption of proper packaging, storage and processing technologies, problems in marketing, like seasonal gluts and distress sales, can also be ameliorated. We have endeavored to deal with most important postharvest aspects of mushrooms – physiological and biochemical changes, packaging and storage of fresh mushrooms, long-term storage and processing.

The production of mushroom is done throughout the year by the environmentally controlled units, but the seasonal growers come into play during the winters and the supply at the local market exceeds causing less profit due to fall in price and spoilage due to market surplus. Mushrooms are highly consumable and get spoiled due to browning, flaccid, liquefaction, loss of consistency, flavor, etc., making it unsalable. Most of the mushrooms, being high in moisture and delicate in consistency, it cannot be stored for more than 24 h at the ambient conditions prevailing in the tropics. Researchers, who studied spoilage of fresh mushrooms, earlier believed the primary cause to be the enzymatic reactions in the living tissue. Later, it was suggested that spoilage might be caused by the action of bacteria on the mushroom tissue and browning of mushrooms was due to a combination of auto enzymatic and microbial action on the tissue. Sound postharvest practices have since been developed to extend the shelf life of fresh mushrooms.

As far as processing technologies are concerned, sun drying of mushrooms is one of the simplest and oldest methods followed by the growers from the time immemorial. Due to the difficulties in drying of some of the mushrooms, new preservation technologies like canning, pickling, mechanical and chemical drying (freeze drying, fluidized bed drying, batch type cabinet drying and osmotic drying) and irradiation treatment of mushrooms have been developed to improve the shelf life and consumption of mushrooms.

During the recent years, there has been an increased emphasis on the quality of fresh vegetables including mushrooms, which is reflected in the price of the produce. In India, the mushroom market is largely the contribution of small and marginal farmers with limited resources, who are dependent on local market for the sale of their produce. The rate of respiration of the harvested mushrooms is high in comparison to the other horticultural crops and this result in a shorter postharvest life.

Many short-term storage measures are followed to retard the deterioration in quality at the level of mushroom grower till it reaches the consumer. By following proper packing, cooling and transportation, the shelf life of mushrooms can be extended.

4.5 WHITE BUTTON MUSHROOM (*Agaricus bisporus*)

White button mushroom still dominates the Indian and International market and a lot of work has been done to minimize the loss in quality of the mushrooms. In case of the button mushroom all the four most deleterious changes namely, color browning, veil-opening, weight loss and microbial spoilage ask for the utmost post-harvest care. Needless to say that these changes are also accompanied by changes in the nutritional and medicinal attributes of these mushrooms.

4.5.1 WASHING

Washing, is normally done to remove soil particles, however, it leads to decline in shelf life and spoilage by bacteria. Small growers wash in solution of reducing agents to retard the browning caused by polyphenol oxidases. Hence, various anti-microbial as well as reductant compounds are used in washing mushrooms to extend the shelf life. Oxine, a stabilized form of chlorine dioxide, was very effective in controlling bacterial growth and color deterioration when used at a level of 50 ppm or higher with a two minute or longer wash period at 12°C. The use of sodium hypochlorite (100 ppm) and calcium chloride (0.55%) with oxine (100 ppm) resulted in increased antibacterial effectiveness. Use of calcium chloride and oxine also resulted in lower cap opening and firmer mushrooms during storage. Washing fresh mushrooms in water containing sodium sulphite solutions results in lower bacterial numbers and an improved initial appearance, but more rapid bacterial growth and browning occurred during subsequent storage compared to unwashed controls. Mushrooms washed in hard water (150 ppm calcium carbonate) reduced bacterial growth and there was less color deterioration during storage. Washing mushrooms in a solution consisting of oxine (50 ppm), sodium erythorbate (0.1%), and calcium chloride (0.5%) resulted in significantly lower bacterial populations and less color deterioration during the storage. Based on experiment done at this organization and its co-coordinating centers, it has been found that washing of mushrooms in 0.05% potassium metabisulphite improved the initial whiteness, which lasted longer during the storage. Even though many farmers are adapting this approach of washing, but selling clean unwashed properly packed mushroom may be a better option, as many people prefer mushroom not just because of health benefit, but also considered it a more chemical free food (Wakchaure, 2011).

4.5.2 PACKING AND PACKAGING

The mushrooms are packed for transporting them to the market. While a good package sells a product, a mediocre package can interfere with sale of an otherwise excellent product. For the local markets in India, mushrooms are packed in retail packs of 200 g or 400 g in simple polythene packs of less than 100 gauge thickness. Large quantity packing of mushrooms is done using polythene or pulp-board punnets, which will withstand long distance transport. Plastic punnets 130 × 130 × 72–0.40, Cardboard chips 305 × 125 × 118–1.82, Plastic tray 330 × 280 × 145–2.30, Expanded polystyrene 400 × 333 × 167–4.56. These punnets are over-wrapped with differentially permeable PVC or polyaccetate films. These over-wrappings help in creating modified atmosphere in punnets with 10% CO_2 and 2% O_2 and mushrooms maintain their fresh look for 3 days at 18°C.

In the recent years controlled atmosphere packaging (CAP) and modified atmosphere packaging (MAP) are catching up fast for all types of fruits and vegetables (Siddiqui et al., 2016; Siddiqui, 2016). These packaging techniques have to be effectively used for packing mushrooms to have improved shelf life. If simple polythene bags are used, it is important to make desired number of holes for proper humidity control.

4.5.2.1 Modified Atmosphere Packaging

Modified atmosphere packaging is a method by which a modified atmosphere is created in a sealed package of a fresh product by respiratory gas exchange, namely oxygen (O_2) intake and carbon dioxide (CO_2) evolution. When the rate of gas permeation through the packaging material equals respiratory gas exchange, equilibrium concentrations of O_2 and CO_2 are consequently established. The equilibrium depends on: temperature, respiration rate of specific product, and product weight O_2 and CO_2 permeability of the packaging material, free volume in the package and film area. Thus MAP helps in extending the shelf life and maintenance of quality of perishable produce by way of creation of appropriate gaseous atmosphere around the produce packed in plastic films. In this technique, the natural process of respiration of the produce in conjunction with the restricted gas exchange through a polymeric film such as low-density polyethylene (LDPE), normal and oriented polypropylene (PP) is used to control the in-pack oxygen and carbon dioxide. Modified atmosphere can be created by two methods: active and passive modifications. In passive modification, the product is just sealed in a polymeric package and due to the respiration of the fresh product and permeation of gases into the package, the atmosphere is modified. In active modification, air is flushed into the package initially, so that the steady state atmosphere is reached quickly after packaging. In passive modification, it takes a long time to reach the steady state conditions within the package.

Pulp board punnet with over wrapped polyethylene sheet MAP of mushrooms has been shown successfully to delay senescence and maintain quality after harvest. Shelf life of mushrooms can be increased by over wrapping them with PVC films. Thickness of 100-gauge polythene bags with 0.5% venting area are recommended for packing mushroom in case of refrigerated storage. For transporting mushroom to long distance, polystyrene or pulp-board punnets should be used instead of using polythene bags. The punnets are over-wrapped with differentially permeable poly vinyl chloride (PVC) or poly acetate films. They create modified atmosphere in punnets producing an atmosphere of about 10% CO_2 and 2% O_2. The optimum atmosphere for storage of mushrooms is found to be 2.5–5% CO_2 and 5–10% O_2. According to Simon et al. (2005) a modified atmosphere containing 2.5% CO_2 and 10–12% O_2 improves improved the appearance of *A. bisporus* and reduces the bacterial count. According to Roy et al. (1995 the storage of mushroom species in a modified atmosphere containing 26% CO_2 and decreases weight loss by 3–4.5%.

In modified atmospheric packaging, the packaging material plays a vital role in modifying the inside atmosphere around the product and the product quality as well. Among the various packaging materials viz., polyvinylidene chloride coated, oriented nylon, anti-fogging, wrap or vacuum packing film, the antifogging film maintained the quality of mushrooms for 24 days. The best packaging material polyethylene extends the shelf life of fresh mushrooms to 15 days under MAP conditions. In modified atmospheric packaging, the shelf life of the product can further be extended by supplementing some chemicals in addition to modifying the atmosphere inside the package. The various supplementary packaging materials viz., activated carbon, sorbitol, chitosan, potassium permanganate can be used in MAP for maintaining the quality of oyster mushrooms at ambient temperature. A study was conducted in Finland about the washing and use of a humidity absorber (Silicagel) in the packages during the modified atmosphere and it was found that washing in chlorinated water and incorporation of dehumidifiers decreased the microbial contamination and increased shelf-life in *A. bisporus*. Sorbital maintained the best color in mushroom when it was packed and stored along with fresh mushroom trays over wrapped with PVC films. Mushrooms treated with honey (0.5 and 1%) for 18 h, air dried to remove surface moisture, packed in 100 gauge polythene pouches with 0.5% venting area, increased the shelf life by more than a week over control at 3–5°C and 2–3 days at ambient temperature.

4.5.2.2 Modified Humidity Packaging

Most polymeric films used in conventional packing have lower water vapor transmission rates relative to transpiration rates of fresh produce.

This leads to nearly saturated conditions within packages. The high in-package-relative-humidity (IPRH) can cause condensation of water vapor within a package and allow microbial growth. This may either increase or decrease the spoilage depending on the product, depending on their transpiration coefficients and water potentials (Siddiqui et al., 2016). To obtain the desired IPRH, there are two possible approaches: perforation of the package, which precludes the possibility of achieving modified atmosphere conditions within the package, and use of in-package water absorbing compounds like calcium chloride, which can maintain the required RH. MAP in combination with MHP further improved the shelf-life of fresh mushrooms. An IPRH of 87–90% is desirable for best color in mushrooms during storage.

4.5.3 STORAGE

4.5.3.1 Optimum Storage Conditions

The storage condition is very important for maintaining the quality of fresh mushroom. The best result for storing fresh mushroom in a cool chamber The most favorable temperature for storage of mushrooms at 0 to 2°C with 95% RH. For the period of 7–9 days upon rapid cooling. Storage at 2°C (35.6°F) shortens storage-life to 3–5 days by accelerating surface browning, stipe elongation, and veil opening (Umiecka, 1986). High RH is essential to prevent desiccation and loss of glossiness. Moisture loss is correlated with stipe blackening and veil opening. Mushrooms should be packed in cartons with a perforated over-wrap of polyethylene film to reduce moisture loss. It is important to avoid water condensation inside packages. There are no chemical treatments to extend storage-life of mushrooms intended for fresh consumption.

4.5.3.2 Controlled Atmospheric Storage

In this method, the oxygen and carbon dioxide concentrations are altered inside the package and respiration rate gets altered. Controlled atmospheric package reduces brown discoloration (enzymatic browning) and the shelf life is extended (Ahmad and Siddiqui, 2015).

4.5.4 DRYING OF MUSHROOMS

Drying is perhaps the oldest technique known to the mankind for preservation of food commodities for long duration. It is the process of removal of moisture from the product to such a low level that microbial and biochemical activities are checked due to reduced water activity, which makes the products suitable for safe storage and protection against the attack by microorganisms during the storage. Mushrooms contain about 90% moisture at the time of harvesting and are dried to a moisture level down below 10–12%. At a drying temperature of 55–60°C, the insects and microbes on the mushrooms will be killed in few hours, which gives us the dehydrated final product of lower moisture content with longer shelf-life. The temperature, moisture of the mushroom and humidity of the air affect the color of the dried product. Dehydrated mushrooms are used as an important ingredient in several food formulations including instant soup, pasta, snack seasonings, casseroles, and meat and rice dishes. Dried mushrooms can be easily powdered and used in soups, bakery products, etc. Mushroom dried at higher temperature loose texture, flavor, color along with reduced rehydrability. Most of the mushrooms except the button mushroom have been traditionally dried for long-term storage, for example, oyster, shiitake, paddy straw, milky mushroom, etc. In case of button mushroom, it is the blackening and irreversible change of texture, which often discourages the use of this otherwise simple technique of preservation. Recently with advances in drying technologies, various drying methods such as solar drying, fluidized bed drying, dehumidified air- cabinet drying, osmo-air drying, freeze-drying cabinet drying and microwave drying are efficiently used for almost all types of mushrooms. There are many drying methods such as sun-drying, freeze-drying, cabinet air drying, osmo-air drying, fluidized-bed drying and microwave drying. Sun-drying is very oldest, cheapest and very simple method among various drying methods. Under this method mushrooms are spread over the trays or sheets and kept in open under the sun with 25°C temperature, with less than 50% relative humidity and high wind velocity. Sun dried product contains more than 10–12% moisture and should therefore be oven dried at 55–60°C for 4–6 h to further reduce the moisture to 7–8% to avoid any spoilage during storage. Under freeze drying; removal of water from a substance by sublimation from the frozen state to the vapor state is known as freeze-drying. The product is frozen at −22°C for one min. The frozen mushrooms are dried to moisture content of 3% in a freeze drier and packed under vacuum (Kannaiyan and Ramsamy, 1980).

4.5.5 COOLING

4.5.5.1 Pre-Cooling or Refrigeration

The temperature of the button mushroom after picking varies between 15 and 18°C, and it rises steadily during the storage due to respiration and atmospheric temperature. This heat causes deterioration in quality. Hence, the heat should be removed immediately after the harvest and the temperature of mushrooms should be brought down to 4–5°C as quickly as possible. The choice of the cooling system depends upon the quantity to be handled, which may be a refrigerator for a small grower to a cold room with all facilities for a commercial grower. Using evaporative cooling, hydro-cooling, forced-chill air, ice bank or vacuum cooling systems, and mushrooms can be cooled up to 2–4°C. The temperature of the mushrooms increases through respiration after picking and the respiratory rate increases with the increase in the storage temperature. It has been estimated that mushrooms at a temperature of 10°C have 3.5 times higher respiratory capacity than those at temperature of 0°C, which necessitates immediate shifting of mushrooms to the refrigerated zone. The size and shape of the packs also play an important role in cooling room. Packs with more than 10 kg mushrooms or with 15 cm thick layers of mushroom causes problem. Vertical flowing of air is more suitable for cooling. The mushrooms should not be stored in the same cooler along with fruits as the gases produced by fruits, for example, ethylene cause discoloration of mushrooms. This forced-chill air-cooling system is time consuming and vacuum cooling is becoming more popular.

4.5.5.2 Vacuum Cooling

In vacuum cooling, the water in cell walls and inter hyphal spaces of mushrooms is evaporated under low pressure and the evaporative cooling lowers the temperature from ambient to 2°C in 15–20 min. Vacuum cooling is a uniform and faster process, where mushrooms are subjected to very low pressure and water evaporates giving off the latent heat of vaporization, thus cooling itself. The vacuum-cooled mushrooms have superior color than conventional-cooled mushrooms. The major drawback of the system is the high capital cost; and an inevitable loss of fresh weight during the process of cooling. Filling and emptying the cooling chamber introduces another operation and expenses into the marketing chain. Air spray moist chillers can also cool the mushrooms rapidly. The temperature can be lowered by 16–18°C in an hour without any moisture loss.

4.5.5.3 Ice-Bank Cooling

With a view to reduce the weight loss during the conventional vacuum cooling, ice bank cooling of mushrooms is now in vogue in some countries wherein a stack of mushrooms is passed through forced draft of chilled but humidified air from the ice bank.

4.5.5.4 Steeping Preservation

This method is simple and economical and the mushrooms can be preserved for short period by steeping them in solution of salt or acids. The common practice is that cleaned mushrooms are washed in water or chemical added water and filled in large plastic containers. Blanching in brine solution for 5 min is generally done before filling them in cans. Brine solution is then added into the cans or containers. Steeping of water blanched mushrooms in 1% potassium meta bisulphite (KMS) along with 2% citric acid (overnight), before drying improves color, texture and reconstitution properties. Solution consisting of 2% sodium chloride, 2% citric acid, 2% sodium bicarbonate and 0.15% KMS is used for steeping preservation of blanched mushrooms for 8–10 days at 21–28°C. Chemical solution of 2% salt, 2% sugar, 0.3% citric acid, 0.1% KMS and 1% ascorbic acid is also used for steeping preservation of mushrooms. It helps to extend shelf life of mushrooms.

4.5.5.5 Radiation Preservation

Low doses of gamma radiation can be used to reduce the contamination and extend the shelf life of mushrooms. Irradiation should be given immediately after harvest for optimum benefits. Irradiation can potentially delay the maturation, for example, development of cap, stalk, gill and spore and also reduces the loss of water, color, flavor texture and delays the quality losses. Cobalt 60 is used as a common source of gamma rays. A dose of 10 KGy (Kilo Gray) will completely destroy microorganisms. An enhancement in shelf-life of *Agaricus bisporus* upto a period of 10 days can be achieved by application of gamma ray close to 2 KGy and storage at 10°C. Irradiation reduces the incident of fungal and bacterial infection. The loss of flavor components is noticed in irradiated mushrooms. The cap opening is also delayed by irradiation. Amino acids in fresh mushrooms are preserved by gamma irradiation. Irradiation at low levels proved better than irradiation levels of 1 and 2 KGy. The permissible doses for such preservation have not been worked out in our country. Even for export, it will be necessary to follow the standards of importing country.

4.5.6 CANNING

Canning is the technique by which the mushrooms can be stored for longer periods up to a year and most of the international trade in mushrooms is done in this form. The caning process can be divided into various unit operations namely cleaning, blanching, filling, sterilization, cooling, labeling and packaging. In order to produce good quality canned mushrooms, these should be processed as soon as possible after the harvest. In case a delay is inevitable; mushrooms should be stored at 4–5°C till processed. The mushrooms with a stem length of one cm are preferred and are canned whole, sliced and stems and pieces as per demand. Well-graded fresh mushrooms white in color, without dark marks on either caps or stems are preferred for canning. Whole mushrooms are washed 3–4 times in cold running water to remove adhering substances. Use of iron free water with 0.1% citric acid prevents discoloration. Thereafter blanching is normally done to inhibit polyphenol oxidase enzymes activity and to inactivate microorganisms. It also removes the gases from the mushroom tissue and reduces bacterial counts. The mushrooms are blanched in stainless steel kettles filled with a boiling solution of 0.1% citric acid and 1% common salt. The blanching time ranges from 5–6 min at 95–100°C. The mushrooms after blanching are filled in sterilized tin cans (A-2½ and A-1 tall can sizes containing approximately 440 and 220 g drained mushroom weight, respectively). Brine solution (2% salt with 0.1% citric acid or 100 ppm ascorbic acid) is added to the mushroom-filled cans after bringing its temperature to 90°C. After filling, the cans are exhausted by passing them in exhaust box for 10–15 min, so that temperature in the center of cans reaches up to 85°C. Then the cans are sealed hermetically with double seamer and kept in upside down position. After exhausting of cans, sterilization of cans is needed. Sterilization is the process of heating the cans up to 118°C to prevent the spoilage by microorganisms during storage. The cans cooled immediately after sterilization process to stop the overcooking and to prevent stack burning. Cooling can be done by placing the cans in a cold-water tank. Thereafter the clean and dry cans are labeled manually or mechanically and packed in strong wooden crates or corrugated cardboard cartons. The cans are stored in cool and dry place before dispatching to market. In a hot country like India, where the ambient temperature is high during the several months in a year, basement stores are useful, especially during the summer months (Figure 4.2).

4.6 OYSTER MUSHROOM (*PLEUROTUS SP.*)

The oyster mushrooms are harvested and the straw adhered to mushroom is removed and are packed in polythene bags of less than 100 gauge thickness with perforations having vent area of about 5%. Though these perforations cause slight reduction in weight during storage, it helps in maintaining the freshness and firmness of the produce. Rough handling should be avoided storage of oyster mushroom at very low temperatures especially in non-perforated polypacks results in condensation of water with increased sliminess and softening of the texture. Cooling with positive ventilation is desirable, for example, cold air should be directed through the packed produce. For transporting 'dhingri,' the fruit bodies are stacked in trays or baskets. Few polypouches containing crushed ice are kept along with mushrooms. The tray is then covered with thin polythene

Postharvest Management and Processing Technology of Mushrooms 173

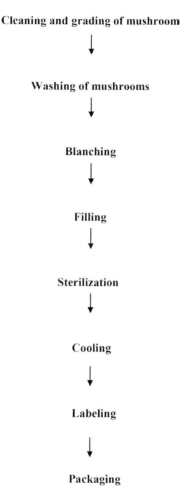

FIGURE 4.2 Flow chart of canning.

sheet with perforation. The prepacked polythene packs with perforations may also be transported in this way.

4.7 MILKY MUSHROOM (*CALOCYBE INDICA*)

Milky mushroom is the new introduction from India to world and its production is catching up fast in different parts of the country during the summer months and the mushroom has revolutionized the so called off-season mushroom growing. Fresh mushroom market is largely catered by seasonal growers who do not have cold-chain storage and transport facilities. They sell the produce in highly localized market. This mushroom has very good shelf life of 3–4 days without loss of color and appearance. Washing, packaging, pre-cooling and refrigeration, transports and storage of fresh milky mushroom, if needed, are almost same as for the button mushroom.

4.8 PADDY STRAW MUSHROOM (*Volvariella volvacea*)

This mushroom is packed in polythene bags. As low temperature storage causes frost injury and deterioration in quality, the best way of storage is at 10–15°C in polythene bags with perforations. Mushrooms packed in bamboo baskets with an aeration channel at the center and dry ice wrapped in paper placed above the mushrooms, is in practice for transportation in Taiwan. Packing in wooden cases for transport by rail or boat is practiced in China. In general the shelf life of this mushroom is very less and mushrooms are sold on the day of harvest.

4.9 TRANSPORTATION

Obviously, fresh mushrooms need to be properly stored to retard postharvest deterioration till these are consumed. Needless to reiterate that the refrigeration or cold-storage is the most essential part of the postharvest care of all the horticultural commodities including mushrooms. Pretreatments, if any, packing and precooling precede the refrigerated storage in most cases. The effect of pre-cooling and packing will be partially negated if the product is later stored and transported in a hot environment. Mushrooms, therefore, need refrigerated transport. To keep the mushrooms cool during transport to short distances, the polypacks of mushrooms can be stacked in small wooden cases or boxes with sufficient crushed ice in polypacks (over-wrapped in paper). For long distance, transport of large quantities in refrigerated trucks is essential though it is costlier.

4.10 DISEASES AND DISORDERS

Disease is generally not an important source of postharvest loss in comparison with physiological senescence and improper handling or bruising.

All diseased caps must be eliminated at harvest. Bacterial blotch or *Pseudomonas* spp. can become a problem during extended storage at elevated temperatures (Suslow and Cantwell, 1998). Low storage temperatures are needed to reduce continued development of mushrooms that occurs after harvest. Common disorders include upward bending of caps and opening of the veil. Mushrooms are easily bruised by rough handling and develop brown discolored tissue.

KEYWORDS

- **Mushroom**
- **Mushroom packaging**
- **Mushroom processing**
- **Postharvest quality**
- **Postharvest Technology**
- **Shelf life and storage**

REFERENCES

Abou-Heilah, A.N., Kasionalsim, M.Y., & Khaliel, A.S. (1987). Chemical composition of the fruiting bodies of *Agaricusbisporus*. *Int. J. Expt. Bot. 47*, 64–68.

Ahmad, M.S., & Siddiqui, M.W. (2015). Postharvest quality assurance of fruits: practical approaches for developing countries. Springer, New York, pp. 265.

Bahl, N. (1983). Medicinal value of edible fungi. In: Proceeding of the International Conference on Science and Cultivation Technology of Edible Fungi. Indian Mushroom Science II, pp. 203–209.

Bano, Z., & Rajarathanam, S. (1982). *Pleurotus* mushrooms as a nutritious food. In: *Tropical Mushrooms–Biological Nature and Cultivation Methods*. Chang, S.T., & Quimio, T.H., (Eds.) The Chinese University Press, Hong Kong, pp. 363–382.

Bano, Z., Bhagya, S., & Srinivasan, K.S. (1981). Essential amino acid composition and proximate analysis of Mushroom, *Pleurotusflorida*. *Mushrooms News Letter Trop., 1*, 6–10.

Chadha, K.L., & Sharma, S.R. (1995). Mushroom Research in India. History, Infrastructure and Achievements. In: *Advances in Horticulture*, Chadha, K.L., & Sharma, S.R. (Eds.), Malhotra Publish House, New Delhi, pp. 1–33.

Chang, S.T., & Buswell, J.A. (1996). Mushroom nutriceuticals. *World J. Microb. Biotech. 12*, 473–476.

Chang, S.T., & Hayes, W.A. (1978). *The Biology and Cultivation of Edible Mushrooms*. Academic Press, New York.
Chang, S.T., & Miles, P.G. (1989). *Edible Mushroom and Their Cultivation*. CRC Press: Florida.
Chu, K.K., Ho, S.S., & Chow, A.H. (2002). *Coriolusversicolor*: A medicinal mushroom with promising immunotherapeutic values. *J. Clin. Pharmacol. 42*, 976–984.
Daugulis, A.J., & Bone, D.H. (1977). Submerged cultivation of edible white rot fungi on tree bark. *Eur. J. Appl. Microbiol. 4*, 159–168.
Esselen, W.B., & Fellers, C.R. (1946). Mushrooms for food and flavor. *Bull. Mass. Agric. Exp. Sta.* pp. 434.
Kannaiyan, S., & Ramaswamy, K. (1980). *A Handbook of Edible Mushrooms*. Todays & Tomorrow's Printers and Publishers, pp. 44–50.
King, T.A. (1993). Mushrooms, the ultimate health food but little research in U.S. to prove it. *Mushroom News, 41*, 29–46.
Li, G.S.F., & Chang, S.T. (1982). Nutritive value of *Volvariellavolvacea*, In: *Tropical Mushrooms–Biological Nature and Cultivation Methods*. Chang, S.T., & Quimio, T.H. (Eds.). Chinese University Press, Hong Kong, pp. 199–219.
Litchfield, J.H. (1964). Nutrient content of morel mushroom mycelium: B vitamin composition. *J. Food Sci. 29*, 690–691.
Manzi, P., Aguzzi, A., & Pizzoferrato, L. (2001). Nutritional value of mushrooms widely consumed in Italy. *Food Chemistry, 73*, 321–325.
Mattila, P.K., Konko, M., Eurola, J., Pihlava, J., Astola, L., Vahteristo, V., Hietaniemi, J., Kumpulainen, N., Valtonen, V., & Piironen, V. (2000). Contents of vitamins, mineral elements and some phenolic compounds in the cultivated mushrooms. *J. Agric. Food Chem. 49*, 2343–2348.
Mehta, B.K., Jain, S.K., Sharma, G.P., Doshil, A., & Jain, H.K. (2011). Cultivation of button mushroom and its processing: an techno-economic feasibility. *International Journal of Advanced Biotechnology and Research, 2*(1), pp. 201–207.
Mizuno, T., Inagaki, R., Kanao, T., Hagiwara, T., Nakamura, T., & Hohshimura, R., (1990). Suniyat; Asukura. Antitumor activity and some properties of water insoluble Heteroglycans from "Himematsutake," the fruiting body of *Agaricusblazei* Murill. *Agricultural and Biological Chemistry, 54*(11), 2889–2896.
Namba, H. (1993). Maitake mushroom the king mushroom. *Mushroom News, 41*, 22–25.
Prakasam, V. (2012). Current scenario of mushroom research in India. *Indian Phytopathol. 65*(1), 1–11.
Rai, D.R. (2009). Quality Assurance and Shelf Life Enhancement of Fruits and Vegetables Through Novel Packaging Technologies, Summer School, CIPHET, Ludhiana.
Rai, R.D. (1995). Nutritional and medicinal values of mushrooms. In: *Advances in Horticulture*. Chadha, K.L., & Sharma, S.R., (Eds.). Malhotra Publishing House, New Delhi, pp. 537–551.
Rai, R.D., & Arumuganathan, T. (2008). *Postharvest Technology of Mushrooms, Technical Bulletin*, NRCM, ICAR, Chambaghat, Solan.
Rai, R.D., & Saxena, S. (1989). Biochemical changes during the post-harvest storage of button mushroom (*Agaricusbisporus*). *Curr. Sci. 58*, 508–510.
Roy, S., Anantheswaran, R.C., & Beelman, R.B. (1995). Fresh mushroom quality affected by modified atmospheres packaging. *J. Food Sci. 60*(2), 334–340.

Sharma, R.K., & Dhar, B.L. (2010). Mushroom cultivation: A highly remunerative crop for Indian farmers. *Indian Farming* (New Direction).

Siddiqui, M.W. (2015). *Postharvest Biology and Technology of Horticultural Crops: Principles and Practices for Quality Maintenance*. CRC Press, Boca Raton, Florida, USA. pp. 550.

Siddiqui, M.W. (2016). *Eco-Friendly Technology for Postharvest Produce Quality*. Academic Press, Elsevier Science, USA. pp. 324.

Siddiqui, M.W., Ayala-Zavala, J.F., & Hwang, C.A. (2016). *Postharvest Management Approaches for Maintaining Quality of Fresh Produce*. Springer, New York. pp. 222.

Simon, A., Gonzalez-Fandos, E., & Tobar, V. (2005). The sensory and microbiological quality control of fresh sliced mushroom (*Agaricus bisporus* L.) packaged in modified atmospheres. *Inter. J. Food Sci. Technol. 40*(9), 943–952.

Singh, H.P. (2011). *Mushrooms, Cultivation, Marketing and Consumption*. RMCU, DMR, Solan.

Sohi, H.S. (1988). Mushroom culture in India, recent research findings. *Indian Phytopath. 41,* 313–326.

Sohi, H.S. (1982). Role of edible mushroom in recycling of agricultural waste and as an alternative protein source: present status of mushroom cultivation in India. *Frontier of Research in Agriculture,* S.K. Roy (Ed.), ISI Calcutta, pp. 565–579.

Swaminathan, M.S. (1995). Agriculture, food security and employment: changing time, uncommon opportunities. *Nature & Resources, 31,* 2–15.

Tewari, R.P., & Pandey, M. (2002). Sizeable income generating venture. *In the Hindu Survey of Indian Agriculture,* pp. 155–167.

Toyama, N., & Ogawa, K. (1974). Comparative studies on cellulolytic and oxidizing enzyme activities of edible and inedible wood rotters. *Mushroom Science, 1,* 745–760.

Verma, R.N., Singh, G.B., & Bilgrami, K.S. (1987). Fleshy fungal flora of N.E.H. India—I. Manipur and Meghalaya. *Indian Mush. Sci. 2,* 414–421.

Wakchaure, G.C. (2011). *Postharvest Handling of Fresh Mushrooms*. RMCU, DMR, Solan.

Wood, D.A. (1989). Mushroom biotechnology. *Inter. Indust. Biotechnol.*

CHAPTER 5

GIBBERELLINS: THE ROLES IN PRE- AND POSTHARVEST QUALITY OF HORTICULTURAL PRODUCE

VENKATA SATISH KUCHI,[1] J. KABIR,[1] and
MOHAMMED WASIM SIDDIQUI[2]

[1]Department of Postharvest Technology of Horticultural Crops, Bidhan Chandra Krishi Viswavidyalaya, Mohanpur, Nadia, West Bengal–741252, India

[2]Department of Food Science and Postharvest Technology, Bihar Agricultural University, Sabour, Bhagalpur, Bihar–813210, India, E-mail: wasim_serene@yahoo.com

CONTENTS

Abstract 181
5.1 Introduction 181
5.2 Chemical Structure 182
5.3 Biosynthesis 182
5.4 Translocation 183
5.5 Functions 184
5.6 Commercial Availability 185
5.7 Role of Gibberellins in Improving and
     Maintaining Postharvest Quality of Fruit Crops 185
     5.7.1 Parthenocarpy 185
            5.7.1.1 Apple 185
            5.7.1.2 Pear 186

5.7.1.3	Mango	187
5.7.1.4	Guava	187
5.7.1.5	Citrus	188
5.7.1.6	Vegetables	189
5.7.2	Fruit Set	189
5.7.3	Control of Fruit Drop	191
5.7.4	Fruit Size	191
5.7.5	Fruit Weight	192
5.7.6	Physiological Loss in Weight	192
5.7.7	Fruit Volume	193
5.7.8	Fruit Color	193
5.7.9	Fruit Firmness	194
5.7.10	Fruit Yield	195
5.7.11	TSS	195
5.7.12	Titrable Acidity	196
5.7.13	Sugars	197
5.7.14	Ascorbic Acid	198
5.7.15	Storage Life	198
5.7.16	Spoilage	199
5.7.17	Organoleptic Rating	200
5.7.18	Carotene Content	200
5.7.19	Total Phenols	200
5.7.20	Enzymatic Activity	201
5.7.21	Mode of Action in Relation to Anti-Ripening and Delaying Senescence	201
5.7.22	Effect of Gibberellins on Vegetables	201
5.7.23	Influence of GA on Plant Height and Internodal Length of Flower Crops	203
5.7.24	Number of Stems	203
5.7.25	Flower Bud Emergence and Flower Harvesting	204
5.7.26	Quality Attributes	204
5.7.27	Length of Stalk	207
5.7.28	Flower Length and Diameter	208

5.7.29	Neck Length	209
5.7.30	Flower Yield	209
5.7.31	Vase Life	213
5.7.32	Bent Neck in Gerbera	217
5.7.33	Disease Resistance in Rose	217
5.7.34	Extra Points	218
Keywords		219
References		219

ABSTRACT

Research in the field of plant hormones is an interesting aspect of physiology where new research findings have been established every year. Plant hormones are synthesized in minute quantities in different plant parts with specific action on target tissues. They are synthesized artificially and applied in the horticultural industry to obtain desired results. Authors among the world had compiled their structure, synthesis, and mode of action. Literature regarding their utilization for quality produces after harvest is scanty. In this chapter, a brief introduction about the discovery, structure, synthesis and functions has been discussed. Interaction of gibberellins with other plant hormones and its role in maintaining postharvest quality of horticultural produce has been highlighted under various headings.

5.1 INTRODUCTION

Gibberellin (GA) was discovered when the attention of plant pathologists of Japan was drawn while rice plants were unable to set seed and had grown excessively tall (named as foolish seedling). They discovered that it was caused by fungus, *Gibberella fujikuroi*. Subsequently, they have extracted in crystal form by culturing the fungus in laboratory and named it as Gibberellin A in 1926. After three decades, Japanese scientists had alienated three various forms of Gibberellins namely gibberellin A_1 (GA_1), gibberellin A_2 (GA_2) and gibberellin A_3 (GA_3) from gibberellins A. So far, 136 gibberellins have been recognized GA_1, GA_2, GA_3, GA_4, GA_5, etc. The number is given based on order of their discovery. Gibberellin usually represents to the total class of hormones and Gibberellic acid particularly to GA_3, which exists most commonly in commercial form (Taiz and Zeiger, 2006). Naturally occurring GA's which are active in higher plants are GA_1 and GA_4. Young leaves, apical buds, elongating shoots, apical regions of roots and developing fruits and seeds are important sites for synthesis of GA. Abundant quantities of gibberellins were found in developing seeds and fruits (Hopkins and Huner, 2008).

5.2 CHEMICAL STRUCTURE

Gibberellins are acidic, tetracyclic, diterepenes with ent-gibberellane structure (Figure 5.1) with 19 or 20 carbon atoms. GAs with 19 carbon atoms are biologically active than C-20 GAs (GA_3). The structure of most widely used GAs in research is given in Figure 5.2.

5.3 BIOSYNTHESIS

Gibberellins are synthesized from geranylgeranyl pyrophosphate (GGPP) which is a multi enzyme pathway (simplified and given in Figure 5.3). In plastids, four isoprene units assemble to give GGPP (20 carbon molecule). Ent-kaurene, a tetracyclic is formed with the help of enzymes ent-copalyl-diphosphate and ent-kaurene synthase. In a multistep process GA_{12} is formed from ent-kaurene which is catalyzed by ent-kaurene oxidase and ent-kaurenoic acid oxidase in endoplasmic reticulum (synthesis of GA up to GA_{12} same in all plant species). $GA-C_{20}$ and $GA-C_{19}$ (including bioactive forms)

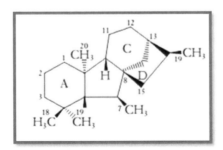

FIGURE 5.1 Ent-gibberellane structure of GA (Source: http://www.planthormones.info/gibberellins.html).

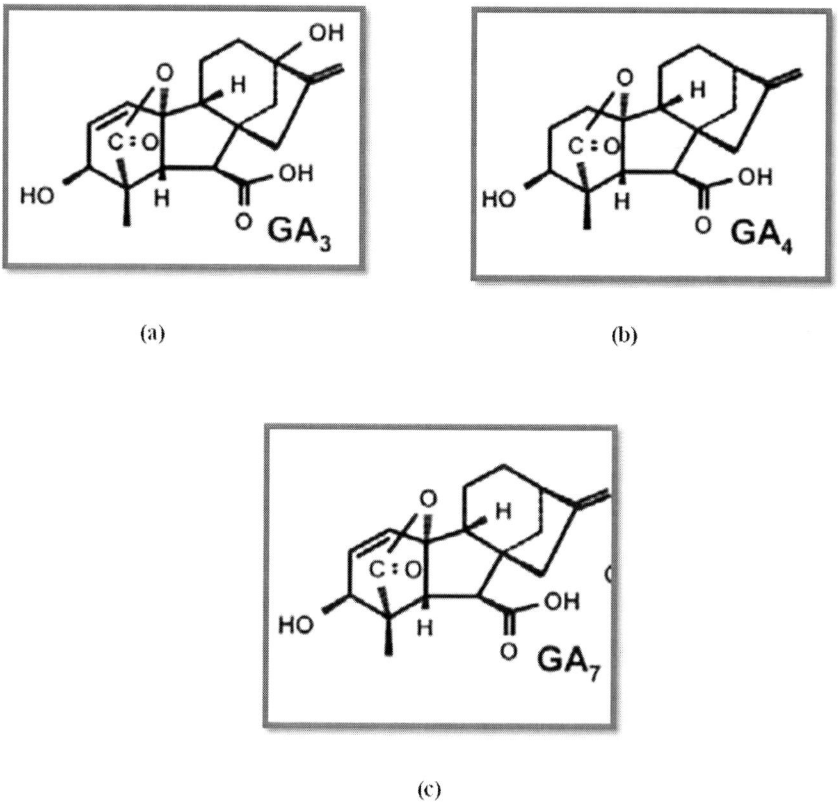

Gibberellins: The Roles in Pre- and Postharvest Quality 183

(a)

(b)

(c)

FIGURE 5.2 (a). GA_3 (b). GA_4 (c). GA_7 (Source; Davies, 2004).

are formed from GA_{12} through a series of oxidative reactions (Krishna, 2012). Gibberellin biosynthesis is influenced by light temperature and ratio of bioactive GA to ABA (abscisic acid) ratio (Taiz and Zeiger, 2006).

It is important to know the biosynthesis of gibberellins to know the mode of action of growth retardants. High GA content in vegetative tissues interferes with flowering of some of the fruit trees (e.g., Mango). Growth retardants such as daminozide (Alar), paclobutrazol (Cultar), CCC (Cycocel), phosphon D block the synthesis of GA's (Figure 5.3). So, flowering is achieved with reduced intermodal elongation in such trees. Foliar application (2000 ppm) or soil drenching (10 g/tree) of paclobutrazol can promote flowering in 'Alphonso' mango trees during off year (Rao and Srihari, 1996).

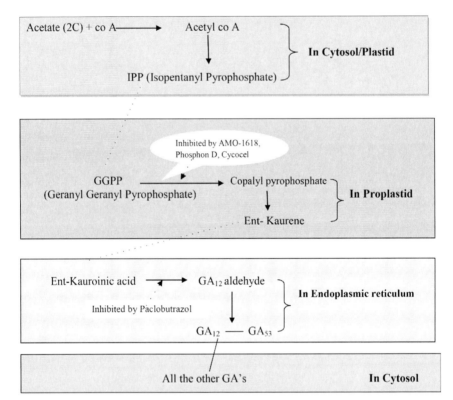

FIGURE 5.3 Simplified diagram of biosynthesis of gibberellins.

5.4 TRANSLOCATION

Gibberellins transport is not polar but a velocity up to 1 mm/h had been reported (Jacobs, 1979). Xylem and phloem can interchange GA while transporting to target tissues (Pessarakli, 2002). Many researchers conducted experiments with radioactively labeled GA but their translocation whether they are transported as free hormone or as conjugated form is not known (Hopkins and Huner, 2008).

5.5 FUNCTIONS

Some physiological processes stimulated by gibberellins are stimulation of stem elongation by stimulating cell division and cell elongation. Breaking seed dormancy and promotion of bolting or flowering in response to long days, breaking of seed dormancy in some plants, which require stratification or light for induction of germination. Stimulation of enzyme production (α-amylase) in germinating seeds for mobilization of seed reserves, induction of maleness in dioecious flowers (sex expression), promotion of seedlessness during development of fruits and delaying senescence of in leaves and fruits (Siddiqui et al., 2016).

5.6 COMMERCIAL AVAILABILITY

Commercially available forms of gibberellins are Gibberellic acid or GA_3 ($C_{19}H_{22}O_6$) in the trade of Pro-Gibb, RyzUp, Release, Relax and Berelex (powdered form). It is soluble in water and highly soluble in ethanol, methanol or acetone. A new formulation of GA_{4+7} is available in commercial form in the trade name 'Provide' which is used in apple fruit growers to minimize physiological disorders and to improve overall appearance (Arteca, 2015).

5.7 ROLE OF GIBBERELLINS IN IMPROVING AND MAINTAINING POSTHARVEST QUALITY OF FRUIT CROPS

5.7.1 PARTHENOCARPY

Seedlessness is regarded as a valuable character for commercial purposes and is considered ideal to the consumer as well as to the producer. In many fruits like banana, grape and pineapple parthenocarpy is essential while in some fruits like pomegranate guava and citrus even reduction in seed number would be an asset. There are some fruits, which normally develop seeds, either by disturbing their genetic construction, by artificial pollination, by growth substances or by mechanical means (girdling of grape vines). These fruits can be made to develop without fertilization resulting into seedless fruits. Consumer preference towards parthenocarpic fruits was increasing as it saves time, labor, easy to eat and improves mouth feel attribute. Seedlessness is appreciated by consumers both in fruits for fresh consumption (e.g., grape, citrus, and banana) as well as in conserved or processed fruits (e.g., frozen eggplant, tomato sauce). Seedlessness can contribute to increase the quality of the fruits when seeds are hard or have a bad taste. Gibberellins promote parthenocarpy by abandoning fertilization. Parthenocarpic fruits are formed under high expression level of GA biosynthetic genes such as SlGAox1 (De Jong et al., 2011). Gibberellic acid is widely employed exogenously in the production of seedless fruits.

5.7.1.1 Apple

Gibberellic acid induced parthenocarpy was reported first time in apple by Davidson (1960). Dennis and Egerton (1962) obtained parthenocarpic fruit of apple cultivar McIntosh with the application of potassium salt of 1000 ppm GA. Bukovac (1963) found parthenocarpic fruit development in delicious apple when emasculated flowers were subjected to GA application. Varga (1966) sprayed different concentrations of gibberellins to emasculated flowers of apple cultivar Golden Delicious and obtained seedless fruits with 100 ppm GA. Parthenocarpy was induced in Cox's orange pippin apple by single spray of gibberellins (Schwabe, 1973). Seedless apple fruits were produced when decapitated blossoms were sprayed with 50 ppm GA_{4+7} (Modlibowska, 1975). Parthenocarpy was induced up to 60% in Jonathan apple cultivar with 200 ppm GA_{4+7} (Taylor, 1975).

5.7.1.2 Pear

Pear fruits set parthenocarpically when flowers were sprayed with potassium salt of GA at 50 ppm (Luckwill, 1953). Gill et al. (1972) developed parthenocarpic fruits in pear cultivar Winter Neils with bloom application of 200 ppm GA_3; 200 and 50 ppm GA_{4+7}. Similarly, Modic and Turk (1978) obtained parthenocarpic fruits in pear sprayed with 45 ppm GA_3 or GA_{4+7} at full bloom stage. Large fruits of Forell pear fruits were found parthenocarpic when treated with Progibb at flowering time (Honeyborne, 1996).

Other temperate fruits

Crane et al. (1960) obtained parthenocarpic fruits in peaches and apricots with 50 and 500 ppm potassium salt of GA. Constanin and Crane (1961) reported parthenocarpic fruit set in cherry with application of 1000 ppm GA at full bloom stage. Parthenocarpic fruit development has been observed in plums (100%) with GA (Webster and Goldwin, 1978). In peaches 500 ppm GA applied to emasculated flowers resulted in greater percentage of parthenocarpic fruits (Stembridge and Gambrell, 1970). Sansavini and Fility (1972) reported parthenocarpy in peaches with the application of 20 and 40 ppm GA_3. Sansavini (1973) further stated that 50% and 80% parthenocarpic fruits were obtained when GA was treated during flowering time at 10 days after flowering. 1000 ppm (Marlangeon and Comes, 1974) and 500 ppm (Bengoa and Marlangeon, 1975) of GA_3 induced higher percentage of parthenocarpic fruits applied to emasculated flowers of peach.

Loquat

GA sprayed at 300 ppm to Thames pride and California Advance of loquat produced seedless fruits (Rao et al., 1960). High percentage of parthenocarpic fruit development in loquat was reported with the application of GA before flowering (Zhang et al., 1999).

Grape

Itakura et al. (1965) observed more than 95% parthenocarpy in Delaware cultivar of grapes when flowers were sprayed with 100 ppm of GA. Seedlessness in grapes have been reported by Volsoueov (1966), and Das & Randhawa (1968, 1970) in Anab-e-Shahi, Bhokri and Pusa Seedless grapes with 75–100 ppm GA. Total seedlessness was achieved with 100 ppm GA in Delaware grapes (Muranishi and Iwagawa, 1972; Prasad and Prasad, 1973). GA was proved to be effective than GA_{4+7} in inducing seedlessness in grapes (Modic and Turk, 1978). Lee et al. (1986) opined that 25 ppm GA when applied at pre-bloom stage induced 90% seedlessness in Koyoho grapes. GA 100–300 ppm sprayed at late bloom stage and one week later induced parthenocarpy in grape berry (Jiang et al., 1997).

5.7.1.3 Mango

Venkataratnam (1949) obtained parthenocarpic fruits in mango cultivar Langra, Banglora and Banganapalli with GA when applied after emasculation. Similarly, parthenocarpic mangoes were obtained by Chacko and

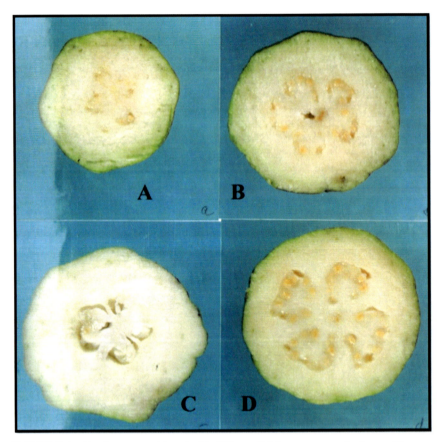

FIGURE 5.4 Seedlessness achieved with different GA3 treatments at bud stage (Source: Kaur, 2003) [A. GA3 (800 ppm); B. GA3 (1000 ppm); C. GA3 (1200 ppm); D. Control].

Singh (1969) with 250 ppm GA; Singh et al. (1977) with 100 ppm of GA in Langra mangoes, and Kulkarni & Rameshwari (1978) in Thambva variety of mango with 200 ppm GA.

5.7.1.4 Guava

Potassium salt of gibberellic acid (10,000 ppm) induced parthenocarpy in guava and reached maturity early (10 days earlier than normal fruits)

(Teaotia et al., 1961). A 100% parthenocarpic fruits were obtained when buds of 2.5 cm size were treated with GA3 (1200 ppm) and had 2–4 seeds but the fruits are misshapen (Figure 5.4) (Kaur, 2003).

5.7.1.5 Citrus

Randhawa et al. (1964) obtained parthenocarpic fruits in grape fruit and mandarin with application of 25–10,000 ppm GA. Parthenocarpy was observed in mandarin cultivar 'Monreal' subjected to 5–200 ppm GA (Garcia and Garcia, 1979). In valencia oranges, Turnbull (1989) obtained 100% seedless fruits with the application of GA and paclobutrazol.

5.7.1.6 Vegetables

Active gibberellins (GA_1 or GA_3) are able to induce fruit set in several horticultural species and in the model species *A. thaliana* (Gillaspy et al., 1993). However, tomato gibberellin induced fruits are smaller than seeded fruits suggesting that other signals are required for tomato fruit growth and development (de Jong et al., 2009). Increased levels of gibberellin have been detected in pollinated ovary together with an increased expression of GA biosynthetic genes (Serrani et al., 2007). The role of gibberellins in fruit set was also supported by the analysis of the parthenocarpic tomato mutans *pat*, *pat2* and *pat3/4*. These mutants show increased level of GA's and an enhanced expression of GA biosynthetic genes (Fos et al., 2000; Olimpieri et al., 2007). Auxin and gibberellin may act in parallel or in a sequential way on fruit set. A synergistic effect of auxin and gibberellin on fruit growth has been observed in pea and tomato suggesting that the two phytohormones interact in regulating fruit development (Ozga, 1999). Data recently obtained in *A. thaliana* and in tomato suggest a hierarchical scheme of interaction where GA acts downstream of auxin. In tomato, auxin-induced fruit initiation is mediated by GAs and in ovaries treated with auxins GA biosynthetic genes are induced (shown in Figure 5.5 and 5.6) (Serrani et al., 2007). Similar results were obtained in *A. thaliana* where fertilization triggers an increase in auxin response in the ovules and the activation of GA synthesis. The depletion of SlDELLA proteins which are negative

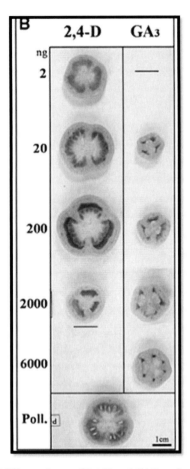

FIGURE 5.5 Effect of different doses of 2,4-D and GA3 on the size of the fruit of tomato (Serrani et al., 2007).

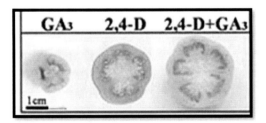

FIGURE 5.6 Cross section of tomato fruit treated with GA3, 2,4-D and the combination (Source: Serrani et al., 2007).

regulators of GA signal transduction pathway, allowed to overcome the growth arrest normally imposed on the ovary at anthesis (Marti et al., 2007). Antisense *SlDELLA* engineered tomato fruits were seedless, but smaller in size and elongated in shape compared with pollinated fruits. Cell number estimations showed that fruit set, resulting from reduced *SlDELLA* expression, arose from activated cell elongation at the longitudinal and lateral axes of the fruit pericarp. Interestingly also the quadruple-DELLA loss-of-function mutant of *A. thaliana* displays the parthenocarpic phenotype (Dorcey et al., 2009).

5.7.2 FRUIT SET

In the investigations of Phuangchik (1994), the application NAA 10 ppm + GA_3 20 ppm + Fulmet 20 ppm (1.63%) resulted in maximum fruit set in mango cv. Nam Dok Mai Tawai. Bankar and Prasad (1990) sprayed GA_3 and NAA 10, 20 and 30 ppm on eight-year-old trees of ber cv. Gola at flowering and again 15 days later at fruit set and revealed that the fruit retention was increased by all the treatments, compared with water-sprayed controls. Pandey (1999) reported that the application of NAA 20 ppm and GA_3 15 ppm resulted in the greatest fruit retention in ber cv. Banarasi Karaka. No beneficial effect of GA_3 on fruit retention was observed (Ghosh et al., 2008). Wangbin et al. (2008) examined the effects of molybdenum (Mo) foliar sprays on fruiting of jujube 'Dongzao' (*Zizyphus jujuba* Mill). Two treatments (Mo 50 ppm + girdling + GA_3 15 ppm and Mo 100 ppm + girdling + GA_3 15 ppm) applied during flowering period increased the fruit set significantly as compared to other treatments.

5.7.3 CONTROL OF FRUIT DROP

Pramanik and Bose (1974) studied the effect of varying doses of GA_3 and NOXA on fruit drop of ber cv. Banarasi Karaka and concluded that both GA_3 and NOXA 10 ppm decreased the fruit drop considerably. GA_3 60 ppm accounted for the lowest fruit drop and highest fruit set in Umran Ber (Singh and Randhawa 2001). Ram et al. (2005) implicated that application of GA_3 15 and 25 ppm decreased the fruit drop and improved the fruit retention in ber cv. Banarasi Karaka.

5.7.4 FRUIT SIZE

Dhillon and Singh (1968) reported that 75 ppm GA_3 increased the fruit diameter significantly in Dandan variety of ber as compared to control. Patil and Patil (1979) concluded that CCC 250 ppm reduced the size of fruit in ber. They further reported that application of 10 ppm NAA and 10 ppm GA_3 increased the fruit size. The results were again confirmed by Patil (1981). Pandey (1999) reported that the application of NAA 20 ppm and GA_3 15 ppm resulted in increased fruit size (length x breadth) in ber cv. Banarasi Karaka. Kale et al. (2000) studied the effect of plant growth regulators on fruit characters of ber and reported that the fruit size increased appreciably with GA_3 20 ppm. Gibberellic acid (Pro-Gibb 4%) applied 3 weeks beforeharvest delayed maturity from 3 to 7 days in cherries (Willemsen, 2000) and gave larger fruit with storage life. They further reported that the lightly cropped trees required lower doses of GA_3.

5.7.5 FRUIT WEIGHT

Banker and Prasad (1990) reported that fruit weight in ber cv. Gola was significantly increased by application of GA_3 (30 ppm) and NAA (30 ppm). Pandey (1999) reported that GA_3 15 ppm and NAA 10, 15 or 20 ppm increased the fruit weight in Banarasi Karaka cultivar of ber. Seven year old Umran ber trees were sprayed with gibberellic acid (GA_3) and NAA (10 and 20 ppm) alone and in combination up to 60 days after full bloom by Kale et al. (2000). The fruit weight was found significantly improved by higher concentration of GA_3 and NAA. However, the combination of GA_3 and NAA were not much effective. In Peach cvs. Springtime and Early Red, Engin et al. (2007) reported that application of GA_3 coupled with irrigation at final swell of the fruits resulted in increase in fruit weight as compared to control.

Dhillon et al. (1982) observed a progressive decrease in loss in weight of Flordasum peach with the increase in Gibberellin, concentration.

5.7.6 PHYSIOLOGICAL LOSS IN WEIGHT

Randhawa et al. (1980) recorded minimum loss in weight of pear treated with 4% $CaCl_2$ followed by GA_3 at 50 ppm concentration. Subtropical peaches showed the minimum loss in weight, less than 1000 ppm GA_3. However the loss in weight in Sharbati peach was not affected with the increasing concentration of GA_3. Parmar and Chundawat (1988) indicated significant variability in cumulative physiological weight loss during storage under various treatments in mango cv. Kesar. Kinetin (75 ppm) in combination with bavistin (1000 ppm) and GA_3 (150 ppm) recorded significantly low percentage of physiological weight loss (PLW). Sindhu and Singhrot (1993) reported minimum PLW in lemon cv. Baramasi when treated with GA_3 (200 ppm) as pre-harvest application. Bhanja and Lenka (1994) found that combined pre-harvest spray with calcium chloride, calcium nitrate and GA_3 one month before harvesting on sapota fruits and postharvest dip in GA_3 @100 ppm revealed reduced PLW. Sudhavani and Shankar (2002) concluded that physiological loss in weight was reduced in GA_3 treated mango cv. 'Baneshan' fruits. It has also been reported for other fruits like Nagpur mandarin, Clementine mandarin and Washington Navel Oranges (El-Otmani and Coggins, 1991; Ladaniya, 1997), and banana (Osman and Abu-Goukh, 2008). The reduced loss in weight in GA_3 treated fruits could be attributed to their increased affinity to water. For that reason fruit retain more water against the force of evapotranspiration, resulting in lesser weight loss during storage. Other contributing factors might be changes in some of the proteinaceous constituents of cell (Yadav and Shukla, 2009), as well as inhibition of respiration.

5.7.7 FRUIT VOLUME

Rani and Brahmachari (2004) reported that mango cv. Amarpalli when sprayed with GA_3 200 ppm produced fruits with the greatest volume. Masalkar and Wavhal (1991) studied the influence of GA_3, NAA, chlormequat and ethephon, alone and in various combinations on fruit volume in ber. The highest fruit volume was obtained with GA_3 10 and 20 ppm. Pandey (1999) conducted trials on ber cv. Banarasi Karaka, by spraying with NAA 5, 10, 15 and 20 ppm and GA_3 5, 10 and 15 ppm at the pea-stage. The GA_3 15 ppm and NAA 10, 15 and 20 ppm increased the fruit volume significantly as compared to control.

Ranjan et al. (2005) reported that the postharvest application of calcium salts and GA_3 during storage resulted in the gradual reduction in specific gravity of Langra mango although the effect of treatments was marginal.

5.7.8 FRUIT COLOR

Color retainment of ber fruit can be achieved by dipping in GA_3 (60 ppm) and calcium chloride (2.0%) for 5 minutes up to 20 days of storage (Jawanda and Randhawa 2008). Balasubramaniam and Agnew (1990) investigated the effect of gibberellic acid sprays on quality of cherry and reported that GA_3 reduced average skin color. The cv. Rainier of cherry did not develop the pink blush, but turned a medium straw color.

In peach cv. Rubiduox, the effects of pre-harvest spraying of gibberellic acid (GA_3) and aminoethoxyvinylglycine (AVG) were investigated by Amarante et al. (2005). They found that the treatment with GA_3 (100 mg/L) and AVG (75 and 150 mg/L) resulted in better retention of skin background color. In addition, GA_3 reduced the number of fruits with skin splitting and decay and reduced the incidence of flesh browning after cold storage.

5.7.9 FRUIT FIRMNESS

Sankhla et al. (2006b) conducted a study to gain an insight into the effect of CEPA, 1-MCP and gibberellic acid (GA_3) on postharvest ripening, quality and shelf life of ber fruits. They implicated that application of GA_3 delayed decrease in fruit firmness towards ripening. In combination, GA_3 and 1-MCP exhibited additive effects on fruit firmness. Ozkaya et al. (2006) opined that the application of GA_3 10 ppm decreased the loss of fruit firmness and maintained the surface brightness in sweet cherry. The preharvest spray of 50 ppm GA_3 with postharvest dipping of calcium chloride along with bavistin recorded maximum firmness in sapota cv. PKM-1 (Sudha et al., 2006). Sudha et al. (2007) reported that preharvest application of potassium chloride (1%) sapota cv. PKM-1 recorded least firmness (2.7 kg/cm^2) as compared to GA_3 (3.45 kg/cm^2) treatment.

Kappel and MacDonald (2007) reported that a single spray of 20 ppm GA_3 at the straw-yellow stage of fruit development of 'Sweetheart' sweet cherry increased the fruit firmness by 15 per cent. The attractiveness of fruit surface responded linearly to the GA_3 applications. Randhawa et al. (1980) reported that the firmness of the pear fruits decreased during storage, but this decrease was minimum under 100 ppm GA_3, followed by the 50 ppm GA_3 treated fruits. Rathore et al. (2009) Mango fruits CV. Chausa treated with GA3 showed uniform texture and maximum firmness of fruits after 19 days of storage.

Delay in softening by GA_3 during on tree-storage of grapefruit and tight skin oranges is reported by Ferguson et al. (1982). It has an antagonistic effect on the biosynthesis of endogenous ethylene, the compound that at threshold level triggers the ripening process in climacteric fruits (Burg and Burg, 1962; Ben-Arie et al., 1996; Ben-Arie et al., 1986). The mechanism of delaying softening and other degradative changes due to GA_3 application is also supported by Lewis et al. (1967) in the studies on Navel Orange peel involving calcium metabolism in cell wall and vegetative tissue, sites responsive to GA_3's effects (Jona et al., 1989; Ben-Arie et al., 1996). Because cellulose micro fibrils provide the structure and support for all the components of plant cell walls (Greve and Labavitch, 1991; Carpita and Gibeaut, 1993), the increase in cellulose would increase fruit firmness. Ben-Arie et al. (1996) suggested that the greater firmness of GA_3 treated fruit accounts for the 37% higher cellulose content in the cell walls of the treated fruit. Both factors, such as cellulose synthesis and hydrolytic activity, probably affect fruit firmness as determined by GA_3 treatment.

5.7.10 FRUIT YIELD

Hassan et al. (2005) applied boron, GA_3 and active dry yeast at different concentrations alone or in their combinations on 'Canino' apricot trees. The results showed that combined application of boric acid 400 ppm, GA_3 40 ppm and active dry yeast 2% at full bloom stage caused a remarked promotion of fruit yield. Webster et al. (2006) reported that the yield of sweet cherry could be improved by foliar sprays of gibberellic acid (GA_3) or aminoethoxyvinylglycine (AVG) applied post blossoming stage.

5.7.11 TSS

In Gola cv. of ber, Bankar and Parsad (1990) found that TSS was appreciably influenced by application of GA_3 but not by NAA. Kale et al. (2000) studied the effect of plant growth regulators on quality of ber and reported the application of GA3–20 ppm was the most effective in inducing highest increase in total soluble solids as compared with other treatments.

Balasubramaniam and Agnew (1990) investigated the effect of gibberellic acid sprays on quality of cherry and concluded that the application of GA3 at the beginning of stage three of fruit growth (approximately three weeks before harvest) significantly improved the total soluble solids of cherries. Kappel and MacDonald (2007) reported that TSS content of 'Sweetheart' sweet cherry fruit could be increased by repeated application of 20 ppm GA3. The response was the greatest in cultivar Bing, followed by cultivars Rainier, Stella and Dawson. GA3–10 ppm was as effective as GA3–20 ppm in improving overall cherry quality. Kumar and Singh (1993) found that the pre-harvest sprays of GA3 (50 or 75 ppm) or ethephon 500 ppm on mango cv. Amarpalli significantly improved fruit total soluble solids content. Sarkar and Ghosh (2005) studied the effect of growth regulators (2,4-D, NAA, GA3 and Planofix) on the biochemical composition of mango cv. Amrapalli. They reported that GA3–20 ppm gave the highest total soluble solids content (21.22 degrees Brix). Sandhu et al. (1983) observed maximum TSS (13%) in Kinnow mandarin treated with 30 ppm GA_3 and covered with sugarcane trash followed by control. According to Ghosh and Sen (1984) the total soluble solids were higher m sweet orange fruits treated with GA and stored m room temperature and lower in fruits kept at 10°C.

5.7.12 TITRABLE ACIDITY

Khader (1991) observed that preharvest application of GA_3 (300 or 400 ppm) increased the total acidity in 'Dashehari' mango fruits. Masalkar and Wavhal (1991) recorded the highest percentage of acidity with foliar application of GA3 coupled with NAA treatments in fruits of ber cv. Umran. Kale et al. (2000) sprayed seven-year-old Ber cv. Umran with gibberellic acid and NAA (10 and 20 ppm) alone and in combination up to 60 days after full bloom. There was significant decrease in acidity of fruits with higher concentration of GA3 and NAA. Kaur et al. (2004) reported that application of GA3–20 and 50 ppm, NAA 10 and 20 ppm, 2,4-D 4 and 8 ppm and 2,4,5-T 10 and 20 ppm decreased acid content of fruits of plum cv. Satluj Purple. Amarante et al. (2005) reported that in peach cv. Rubiduox, the treatments with GA_3 (100 mg/L) and AVG (75 and 150 mg/L) resulted in the least increase of acidity. Benjawan et al. (2006) reported that application of GA$_3$ increased the acidity of the fruits of Kaew mango cv. Srisaket. Shrivastava and Jain (2006) reported that the acidity in mango cv. Langra decreased with foliar applications of urea and GA$_3$. The minimum acidity was observed with urea 2% and GA$_3$ 100 ppm (0.229%) whereas, the unsprayed control trees had maximum acidity (0.289%).

Kotecha et al. (1993) reported that total sugar content in banana fruits treated with either GA$_3$ or Kinetin increased progressively through the ripening, but had significantly lower values of total sugars than control.

5.7.13 SUGARS

Masalkar and Wavhal (1991) sprayed ber trees with GA3 (10 and 20 ppm), NAA (10 and 20 ppm), chlormequat (250 ppm) and ethephon (400 ppm), alone and in various combinations. The maximum increase in non-reducing sugars was obtained with application of GA3 alone. Kale et al. (1999 and 2000) reported that reducing sugar and total sugars of Umran ber increased with GA3 and NAA 10 and 20 ppm alone. The combination of GA3 and NAA were not much effective. Kumar and Singh (1993) found that the preharvest sprays of GA3 (50 or 75 ppm) or ethephon 500 ppm on mango cv. Amarpalli significantly improved the total sugar content of the fruits. In Flordasun cv. of peach, Babu and Yadav (2004) compared efficacy of various thinning agents and found that the application of GA3–100 ppm as thinning agent resulted in maximum total sugars (6.15%). The application of GA3–20 ppm produced Amrapalli mango fruits with maximum total sugars (16.77%) and reducing sugars (7.00%) in the studies of Sarkar and Ghosh (2005). Shrivastava and Jain (2006) found that application of GA3–100 ppm to mango cv. Langra produced the maximum reducing (12.24%) and total sugars (17.90%). Singh (1988) observed that in both cultivars Zardalu and Langra of mango reducing sugar of the fruits increased gradually in all the treatments as the storage period advanced. Treatment with 2000 ppm captan followed by wrapped in perforated polyethylene bag effectively increased and maintained maximum reducing sugar (3.74%) in Zardalu cultivar whereas 50 ppm GA$_3$ in combination with perforated polyethylene bags effectively increased and maintained maximum (3.31%) reducing sugar in Langra cultivar. Randhawa et al. (1980) obtained the maximum amount of total sugar in pear under GA 100 ppm and CCC 1000 ppm closely followed by GA_3 50 ppm and untreated fruits during storage. Saini et al. (1982) noticed that the total sugars in peaches were also increased during storage but the maximum total sugars were recorded (4.80–5.10%) in fruits treated with 100 ppm GA_3 as compared to (4.55–4.75%) fruits under control.

5.7.14 ASCORBIC ACID

Dhillon and Singh (1968) found that GA_3 (25 and 75 ppm) and 2,4,5-T (5 and 15 ppm) increased the ascorbic acid content in Dandan cultivar of ber. In Kaithali cultivar both GA_3 and 2,4,5-T were equally effective. Wavhal (1991) applied chemicals like GA_3 (10 and 20 ppm), NAA (10 and 20 ppm), Chlormequat (250 ppm) and ethephon (400 ppm) alone and in various combinations. The highest ascorbic acid content was obtained with GA_3 alone. Pandey (1999) concluded that the application of NAA 20 ppm and GA_{3-15} ppm resulted in enhancing the ascorbic acid content of ber cv. Banarasi Karaka. Jiang et al. (2004) studied the role of 1-methylcyclopropene (1-MCP) and GA_3 in fruit ripening of Chinese jujube during storage in relation to quality. The treatment with 1-MCP or GA_3 delayed the decrease in vitamin C content of fruits. Kumar and Singh (1993) found that the pre-harvest sprays of GA_3 (50 or 75 ppm) or ethephon 500 ppm on mango cv. Amarpalli significantly improved the ascorbic acid content of fruits. Babu and Yadav (2004) implicated that in Flordasun cv. of peach, the application of GA_{3-100} ppm as thinning agent resulted in maximum vitamin C content in fruits (212.30 mg/100 g). Singh (1988) observed that the ascorbic acid content of fruits decreased gradually in both Zardalu and Langra mango fruits during storage under all the treatments. Treatment with 50 ppm GA_3 and kept in perforated polyethylene was found most effective in minimizing the loss of ascorbic acid in both the cultivars, for example, Zardalu and Langra. Singh (1996) also recorded similar results in case of mango fruits under storage.

5.7.15 STORAGE LIFE

Use of growth regulators became an important orchard practices to enhance quality and shelf life of fruits. In this aspect needs a precise knowledge of growth regulators, their action and practical application. Growth regulators like gibberellic acid are known to promote the shelf life of fruits. The prolongation of fruit life due to growth regulators was probably due to effectiveness of these chemicals in retaining of green pigments, retardation of ripening and senescence (Huang, 1974). The preharvest spray of GA_{3-100} ppm significantly suppresses the succinate activity of malate-dehydrogenase during postharvest ripening of papaya and thus delay ripening. Ahmed and Singh (1999) stated that 'Amrapali' mango fruits recorded maximum economic life of 11 days when treated with GA_3 (50 ppm). Preharvest application of GA_3 (40 ppm) has enhanced the postharvest life of peaches when compared to untreated fruits when stored at 0–1°C and 90–95% relative humidity Figure 5.7 (Kaur, 2012).

Investigations conducted by Singh and Singh (1993) to ascertain the efficiency of various treatments on extending the post-harvest life of Zardalu and Langra cultivar of mango fruits at room temperature showed that spoilage reduced to greater extent when fruits were treated with 50 ppm GA_3. This treatment extended the economic storage life of fruits by 3–5 days. Treatment with 100 ppm GA_3 could be a useful method to extend postharvest life and availability of 'Himsagar' mango with appreciable quality (Siddiqui et al., 2013).

FIGURE 5.7 Pre-harvest application of GA_3 on Peach (cv. Shan-I-Punjab) fruits stored at 0–1°C and 90–95% RH after 35 days of storage (Source: Kaur, 2012).

5.7.16 SPOILAGE

Preharvest application of calcium nitrate (2%) + NAA (100 ppm) + GA_3 (50 ppm) has reduced rot percentage of Allahabad Safeda guava during storage (Singh, 1988). Khader (1991) reported that pre-harvest application of GA (300 or 400 ppm) has reduced the decay loss in mango fruits cv. Dashehari. Application of GA_3 (50 ppm) has reduced decay loss in sapota by 10, 30 and 70% on 3rd, 6th and 9th day, respectively (Banik et al., 1988). Fruits of ber cv. Umran treated with $CaCl_2$ (2%) registered minimum rotting followed by GA_{3-60} ppm treatment (Jawandha et al., 2009).

Singh (1988) observed that the postharvest application of calcium nitrate (1%) and GA_3 (40 ppm) was effective in reducing rot percentage and finally maintained the edible quality and marketability of fruits of guva cv. Allahabad Safeda for more than 6 days during storage.

5.7.17 ORGANOLEPTIC RATING

Fruits of Umran ber after 30 days of storage, the maximum palatability rating was recorded in $CaCl_2$ (2%) followed by GA_3 60 ppm treatment (Jawandha et al., 2009). Similar results were obtained by Kumar et al. (1992) in 'Baramasi Pawandi' fruits. Kumar (1998) reported that mango fruits cv. Sipia showed maximum organoleptic score (7.8 out of 9) when treated with GA_3 (250 ppm) alone and along with 500 ppm bavistin retained the consumer acceptability for a long time.

5.7.18 CAROTENE CONTENT

Kumar and Singh (1993) studied the effect of GA_3 and ethrel on quality of mango cv. Amrapali and they found that pre-harvest spray of GA_3 (50 and 75 ppm) or ethrel (500 ppm) significantly improved carotene concentration in fruit.

5.7.19 TOTAL PHENOLS

Phenols are one of the most important stable secondary metabolites (Dewich and Haslam, 1969). They are synthesized through Shikimmic pathway (Kefeli et al., 2003). Phenols help in determining color and flavor (especially astringent flavor) in most of the fruits (Van Buren, 1970). Aly and Ismail (2000) reported that the pre-harvest treatment of 'Balady' guava with GA_3 (150 ppm) decreased fruit skin browning and polyphenol oxidase activity compared with control and boron treated fruits.

5.7.20 ENZYMATIC ACTIVITY

Pectin methyl esterase (PME) is an endogenous enzyme in plants that catalyzes desertification of carboxyl groups in pectin molecules, yielding galacturonic acid and methanol. The resulting galacturonic units in pectin molecules are potentially substrates for polygalacturonase (PG). PG is also an endogenous enzyme that catalyzes hydrolytic cleavage of α-1,4 glycoside bonds between galacturonic acid residues, resulting in tissue softening of fruits.

The changes in cell wall composition which accompany softening of ripening fruit apparently result from the action of enzymes produced by the fruit. Prominent enzymes such as PME and PG bring about striking changes in cell wall pectin content in ripening fruit (Seymour et al., 1993). Gibberellins treatment proved to be highly effective in reducing enzymatic activity. Pre-harvest spray of GA_{3-50} ppm reduced the activities of softening enzymes (PME and PG) in the fruits of Cape gooseberry by protecting stiff pectin macromolecules against demethylation or reduced depolymarization of 2 olygalacturonase (Majumder and Majumder, 2001). Mehta et al. (1986) reported that GA_3 (100 ppm) significantly suppresses the succinate activities of malate-dehydrogenase in papaya and cellulose activity in ber (Jawandha et al., 2009) during post-harvest ripening thus increasing shelf life.

5.7.21 MODE OF ACTION IN RELATION TO ANTI-RIPENING AND DELAYING SENESCENCE

Gibberellin is generally regarded as a stimulant of vegetative growth, yet it can have the reverse effect on reproductive tissue and it may even inhibit fruit growth (Ben-Arie et al., 1986). This, however, could be reconciled with its effect of delaying fruit senescence, thereby maintaining fruit at a more juvenile stage.

5.7.22 EFFECT OF GIBBERELLINS ON VEGETABLES

There are various effects of gibberellins on vegetables, which are as follows:

Use in induction of flowering:

Flowering was induced by GA_3 treatments replacing the cold requirement for flowering. In pepper, spraying of GA_3 at 25 ppm 3 times during vegetative growth caused flowering 18 days later than control (La Red and Cucchi, 1966).

Regulation of sex expression:

In cucumber, GA_3 at 5–10 ppm promote greater number of female flowers (Choudhury, 1966 and 1967). GA_3 at 25 ppm to watermelon were most effective in lowering the male: female ratio setting additional fruit and higher yield (Arora et al., 1985). In Momordica charantia at low concentration of GA_3 increased the number of female flowers and the ratio female: male flowers (Wang and Zeng 1997). In pumpkin it however tended to increase the ratio of staminate to pistillate flower (Das and Das, 1995).

Use in fruit set and development:

Higher concentration of GA_3 (100–1000 ppm) has been found to increase the fruit number and size of tomato (Kaushik et al., 1974, Shittu and Adeleke, 1999).

Development of seedless fruits (parthenocarpy):

GA_3 promote parthenocarpic fruit development. It inhibits abscission and also retains fruit for longer. GA_3 has been found to be effective in tomato and brinjal (Mukherjee and Dutta, 1962) in production of parthenocarpic fruits.

Postharvest treatment:

Postharvest treatment of GA_3 at 200 ppm reduced the physiological loss in weight and significantly increased the shelf-life of chili and retained higher oleoresin content (Arora et al., 1998).

Other Uses:

GA_3 is useful in increasing the yield. Multiple application of mixture of gibberellins i.e., GA_{4+7} (80 mg/L) resulted in larger diameter of cauliflower curd (Booij, 1989). GA_3 at 10 ppm and 20 ppm improved vine growth and seed content in pumpkin.(Arora et al., 1989). At 100 ppm it produced most compact plants and highest fruit yields in pumpkin (Das and Das, 1996).

Gibberellic acid improves yield and quality of ornamental plants via plant growth incitation and stem elongation (Fathipour and Esmaellpour, 2000). It enhances plant growth and internode length by increasing the cell division and enlargement. It also increases cell size, stem height, stem thickness and number of leaves. Other studies on the effect of GA_3 on ornamental plants showed that, GA_3 accelerated flowering and enhanced plant height (Gul et al., 2006).

5.7.23 INFLUENCE OF GA ON PLANT HEIGHT AND INTERNODAL LENGTH OF FLOWER CROPS

Mastalerz (1960) stated that 3 to 13 weekly application of GA_3 at concentration upto 100 ppm may result in marked increase in stem length in various plants such as rose, dwarf dahlia, chrysanthemum, snap dragon and datura. Internodal length was significantly increased due to GA sprays. Maximum length of internode in rose was recorded at GA 500 ppm (4.83 cm), whereas; control recorded least (2.50 cm) (Nanjan and Muthuswamy, 1975). Gowda (1980) noticed increased stem length in rose cultivar Super Star with GA_3 at 100–250 ppm when applied in rose cultivar Queen Elizabeth, all the applications increased stem length and intermodal length.

Maharana and Pani (1982) reported increased plant height in rose cultivar. Celebration when the plants were sprayed with GA_3 at 200 ppm one month after pruning. Banker and Mukhopadhya (1982) reported that applying GA at 250 ppm recorded maximum intermodal length (6.16 m) compared to control (3.49 cm) in rose cv. 'Queen Elizabeth.' Venkatesh and Nagarajaiah (1986) reported that GA at 50 ppm produced maximum intermodal length (4.10 cm) in rose cv. 'Queen Elizabeth' whereas; least was recorded by control (2.53 cm). Gowda (1988) observed increased stem length and neck length in rose cv. American Heritage with GA_3 spray at 300 and 350 ppm. The length and diameter of shoot were also increased with the application of GA_3 at 45 ppm in rose cv. Super Star. Sadanand et al. (2000) reported increased plant height, shoot length and maximum number of leaves per plant were recorded with the application of GA_3 at 200 ppm in rose cv. First Red.

5.7.24 NUMBER OF STEMS

Gowda (1985) reported increase number of primary and secondary shoots with GA_3 at 100 and 200 ppm in rose plant with the spray of GA_3 at 45 ppm compared to control (10.50) in rose cv. Super Star. There was a marked increase in lateral branching by spraying of NAA at 500 ppm or GA_3 at 100 ppm in carnation cv. Margherite Crimson (Mukhopadhyay, 1990). Geetha et al. (2000) reported that in China aster GA and IAA were responsible for more number of branches per plant. The effect of plant growth regulators on vegetative growth and flower earliness of damask rose (*Rosa damascena*) was investigated. Treatments comprised three levels of gibberellic acid (GA_3; 50, 100 and 200 ppm), CCC (Chlormequat) (500, 1000 and 2000 ppm), and Ethrel (Ethephon) (0.02, 0.04 and 0.06%), and distilled water spray as control. Plants sprayed with GA_3 at 200 ppm recorded the maximum vegetative growth and earliest flowering. The application of GA3 at 200 ppm resulted in the maximum number of branches in China aster cv. Shashank (Kumar et al., 2003).

5.7.25 FLOWER BUD EMERGENCE AND FLOWER HARVESTING

Shaul et al. (1995) proposed a hypothesis of a dual effect of GA_3 in suppression of *Botrytis* in rose cut flowers. Firstly, it may inhibit senescence related malfunction of cell membranes. Secondly, GA_3 may stimulate formation of endogenous compounds inhibiting *Botrytis* blight development in the petals. Nanjan and Muthuswamy (1975) reported that GA at 200 ppm significantly minimized the days for flower bud emergence and flower harvesting in Edward rose (*Rosa bourboniana*). Parwal et al. (1994) reported that application of GA_3 at 200 ppm recorded early flowering in Damask rose.

Roberts et al. (1999) reported that in rose cv. Felicite Perpetue, floral initiation occurred when concentrations of GA_3 were low and was inhibited when concentrations of GA_3 were high. Padmapriya and Chezhilyan (2003) reported that in chrysanthemum GA_3 was found to increase the IAA oxidase activity, which is responsible for early flowering. Similarly earliness in flowering and harvesting could be attributed to the fact that, GA_3 increased the cell division and cell elongation influencing floral morphogenesis. Hence, rendering early maturity in plants and also could be due to quick availability of optimum level of nutrients.

5.7.26 QUALITY ATTRIBUTES

GA$_3$ decreased flower oil content compared to the control and also showed the shortest flowering period among treatments (Saffari et al., 2004).

Senescence is one of the most puzzling events in plant life. From seed germination to maturation, the plant undergoes several changes. Senescence shows accumulation of hazardous substance. Growth promoters delay the deteriorative process as well as accumulation of hazardous substance. Studies were carried out to determine the effect of plant growth regulators like gibberellic acid (GA$_3$) at 10, 20, 50 and 100 ppm, on abscission and senescence of leaves of (*Rosa indica*) and (*Rosa chinensis*) (Sharma and Tomar, 2008). Bio-chemical contents of leaf *viz*. chlorophyll-α, chlorophyll-β, reducing and non-reducing sugar and protein reduced during the course of development. Exogenous application of GA$_3$ reduced the reduction of this content in attached and deattached condition while Abscisic Acid (ABA) and Etheophon (ETH) enhance the reduction. Perhaps, GA$_3$ delay the production of hydrolytic enzyme, where as ABA and ETH not only promote the production of hydrolytic enzyme but also reduced the production of growth promoter (Sharma and Tomar, 2009). Pre-harvest spraying of plants with gibberellic acid at a concentration of 100 mg x dm^{-3} has a positive effect on the content of photosynthetically active pigments in the leaves of *A. europaeum* cultivated in an unheated plastic tunncl (Pogroszewska et al., 2014). Exogenously applied GA$_3$ accumulate endogenous gibberellic acid which stimulates the production pigments. The growth of chlorophyll content is influenced by the phytohormone during the transformation of etioplasts into chloroplasts on the (Ouzounidou and Illias, 2005). Gibberellic acid also inhibits degradation of the discussed pigments by controlling the process of starch and sucrose hydrolysis into fructose and glucose, which indirectly affects the content of chlorophyll *a* and *b* in leaves (Emongor, 2004). GA$_3$ stimulates the vegetative growth, as mentioned previously and hence high accumulation rate of metabolic components especially carbohydrates such as *chlorophyll* and *carotene*. Gibberellin reduce the occurrence of leaf chlorosis all three cultivars (Table 5.1). Although both GA$_3$ and GA$_{4+7}$ prevents leaf yellowing, at a given concentration (100 ppm), GA4+7 was more effective than GA$_3$ in lilies of cold-stored stems and non- cold-stored stems (Ranwala and

TABLE 5.1 Influence of Different Concentrations of GA_3 and GA_{4+7} on Longevity of Different Varieties of Lily

Preharvest spray application	Vermeer	Vivaldi	Marselli
GA_3 500 ppm	11.4	14.1	13.8
GA_{4+7} 100 ppm	12.8	14.1	14.4
Control	8.5	10.9	10.9

Source: Ranwala and Miller, 2002.

Miller, 2002). GA_3 at 100 ppm recorded the highest total chlorophyll content (1.826 mg g^{-1}) in case of cut rose cv. 'First Red' (Kumar et al., 2012).

The reduction rates of *anthocyanin* contents were alleviated by GA treated tulip flowers (Mohammadi et al., 2012). GA_3 application at a concentration of 600 mg per dm^3 led to the accumulation of the greatest amount of anthocyanins in the leaves of *A. europaeum* cultivated both in an unheated plastic tunnel and in the field (Pogroszewska et al., 2014). There will be an increase in the content of anthocyanin pigments after treatment of plants with gibberellic acid. The phenomenon behind this is the effect that gibberellins had on the synthesis of phenylalanine ammonia lyase (PAL) and tyrosine ammonia lyase (TAL), the key enzymes of flavonoid and anthocyanin synthesis. Phenylalanine ammonialyase catalyzes spontaneous, non- oxidative deamination of L-phenylalanine to trans-cinnamic acid which, in the process of further metabolic changes, can be transformed into flavonoids (Kwack et al., 1997). After foliar application of GA_3 solutions, an increased amount of anthocyanins was also found in the leaves of *Hibiscus sabdariffa* (Rafia et al., 2005) and *Ajuga reptans* (Kwack et al., 1997).

Improved *anthocyanin* (1.72 mg/g) and total carotenoid (4.25 mg/g) content was observed in pre-soaking and foliar spray of gibberellic acid 100 ppm in case of tuberose cv. Jessica (Kumar and Gupta, 2014). Enhanced pigmentation in petal might be due to higher concentration of pelargonidin (Davies et al., 1993). The induction of anthocyanin synthesis requires the presence of sugar in the medium and sugars may serve as specific signals for the activation of specific genes, as cellular osmotic regulators, or as general energy source of carbon metabolism in the developing flower (Delila et al., 1997). Gibberellic acid application at 100 ppm recorded the maximum anthocyanin content findings are in agreement with reports of Dahab et al. (1987) in chrysanthemum, Goyal and Gupta (1994); Arun (1999) and Ramalingam (2008) in rose.

The induction of anthocyanin synthesis and anthocyanin biosynthetic gene expression in detached petunia (Petunia hybrida) corollas by gibberellic acid (GA_3) requires sucrose for activation of anthocyanin biosynthetic gene (Moalum-Beno et al., 1997).

Senescence is considered as a genetically programmed process with culminates the development and differentiations of plant structure and which serve to specific function in the plant. Physiological senescence is definitely followed by death while non-physiological not be followed. It is due to deficiency of any essential mineral. The senescence process allows for the termination of cells, tissues, organs or even organisms in a controlled process. Senescence is age dependent and under the control by hormonal, molecular, and genetical processes. It is controlled by several factor, hormones are one of them. The hormonal regulation is an important aspect of the mechanism of the senescence but not an isolated aspect. Hormones act by controlling the development of the senescence program. Gibberellins delay the senescence. Gibberellin is a powerful retardant (Sharma et al., 2011).

5.7.27 LENGTH OF STALK

In celebration Rose, GA_3 sprayed at 200 ppm accelerated the stem length (Maharana and Pani, 1982). Foliar applications of GA3–200 ppm twice (20 and 30 days after pruning) on 'Queen Elizabeth' cut roses increased length of the internode and shoot (Nagarajaiah and Reddy, 1986). Spraying with gibersol (GA_3) every two weeks throughout the season increased length of flower stalk when flowers were harvested either continuously or in the autumn (Wisniewska and Treder, 1989).

GA_3 sprayed @ 100 and 200 ppm concentration of each on *Rosa hybrida* cv. Super Star three weeks after pruning increased stalk length BA (Anon, 1993). Foliar spray of GA_3 @ 50 to 500 ppm in field increased plant growth in Gladiolus (Bhattacharjee, 1984). In the rose cv. Super Star shoot length was maximum (37.71 cm) when compared to control (25.25) by applying GA at 45 ppm, Goyal and Gupta (1994). Patel and Patil (1998) reported that application of 300 ppm GA increases the length of cut flower stalk in rose. Arun et al. (2000) observed that rose cut flower stalk length was greatest by applying GA_3 (300 ppm). Maximum increase in stem length (55.22 cm) compared to control (32.79 cm) was found in rose cv. Queen Elizabeth by spraying GA (250 ppm) (Banker and Mukhopadhya, 1982). The effect of gibberellic acid at 100 ppm showed improved plant heights (70.50 cm) as compared to untreated ones (62.33 cm) in carnation cv. Improved Margherite (Jana and Jahangir, 1987). Application of GA_3 twice in carnation significantly increased plant height (65.94 cm) and stem length (58.25 cm) (Verma et al., 2000). Maximum stalk length (74.60 cm) was produced in rose cv. First Red by applying GA_3 at 300 ppm, whereas; control recorded least (45.92 cm) (Dhekney et al., 2000). Highest flower stalk length (57.60 cm) was recorded by application of GA_3 (200 ppm) along with water-soluble fertilizer @ 75% in chinaster (Kore et al., 2003). Application of 200 ppm of GA_3 in carnation significantly increased plant height (88.24 cm) (Ramesh and Singh, 2003).

5.7.28 FLOWER LENGTH AND DIAMETER

Garrod and Harris (1974) reported that additional petals could be promoted by application of GA_3 to the shoot tip during flower initiation. Banker and Mukhopadhyay (1982) stated that GA_3 100 ppm produced good diameter of bud in rose cv. Queen Elizabeth. Gowda (1988) noted more number of petals per flower with GA spray at 250 to 350 ppm in rose cv. American Heritage. Bud diameter improved with all the concentrations of GA_3 spray used in carnation (Amitabh, 1990). The maximum size of flower (6.4 cm) associated with longer stalk length (14.84 cm) was produced by applying GA @ 150 ppm. The minimum size was recorded in MH-1000 ppm (4.13 cm) whereas, control recorded 4.85 cm in case of chrysanthemum (Jyoti, et al., 1995). Goyal and Gupta (1996) recorded the highest flower diameter (9.55 cm) and weight of flower (13.91 g) with the foliar spray of GA_3 at 45 ppm compared with control (9.05 cm and 10.15 g, respectively) in rose cv. Super Star. Sadanand et al. (2000) recorded that increased bud length and bud diameter was reported with GA_3 application and the highest length and diameter were obtained in case of bud with GA_3 at 200 ppm in rose cv. First Red. Dhekney et al. (2000) reported that GA at 200 ppm resulted in increase in bud length, bud circumference and flower diameter. Verma et al. (2000) stated that applying Nitrogen (500 ppm) and GA_3 (50 ppm) per week were found to be effective in increasing the flower diameter and flower length in carnation. Chakradhar (2002) reported that the flower quality attributes such as length and diameter of flower bud, number of petals per flower, flower longevity and weight of flower improved with the application of GA_3 60 ppm in rose cv. Gladiator. Chakradhar et al. (2003) reported that flower bud length and diameter were maximum with application of GA_3 at 60 ppm and minimum with BA at 100 ppm in rose cv. Gladiator. Significantly maximum flower diameter (7.09 cm) was recorded at GA_3 100 ppm in China aster as compared to control (4.87 cm) (Kore et al., 2003). Increase in flower diameter was observed by GA_3 (100 ppm) in case of cut rose cv. 'First Red' (Kumar et al., 2012). Similar increase in flower diameter was obtained by Baskaran and Misra (2007) in gladiolus and Sainath (2009) in chrysanthemum. GA_3 seems to affect the flower diameter by forming sink in a position where it accumulates and draws the available photosynthates to this site.

The petal development in *Gerbera hybrida*, the petal size is mainly determined by cell expansion, and not by cell division (Zhang et al., 2012). GA and ABA have antagonistic effects on petal growth by modulation of cell elongation at the basal region and petal/cell elongation is enhanced by GA but repressed by ABA when each phytohormone is applied alone (Figures 5.8 and 5.9). The increase in petal length by GA and the reduction in petal length by ABA are attenuated by the co-application of ABA and GA, respectively (Figure 5.10; Li et al., 2015).

5.7.29 NECK LENGTH

The GA has been reported to increase the neck length in cut roses (Figure 5.11). The length of the pedicle (neck) was increased markedly by GA values than untreated plants (Nanjan and Muthuswamy, 1975). Banker and Mukhopadhya (1982) reported that applying GA at 250 ppm produced maximum neck length (14.93 cm) compared to control (9.00 cm) in rose cv. Queen Elizabeth. Maximum flower neck length (9.073 cm) was produced in rose cv. First Red by application of 300 ppm GA_3 (Dhekney et al., 2000).

FIGURE 5.8 Gerbera flower treated with Gibberellins and ABA (Source: Li et al., 2015).

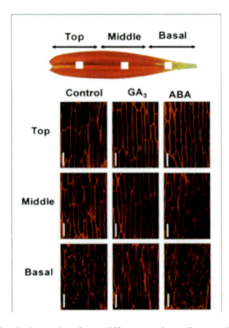

FIGURE 5.9 Petal cell elongation from different regions (Source: Li et al., 2015).

Gibberellins: The Roles in Pre- and Postharvest Quality 211

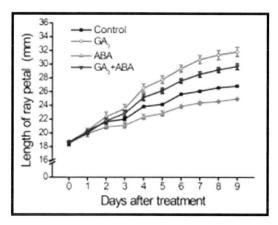

FIGURE 5.10 Graphical illustration showing the influence of GA_3, ABA and (GA_3 + ABA) on length of the ray petal (Source: Li et al., 2015).

First Red   Gold Strike

FIGURE 5.11 GA_3 treated roses in vase solution (Source: Rajesh, 2012).

5.7.30 FLOWER YIELD

Obtaining higher yield with enhanced quality is the final objective of any crop regulation practice. Increased yield with high quality finally contributes to the net returns. When flower crops are grown under protected conditions, increased yield per plant and improved quality are very critical to justify the cultivation of the crop under protected condition since a great amount of investment is involved in this type of cultivation. The number of flowers per plant is the major yield-contributing factor in cut flowers.

GA @ 200 ppm recorded maximum yield (205 flowers) in Edward rose compared to control (168 flowers) (Nanjan and Muthuswamy, 1975). El-Shafie et al. (1980) obtained highest number of flowers with GA_3 at 250 ppm in Queen Elizabeth and Baccara rose varieties. Application of GA_3 (10–100 ppm) twice after pruning in 'Queen Elizabeth' rose cultivar increased number of flowers (Bankar and Mukhopadhyay, 1982). In 'Celebration' rose, GA_3 spray (200 ppm) advanced flowering and accelerated stem (Maharana and Pani, 1982). Foliar spray of GA_3 at 50 to 500 ppm in field grown 'Raktgandha' roses increased number of flowers per plant and petals per flower (Bhattacharjee and Singh, 1983). Gowda (1985) enhanced the flower yield in rose cv. Super Star with the foliar spray of GA at 200 ppm. Nagarajaiah and Reddy (1986) obtained similar results in rose cv. Queen Elizabeth with GA_3 at 10 to 100 ppm applied twice after pruning. Dhekney et al. (2000) reported that applying GA_3 at 200 ppm produced maximum number of flowers (38.610) per m² compared to control (18.22). Kewte (1991) obtained highest number of flowers per plant with GA at 300 and 200 ppm (28.75 and 25.25, respectively) compared to control (14.50) in rose cv. Paradise. Goyal and Gupta (1996) recorded 19.50 flowers per plant with GA at 30 ppm compared to 16.00 flowers per plant with control in Rose cv. Super Star. Kewte and Sable (1997) recorded highest number of A and B grade cut flowers from plants treated with GA_3 at 300 ppm compared with control in rose cv. Paradise. Gowda (1988) reported that the numbers of marketable flowers were increased in rose cv. American Heritage with GA_3 at 300 and 350 ppm. Sadanand et al. (2000) recorded more number of flowers per meter square with GA_3 @ 200 ppm as compared with other treatments in rose cv. First Red. Arun et al. (2000) reported that applying GA_3 at 200 ppm recorded more number of flowers/

m² in rose cv. First Red. *Rosa hybrida* cv. Sntrix plants grown under greenhouse conditions were sprayed with 250 ppm gibberellic acid (GA$_3$) alone or combined with foliar fertilizer Sangral. The highest values of vegetative and flowering parameters were obtained with spraying the plants with 0.40% foliar fertilizer and 250 ppm GA$_3$. Moreover, the total carbohydrate and mineral contents in the leaves were increased as a result of spraying the plants with GA$_3$ and foliar fertilizer (Al-Humaid, 2001). The number of flowers and yield (67.33, 192.59 g) per plant per m² per annum were highest with GA @ 200 ppm while lowest with control (26.00 flowers and 76.8 g yield of flowers) in China aster cv. Kamini (Kumar et al., 2003). Maximum yield (49.60 q ha⁻¹) was recorded by application of GA$_3$ (200 ppm) along with water-soluble fertilizer in China aster (Kore et al., 2003). Increase in yield of the flowers is due to the effect of GA$_3$ on increase in the internodal length (Roberts et al., 1999) in Roses. But there was a decrease in yield with higher concentrations of GA$_3$ (300 ppm) up to 11–20% reported in 'Raktagandha' rose by Bhattacharjee and Singh (1995) and reduced yield was also observed in cut rose 'Poison.'

Application of 0, 10, 25 and 50 mg l-1 GA3 on Aquilegia × hybrida showed that the most yield was obtained in 10 mg l-1 GA3, while 50 mg l-1 GA3 caused to diminishing of cut flowers (Barzegarfallah, 2006). Also, showed that 300 mg l-1 GA3 decreased yield of rose 'Raktagandha' to 11–20%. In current study, 300 mg l-1 GA3 reduced yield of cut rose 'Poison.'

5.7.31 VASE LIFE

Postharvest pulsing for 20 h with solutions containing 20 to 40 mg GA$_3$ per liter extended the vase life and promoted bud opening of unstored and stored flowers of the rose cv. Mercedes. While continuous treatment with GA$_3$ was detrimental (Goszczynska et al., 1990). GA$_3$ stimulates active sucrose uptake in GA$_3$ per sucrose-dependent rose petals (Kuiper et al., 1991). Barthe et al. (1991) reported foliar spraying with GA$_3$ in *Rosa hybrida* cv. Royalty the petal area progressively increased. Goyal and Gupta (1994) observed increased vase life of 141.62 hours with GA$_3$ @ 45 ppm compared to 125.25 hours with control in rose cv. Super Star. Lee and Kim

(1995) reported that flowers kept in distilled water with GA showed high chlorophyll contents (*Rosa hybrida* L. cv. Red Sandra). Foliar application of GA_3 (300 ppm) on the rose cv. Paradise gave highest number of A and B grade cut flowers and longest vase life (Kewte and Sable, 1997).

An investigation was carried out on "Eiffel Tower" cut roses to determine the most effective growth regulating chemical and its concentration in the holding solution for improving postharvest life and quality. GA showed beneficial effects on vase life, flower diameter, water uptake and fresh weight. Among the chemicals, GA_3 (150 ppm) was effective in terms of effectiveness in increasing postharvest life (Bhattacharjee, 2000).

The longevity of harvested rose (*Rosa hybrida*) flowers of cv. Mercedes, has been promoted by application of GA_3. The leakage of electrolytes from the GA treated petals of was lower in comparison with untreated ones (Agbaria et al., 2001). Chakradhar (2002) recorded that maximum vase life of cut flowers (8.90 days) was recorded with GA_3 (60 ppm) compared to other treatments in rose cv. Gladiator. GA_3 treated petal discs of 'Febesa' roses could increase the activity of cell wall bound and vacuolar invertase (Horibe et al., 2010). Pulsing with a solution of sucrose at lower concentrations along with GA_3 (10 mg/L) was promising in increasing vase life of cut roses 'Red One' (Gholami et al., 2011).

Gibberellic acid delayed the flower senescence in gerbera cut flowers and postponed bent neck disorder in gerbera cut flowers by supplying respirable substrates, especially in petals, thus promoting respiration and extending the vase life of gerbera flowers. It has been reported that the main effect of applied sugar in extending cut flower vase life was to maintain mitochondrial structure and functions. However, the effect of sugars on mitochondria may not be a specific effect and may stem from its general protective effect on membrane integrity (Emongor, 2004). GA signaling is crucial to petal growth. As a versatile regulator, abscisic acid (ABA) has been shown to act antagonistically to the function of GA in a variety of developmental processes, including florral transition. Li et al. (2015), applied RNA sequencing technique and established that GA and ABA act antagonistically in the petal growth of gerbera. The possible reasons they suggested are that both GA and ABA target the same genes thus biosynthesis or signaling pathways were affected.

GA$_3$ at 100 ppm recorded the maximum vase life of 2.6 days in cut rose cv. First Red in distilled water. An increase in vase life of cut roses due to spray of GA$_3$ may be attributed to the fact that retardation of senescence by GA$_3$ is associated with the maintenance of a higher level of RNA in petals and leaves (Goszczynska and Rudnicki, 1988). Similar findings of increase in the vase life of flowers with GA$_3$ application was reported by Delvadia et al. (2009) in gaillardia, Kazaz and Karaguzel (2010) in golden rod, cut flowers of anthurium cv. Xavia by 15 days (Sahare and Singh, 2015) and Rao (2010) in chrysanthemum. Spraying cut flowers of *Anemone coronaria* with GA$_3$ (50 ppm) a day beforeharvest or spraying cut flowers with 50 ppm GA$_3$ after harvest extended vase life 8 days and had better preservation of cut flower quality (Sharifani et al., 2005). The GA$_3$ application pre-harvest significantly increased vase life of Solidago inflorescence in comparison with the control and the peak at 100 ppm GA$_3$ (11 days) then declined at higher concentration and the longest vase life was observed (12 days) by combination GA$_3$ at 100 ppm (Osman, A.R. and Sewedan, 2014). Improving the postharvest quality of Solidago inflorescence by using GA$_3$ could be explained through the role of GA3 Improving water balance, fresh weight (EL-Saka et al., 2002) and hence high accumulation of carbohydrates in stem and leaves which consequently increased the vase life (Hassan et al., 2003).

Longevity of the cut-flowers of tuberose cv. Jessica was significantly increased by pre-soaking and foliar spray of gibberellic acid 100 ppm (16.70 days) over control (12.28 days) (Kumar and Gupta, 2014). Saeed et al. (2014) concluded that gibberellic acid applied at lower concentrations renders greater beneficial effects on vase life quality, membrane stability and antioxidant activities in gladiolus cut spike and higher application rates cause no improvement in the flower longevity. Bharathi and Kumar (2009) reported the similar findings for prolonging vase life of cut tuberose spikes and Umrao et al. (2007) for spike durability in gladiolus. Thus, Waters (1966) indicated that gladiolus longevity and other quality can be measured as and the effect of gibberellic acid on senescence as well as extended vase life and delayed senescence may result from direct effect of gibberellic acid on cell senescence associated process which may prevent membrane permeability and subsequent leakage of electrolytes.

TABLE 5.2 Influence of GA_3 at Different Concentrations on Vase Life in Floricultural Crops

Name of the ornamental plant	Concentration of GA_3	VVase life (Days)/ LLongevity (Days)	Source
Gladiolus grandiflorus L. cv. Jessica	Pre-soaking with $GA_{3\_}$ 100 ppm	15.32[L]	Kumar and Gupta (2014)
	Pre-Soaking and Foliar Spray of $GA_{3\_}$ 100 ppm	16.34[L]	Kumar and Gupta (2014)
	Foliar Spray of $GA_{3\_}$ 100 ppm	16.70[L]	Kumar and Gupta (2014)
Rosa hybrida cv. First Red	Pre-harvest Spray application of $GA_{3\_}$ 250 ppm	6.94[v]	Rakesh, 2012 (Figure 5.11)
	Pre-harvest Spray application of $GA_{3\_}$ 100 ppm	2.6[v]	Kumar et al. (2012)
Rosa hybrida cv. Noblesse	Pre-harvest Spray application of $GA_{3\_}$ 250 ppm	7.61[v]	Rakesh (2012)
Rosa hybrida cv. Gold Strike	Pre-harvest Spray application of $GA_{3\_}$ 250 ppm	7.5[v]	Rakesh (2012) (Figure 5.11)
Rosa hybrida cv. First Red cv. Grand Gala	Pre-harvest Spray application of $GA_{3\_}$ 250 ppm	6.54[v]	Rakesh (2012)
Anthurium cv. Xavia	$GA_{3\_}$ 10 ppm in vase solution	15.67[v]	Sahare and Singh (2015)
Solidago canadensis cv. Tara	$GA_{3\_}$ 100 ppm as foliar spray	11.00[v]	Osman and Sewedan (2014)
	$GA_{3\_}$ 200 ppm as foliar spray	10.00[v]	Osman and Sewedan (2014)
Croton (*Codiaeum variegatum*) cv. Mariana	$GA_{3\_}$ 100 ppm in vase solution	16.00[v]	Kumara et al. (2007)
Croton (*Codiaeum variegatum*) cv. Batik	$GA_{3\_}$ 100 ppm in vase solution	18.00[v]	Kumara et al. (2007)
Croton (*Codiaeum variegatum*) cv. Aucubaefolia	$GA_{3\_}$ 100 ppm in vase solution	17.00[v]	Kumara et al. (2007)

Gibberelic acid at 100 ppm significantly increased the vase-life of vars. Mariana, Batik and Aucubaefolia of croton (Codiaeum variegatum) upto 16, 18 and 17 days, respectively (while the control was 7, 11 and 14 days, respectively) (Kumara et al., 2007). GA_{4+7} was effective than GA3 in prolonging inflorescence longevity at a given concentration, especially in cold-stored stems of lilies. The increased inflorescence longevity was because of both by the delayed bud opening and increased longevity of individual flowers (Ranwala and Miller, 2002).

Postharvest spray treatment of GA (100 ppm) effectively increased water uptake and retained fresh weight of cut inflorescence, there by increasing (by almost twofold) vase life stabilizing absolute integrity of cell membrane leading to a delay in bract cell death in heliconia cv. Golden Torch inflorescence (Mangave, 2013). The effect of GA_3 on the vase life of different flowers was given in Table 5.2.

5.7.32 BENT NECK IN GERBERA

The reduction in water uptake, coupled with continuous transportation, leads to water deficit and reduced turgidity in the cut flowers. This may cause the stem to bend under the weight of the flower. The bending occurs often below the flower, a phenomenon known as bent neck (Burdett, 1970). Bending resistance depends on the development of secondary thickening and lignification of the vascular elements in the peduncle area subtending the flower head. GA_3 significantly reduces the number of gerbere cut flower stems with bent neck because GA_3 minimizes water loss and increases water uptake, therefore improving the water balance and mechanical strength of the stems due to turgidity Table 5.3 (Emongor, 2004).

TABLE 5.3 Effect of GA_3 on the Bent of Gerbera (*Gerbera jamesonii* cv. Ida Red) Stems

GA_3 concentration (ppm)	No. of stems with bent neck after	
	7 days	14 days
0.0	3.25	6.00
2.5	1.00	4.25
5.0	2.00	3.25
7.5	1.50	3.00

Source: Emongor (2004).

5.7.33 DISEASE RESISTANCE IN ROSE

The development of Botrytis blight was suppressed by spraying flower buds with a 1 mM solution of GA_3, although the effect of GA_3 was limited by flower petal senescence. Application of GA_3 either prior to or after conidial inoculation suppressed development of Botrytis blight. GA_3 application suppressed Botrytis blight development even after the flowers were kept in cold-storage conditions (Shaul et al., 1995). GA_3 suppression of blight was due to protective function associated with the GA imposed inhibition of cell membrane disruption in rose petals (Sabehat and Zieslin, 1995) and formation of endogenous compounds, such as various phenolics (Verhoeff, 1980), which could be inhibitory to the development of B. cinerea. This possibility is inferred from the inhibition of Botrytis blight with GAs applied after the fungal infection (Figure 5.12). GA_3 could be a more environmentally-friendly alternative to conventional fungicides (Shaul et al., 1995).

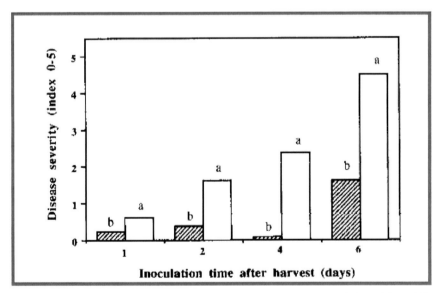

FIGURE 5.12 Effect of time between GAs application and the inoculation with conidia suspension on the development of Botrytis blight in harvested flowers of rose cv. Mercedes. The extent of the disease was evaluated six days after inoculation (Source: Shaul et al. (1995) a: Control; b: GA_3 Treated flowers).

5.7.34 EXTRA POINTS

- Highest amounts of superoxide dismutase enzyme and decreasing rates of relative fresh weight of tulip cut flowers were found in GA_3 treated flowers (Mohammadi et al., 2012).
- No effect of foliar GA_3 application on dry mass content in the leaves of *Iris nigricans* was found by (Al-Khassawneh et al., 2006), *A. Europaeum* (Pogroszewska et al., 2014).
- Flowering branches length inflorescence increases significantly by increasing GA_3 concentration compared to control then decreased at higher concentration (Roberts et al., 1999).
- GA_3 enhances plant growth by increasing the cell division, cell elongation and cell size which agreement with Gul et al. (2006).
- Gibberellic acid prevented leaf chlorosis, which was the major postharvest disorder in many cut flowers such as Santonia cv. Golden light flowers (Eason et al., 2001).
- Gibberellic acid did not delay leaf senescence in most plant species and its content in tissues was not correlated with senescence (Burdett, 1970).
- GA accelerates flower formation. Changes in GA levels occur when plant switches from vegetative phase to reproductive phase.

KEYWORDS

- Gibberellic acid
- Postharvest quality
- Postharvest shelf life
- Preharvest spray
- Senescence regulation
- Storage and packaging
- Vase life

REFERENCES

Agbaria, H., Zamski, E., Zieslin, N., & Zieslin, N. (Ed.). (2001). Agbaria, H. Effects of gibberellin on senescence of rose flower petals. *Acta Hort.*, *547*, 269–279.

Ahmed, M.S., & Singh, S. (1999). Effect of various postharvest treatments on shelf life of Amrapali mango. *Orissa J. Hort.*, *27*(10), 29–33.

Al-Humaid, A.I. (2001). The influence of foliar nutrition and gibberellic acid application on the growth and flowering of 'Sntrix' rose plants. *Alexandria J. Agric. Res.*, *46*(2), 83–88.

Al-Khassawneh, N.M., Karam, N.S., & Shibli, R.A. (2006). Growth and flowering of black iris (*Iris nigricans* Dinsm.) following treatment with plant growth regulators. *Scientia horticulture*, *107*(2), 187–193.

Aly, E.Z., & Ismail, H.A. (2000). Effect of preharvest GA3, CaCl$_2$ and boron treatments on quality and enzymatic browning in Balady guava fruits. *Annals of Agricultural Science*, *38*(2), 1101–1108.

Amarante, C.V.T., Drehmer, A.M.F., Souza, F., & Francescatto, P. (2005). Pre-harvest spraying with gibberellic acid (GA3) and aminoethoxyvinylglycine (AVG) delays fruit maturity and reduces fruit losses on peaches. *Revista Brasileira de Fruticultura*, *27*, 1–5 (Original not seen. Abstract in CAB Abstracts, 100, Accession No. 20053104566, 2005).

Amitabh, M. (1990). Responses of carnation to spray application of NAA and Gibberellic acid. *Haryana J. Hort. Sci.*, *19*(3–4), 280–283.

Anonymous (1993). Annual Report, Division of Horticulture and Landscaping, IARI, New Delhi.

Arteca, R.N. (2015). *Introduction to Horticultural Science, 2nd Ed.* Cengage Learning, Stanford, USA, 186–207.

Arun, D.S., Ashok, A.D., & Rangasamy, P. (2000). Effect of some growth regulating chemicals on growth and flowering of rose cv. First Red under greenhouse condition. *J. Ornamental Hort.*, New Series, *3*(1), 51–53.

Babu, K.D., & Yadav, D.S. (2004). Physical and chemical thinning of peach in subtropical North Eastern India. *Acta Horticulturae*, *662*, 327–331.

Balasubramaniam, R., & Agnew, R. (1990). *The effect of gibberellic acid on post-harvest handling and storage quality of methyl bromide fumigated and artificially covered cherries of cultivars Dawson, Stella, Bing and Rainier.* Report for the Marlborough Fruit Producers Limited. MAF Technology, Marlborough Research Centre.

Bankar, G.J., & Prasad, R.N. (1990). Effect of gibberellic acid and NAA on fruit set and quality of ber (*Zizyphus mauritiana* Lamk.) cv. Gola. *Progressive Horticulture*, *22*, 60–62.

Banker, G.J., & Mukhopadhya, A. (1982). Gibberellic acid influences growth and flowering of rose cv. Queen Elizabeth. *Indian J. Hort.*, *39*, 130–133.

Barthe, P., Vaillant, V., & Gudin, S. (1991). Definition of indicators of senescence in the rose: effect of the application of plant hormones. *Acta Hort.*, *298*, 61–68.

Baskaran, V., & Misra, R.L. (2007). Effect of plant growth regulators on growth and flowering of gladiolus. *Indian J. Hort.* *64*(4), 479–482.

Ben-Arie, R., Bazak, H., & Blumenfeld, A. (1986). Gibberellin delays harvest and prolongs storage life of persimmon fruit. *Acta Hortic.*, *179*, 807–813.

Ben-Arie, R., Saks, Y., Sonego, L., & Frank, A. (1996). Cell Wall metabolism in gibberellin treated persimmon fruit. *Plant Growth Regul.*, *19*, 25–33.

Bengoa, R.E., & Marlangeon, R.C. (1975). A reappraisal of the growth and development of peach fruit. *Phyton* (Argentina), *33*(2), 113–117.

Benjawan, C., Chutichudat, P., Boontiang, K., & Chanaboon, T. (2006). Effect of chemical paclobutrazol on fruit development, quality and fruit yield of Kaew mango (*Mangifera indica* L.) in Northeast Thailand. *Pakistan Journal of Biological Sciences*, *9*, 717–722.

Bhanja, P.K., & Lenka, P.C. (1994). Effect of pre- and postharvest treatments on storage life of sapota fruits cv. Oval. *Orissa Journal of Horticulture*, *22*(1/2), 54–57.

Bharathi, T.U., & Kumar, S. (2009). Effect of growth regulators and chemical on postharvest parameters of tuberose cv. Suvasini. *Advances in Plant Sciences*, *22*(1), 107–109.

Bhattacharjee, S.K. (1984). The effect of growth regulating chemicals on Gladiolus *Gartenbavwissenschaft*, *49*(3), 103–106.

Bhattacharjee, S.K. (2000). Postharvest life of "Eiffel Tower" cut roses and biochemical constituents of petal tissues as influenced by growth regulating chemicals in the holding solution. *Haryana Journal of Horticultural Sciences*, *29*(1/2), 66–68.

Bhattacharjee, S.K., & Singh, U.C. (1983). Growth and flowering response of *Rosa hybrida* cv. 'Raktagandha' to certain growth regulator sprays. *Orissa J. Hort.*, *23*(1–2), 21–25.

Bukovac, M.J. (1963). Induction of parthenocarpic growth of apple fruit with gibberellin A3 and A4. *Bot. Gaz.*, *124*, 191–195.

Burdett, A.N. (1970). The cause of bent neck in cut roses. *J. Am. Soc. Hortic. Sci.*, *95*, 427–431.

Burg, S.P., & Burg, E.A. (1962). Role of ethylene in fruit ripening. *Plant Physiol.*, *37*, 179–189.

Carpita, N.C., & Gibeaut, D.M. (1993). Structural models of primary cell walls in flowering plant: consistency of molecular structure with the physical properties of the walls during growth. *Plant J.*, *3*, 1–30.

Chakradhar, M. (2002). Effect of growth regulators on growth, flowering and vase life of Rose cv. Gladiator. MSc (Agri.) Thesis, Dr. PDKV, Akola.

Chakradhar, M., Khirtkar, S.D., & Rosh, K. (2003). Effect of growth regulators on flower quality and vase life of rose cv. Gladiator. *J. Soil and Crops*, *13*(2), 374–377.

Dahab, A.M., El-dahab, R.S., & Salem, M.A. (1987). Effect of gibberellic acid on growth, flowering and constituents of chrysanthemum. *Acta Hort.*, *205*, 129–135.

Davies, K.M., Bradley, J.M., Schwinn, K.E. Markham, K.R., & Prodivinsky, E. (1993). Flavonoid biosynthesis in flower petals of five lines of lisianthes. Plant Science, *95*, 67–77.

Davies, P.J. (2010). *Plant Hormones: Biosynthesis, Signal Transduction, Action 801 pp.* Springer, Dordrecht, The Netherlands.

Davison, R.M. (1960). Fruit-setting of apples using gibberellic acid. Nature, *188*, 681–682.

De Jong, M., Wolters-Arts, M., Feron, R., Mariani, C., & Vriezen, W.H. (2009). The *Solanum lycopersicum* auxin response factor 7 (*SlARF7*) regulates auxin signaling during tomato fruit set and development. *Plant J.*, *5*, 160–170.

De Jong, M., Wolters-Arts, M., García-Martínez, J.L., Mariani, C., & Vriezen, W.H. (2011). The Solanum Lycopersicum Auxin Response Factor 7 (SlARF7) mediates cross-talk between auxin and gibberellin signaling during tomato fruit set and development. *Journal of Experimental Botany*, 62(2), 617–626.

Delila, M.B., Tamari, G., Yael, L.D., Borochov, A., & Weiss, D. (1997). Sugar dependent gibberellin induced chalcone synthase gene expression in petunia corollas. *Plant Physiology*, 113, 419–424.

Delvadia, D.V., Ahlawat, T.R., & Meena, B.J. (2009). Effect of different GA3 concentration and frequency on growth, flowering and yield in gaillardia (*Gaillardia pulchella* Foug.) cv. Lorenziana. *J. Hortl. Sci.*, 4(1), 81–84.

Dennis, F.G., Jr., & Edgerton, L.J. (1962). Induction of parthenocarpy in the apple with gibberellin and the effects of supplementary auxin application. *Proc. Am. Soc. Hort. Sci.*, 80, 58–63.

Dewich, A.S., Haslam. (1969). Phenol biosynthesis in higher plants. *Biochem J.*, 113, 537–542.

Dhekney, S.A., Ashok, A.D., & Rangaswamy, P. (2000). Effect of some growth regulating chemicals on growth and flowering of rose cv. First Red under greenhouse conditions. *J. Ornamental Hort*. New Series, 3(1), 51–53.

Dhillon, B.S., & Singh, K. (1968). Effect of some plant regulators on fruit set and fruit drop in *Zizyphus jujuba* L. *Journal of Research PAU*, 5, 392–394.

Dhillon, B.S., Saini, H.S., & Singh, S. (1982). Application of GA3 on the (cold) storage life of peach. *Indian Fd. Pckr*, 36(3), 31–33.

Dorcey, E., Urbez, C., Blazquez, M.A., Carbonell, J., & Perez-Amador, A. (2009). Fertilization-dependent auxin response in ovules triggers fruit development through modulation of gibberellin metabolism in Arabidopsis. *Plant J.*, 58, 318–332.

Eason, J.R., Morgan, E.R., A.C. Mullan, A.C., & Burge, G.K. (2001). Postharvest characteristics of santonia cv. Golden light a new hybrid cut flower from *Sandersonia aurantiaca* flowers. *Postharvest Biology and Technology*, 22(1), 93–97.

El-Otmani, M., & Coggins, C.W. Jr. (1991). Growth regulator effects on retention of quality of stored citrus fruit. *Sci. Hortic.*, 45, 261–272.

EL-Saka, M., Auda, M.S., & Abou-Dahab, T. (2002). Effect of nutrition with NPK and calcium chloride as preharvest treatments on flowers quality of *Hippeastrum vittatum* during postharvest handling. *Zagazig J. Agric. Res.*, 29, 1143–1167.

EL-Shafie, S.A., EL-Kholy, S.A., & Afify, M.M. (1980). Effect of gibberellic acid on growth and flowering of Queen Elizabeth and Baccara rose varieties. *Monoufeia J. Agric. Res.*, 3, 291–310.

Emongor, V.E. (2004). Effects of Gibberellic Acid on Postharvest Quality and Vaselife Life of Gerbera Cut Flowers (*Gerbera jamesonii*). *Journal of Agronomy*, 3, 191–195.

Engin, H., Sen, F., & Mengu, G.P. (2007). Effects of irrigation, gibberellic acid and nitrogen on some physiological disorders and fruit quality in 'springtime' and 'early red' peach. *Anadolu*, 17, 58–70 (Original not seen. Abstr in CAB Abstracts, 30, Accession No. 20083067807, 2007).

Fathipour, B., & Esmaellpour, B. (2000). *Plant Growth Substances (Principles and Application)*. Mashhad University Publication, Mashhad, Iran.

Ferguson, L., Iswail, M.A., Davies, F.S., Wheaton, T.A. (1982). Pre- and postharvest gibberellic acid and 2, 4-D application for increasing storage life of grape fruit. *Proc Fla State Hort Soc*, 95, 242–245.

Fos, M., Nuez, F., & Garcıa-Martınez, J.L. (2000). The gene pat-2, which induces natural parthenocarpy, alters the gibberellin content in unpollinated tomato ovaries. *Plant Physiol.*, *122*, 471–480.
García-Martínez, J.L., & García-Papí, M.A. (1979). The influence of gibberellic acid, 2,4-dichlorophenoxyacetic acid and 6-benzylaminopurine on fruit-set of Clementine mandarin. *Scientia Horticulturae*, *10*(3), 285–293.
Garcia-Papi, M.A., & Garcia-Martinez, J.L. (1984). Endogenous plant growth substances content in young fruits of seeded and seedless Clementine mandarin as related to fruit set and development. *Scientia Horticulturae*, *22*(3), 265–274.
Garrod, J.F., & Harris, G.P. (1974). Studies on the glass house carnation: Effect of temperature and growth substances on petal number. *Ann. Bot.*, *38*, 1025–1031.
Geetha, K., Sadawarta, K.T., Maharokar, V.K., Joshi, P.S., Das, D.D. (2000). A note of the effect of foliar application of plant growth regulators on seed yield in China aster. *Orissa J. Hort.*, *28*(2), 113.
Gholami, M., Rahemi, M., & Rastegar, S. (2011). Effect of pulse treatment with sucrose, exogenous benzyl adenine and gibberellic acid on vase life of cut rose 'Red One.' *Horticulture, Environment and Biotechnology.* *52*(5), 482–487.
Ghosh, S.K., & Sen, S.K. (1984). Extension of storage life of sweet orange. *South Ind. Hort.*, *32*(1), 16–22.
Ghosh, S.N., Bera, B., Kundu, A., & Roy, S. (2008). Effect of Plant Growth Regulators on Fruit Retention, Yield and Physico-Chemical Characteristics of Fruits in Ber cv. Banarasi Karka Grown in Close Spacing. *Proceedings of 1st International Jujube Symposium*, pp. 18. Agricultural University of Hebei, Baoding, China.
Gillaspy, G., Ben-David, H., & Gruissem, W. (1993). Fruits: a developmental perspective. *Plant Cell*, *5*, 1439–1451.
Goszczynska, D.M., Zieslin, N., Mor, Y., & Halevy, A.H. (1990). Improvement of postharvest keeping quality of Mercedes roses by gibberellin. *Plant Growth Regulation*, *9*(4), 293–303.
Gowda, J.V.N. (1985). Effect of gibberelic acid on growth and flowering of rose cv. Super Star. *Indian Rose Annual*, *4*, 185–187.
Gowda, J.V.N. (1988). Effect of gibberellic acid on growth and flowering of rose cv. American Heritage. *Indian Rose Ann.*, *7*, 155–157.
Goyal, R.K., & Gupta, A.K. (1994). Effect of growth regulators on nutritional status, Anthocyanin content and vase life of rose cv. Super Star. *Haryana J. Hort. Sci.*, *23*(2), 118–121.
Goyal, R.K., & Gupta, A.K. (1996). Effect of growth regulators on growth and flowering of rose cultivar Super Star. *Haryana J. Hort. Sci.*, *25*(4), 183–186.
Greve, L.C., & Labavitch, J.M. (1991). Cell wall metabolism in ripening fruit. Analysis of cell wall synthesis in ripening tomato pericarp tissue using a D-[U-13C]-glucose tracer and gas-chromatography-mass spectrometry. *Plant Physiol.*, *97*, 1455–1461.
Gul, H., Khattak, A.M., & Amin, N. (2006). Accelerating the growth of *Araucaria heterophylla* seedlings through different gibberellic acid concentrations and nitrogen levels. *J. Agric. Biol. Sci.*, *1*, 25–29.
Hare Krishna. (2012). *Physiology of Fruit Production*. Studium Press India Pvt. Ltd. India.
Hashemabadi, D. (2012). Pre-Harvest Application of Plant Growth Regulators on Rose 'Poison,' *Acta Hort.*, *937*, 883–887.

Hassan, F.A.S., Tar, T., & Dorogi, Z. (2003). Extending the vase life of *Solidago canadensis* cut flowers by using different chemical treatments. *Int. J. Hortic. Sci.*, *92*, 83–86.
Hassan, H.S.A., Mostafa, E.A.M., & Dorria, M.A. (2005). Improving canino apricot trees productivity by foliar spray with boron, GA3 and active dry yeast. *Arab Univ. J. Agric. Sci.*, *13*(2), 471–480.
Hopkins, W.G., & Huner, N.P.A. (2008). *Introduction to Plant Physiology. 4th ed.*, Danvers, MA: John Wiley & Sons; Hormones II Gibberellins, pp. 323–337.
Horibe, T., Ito, M., & Yamada, K. (2010). Effect of plant hormones on invertase activity and on petal growth of cut rose. *Acta Hort.*, *870*, 279–284.
http://www.planthormones.info/gibberellins.html.
Huang, C.C. (1974). Maintaining freshness of pineapple fruits for export. *Taiwan Agricultural quarterly*, *10*, 103–111.
Itakura, T., Kosaki, I., & Machida, Y. (1965). Studies with gibberellin application in relation to response of certain grape varieties. Bull. Hort. Res. Sta. Japan Ser. A *4*, 67–95.
Jacobs, W.P. (1979). *Plant Hormones and Plant Development.* Cambridge University Press, Cambridge, New York.
Jana, B.K., & Jahangir, K. (1987). Responses of growth regulators on growth and flowering of carnation of cv. Improved Margurite. *Prog. Hort.*, *19*(1–2), 125–127.
Jawandha, S.K., Mahajan, B.V.C., & Gill, P.S. (2009). Effect of preharvest treatments on the cellulose activity and quality of ber fruit under cold storage conditions. *Not Sci Biol.*, *1*, 88–91.
Jiang, W., Sheng, Q., Jiang, Y., & Zhou, X. (2004). Effects of 1-methylcyclopropene and gibberellic acid on ripening of Chinese jujube (*Zizyphus jujuba* M) in relation to quality. *Journal of Science Food and Agriculture*, *84*, 31–35.
Kale, V.S., Dod, V.N., Adpawar, R.M., & Bharad, S.G. (2000). Effect of plant growth regulators on fruit characters and quality of ber (*Zizyphus mauritiana* L.). *Crop Research*, *20*, 327–333.
Kale, V.S., Kale, P.B., & Adpawar, R.W. (1999). Effect of plant growth regulators on fruit yield and quality of ber cv. Umran. *Annals of Plant Physiology*, *13*, 69–72.
Kappel, F., & MacDonald, R. (2007). Early gibberellic acid sprays increase firmness and fruit size of 'Sweetheart' cherry. *Journal of the American Pomological Society*, *61*, 38–43.
Kaur, A.P. (2012). Effect of pre-harvest application of nutrients and growth regulators on storage life and quality of peach cv. Shan-I-Punjab. A thesis Submitted to PAU, Ludhiana.
Kaur, H., Randhawa, J.S., & Kaundal, G.S. (2004). Effect of growth regulators on preharvest fruit drop in subtropical plum cv. Satluj Purple. *Acta Horticulturae*, *662*, 341–343.
Kazaz, S., & Karaguzel, O. (2010). Influence of growth regulators on the growth and flowering characteristics of goldenrod (Solidago x hybrida). *European J. Scientific Research*, *45*(3), 498–507.
Kefeli, V.I., Kalevitch, M.V., & Borsari, B. (2003). Phenolic cycle in plants and environment. *J. Cell Mol. Biol.*, *2*, 13–18.
Kewte, M.G. (1991). Influence of foliar spray of nutrients and growth promoters on growth and flowering of rose cv. Paradise (H.T.). MSc (Agri.) Thesis, D. PDKV, Akola.

Kewte, M.G., & Sable, A.S. (1997). Influence of growth regulators and foliar nutrients on vase life, grading and economics of rose cv. Paradise. *J. Soils and Crops, 7*(1), 96–98.

Khader, S.E.S.A. (1991). Effect of preharvest application of GA$_3$ on postharvest behavior of mango fruits. *Scientia Horticulturae, 47*(3–4), 317–321.

Kore, V.N., Meman, S.L., & Burondkar, M.M. (2003). Effect of GA3 and fertigation on flower quality and yield of chinensis (*Callistephus chinesis* L. Meees) var. "Oystrich Plume Mixed" under Konkan agro-climatic conditions. *Orissa J. Hort., 31*(91), 58–60.

Kumar, D. (1998). Post-harvest studies of mango *(Mangifera indica L.)*. PhD Thesis, R.A.U. Bihar.

Kuiper, D., Reenen, H.S., & van Ribot, S.A. (1991). Van-Reenen, H.S. Effect of gibberellic acid and sugar transport into petals of 'Madelon' rose flowers during bud opening. *Acta Hort., 298*, 93–98.

Kumar, P., & Singh, S. (1993). Effect of GA3 and ethrel on ripening and quality of mango cv. Amrapalli. *Horticultural Journal, 6*, 19–23.

Kumar, P., Raghava, S.P.S., Mishra, R.L., & Krishna Singh, P. (2003). Effect of GA3 on growth and yield of China aster. *J. Ornamental Hort., 6*(2), 110–112.

Kumar, R., Kaushik, R.A., & Chharia, A.S. (1992). Effect of post-harvest treatment on the quality of mango during storage. *Haryana J. Hort. Sci., 21*(1–2), 46–55.

Kumar, S., & Gupta, A.K. (2014). Postharvest Life of Gladiolus grandiflorus L. cv. Jessica as Influenced by Pre-harvest Application of Gibberellic Acid and Kinetin. *Journal of Post-Harvest Technology, 2*(3), 169–176.

Kumar, S., Muthu, V.P., Jawaharlal, M., & Kumar, A.R. (2012). Effect of plant growth regulators on growth, yield and exportable quality of cut roses. *The Bioscan, 7*(4), 733–738.

Kumara, G.K.K.P., Angunawela, R., & Weerakkody, W.A.P. (2007). Plant selection, preharvest treatments and post-harvest management to prolong the vase-life of shoot cuttings of Codiaeum variegatum (croton). *Journal of the National Science Foundation of Sri Lanka, 35*(3), 153–159.

Kwack, H.R., & Lee, J.S. (1997). Effects of uniconazole and gibberellin on leaf-variegation of ornamental plants under different light conditions. *Journal of the Korean Society for Horticultural Science (Korea Republic)*.

Ladaniya, M.S. (1997). Response of Nagpur mandarin fruit to preharvest sprays of gibberellic acid and carbandazim. *Indian J. Hort., 54*, 205–12.

Lee, J.S., & Kim, O.S. (1995). Effect of plant growth regulators on change of petal colors in cut flowers of *Rosa hybrida* L. cv. Red Sandra. *J. Korean Soc. Hort. Sci., 36*(1), 107–112.

Lewis, L.N., Coggins, C.W., Labanauskas, C.K., & Dugger, W.M. (1967). Bio-chemical changes associated with natural Gibberellin and GA3 delayed senescence in the Navel orange rind. *Plant Cell Physiol., 8*, 151–160.

Li, L., Zhang, W., Zhang, L., Li, N., Peng, J., Wang, Y., & Zhong, C. (2015). Transcriptomic insights into antagonistic effects of gibberellin and abscisic acid on petal growth in Gerbera hybrida. *Frontiers in Plant Science, 6*, doi: 10.3389/fpls.2015.00168.

Maharana, T., & Pani, A. (1982). Effect of post pruning and spraying different growth regulators on growth and flowering of hybrid roses. *Bangladesh Hort., 10*, 1–4.

Majumder, K., & Mazumdar, B.C. (2002). Changes of pectic substances in developing fruits of cape-gooseberry (*Physalis peruviana* L.) in relation to the enzyme activity and evolution of ethylene. *Scientia Horticulturae, 96*(1), 91–101.

Mangave, B., Singh, A., & Mahatma, M. (2012). Effects of different plant growth regulators and chemicals spray on postharvest physiology and vase life of heliconia inflorescence cv. Golden Torch. *Plant Growth Regulation, 69*(3), 259–264.

Marti, C., Orzaez, D., Ellul, P., Moreno, V., Carbonell, J., & Granell, A. (2007). Silencing of DELLA induces facultative parthenocarpy in tomato fruits. *Plant J., 52*, 865–876.

Masalkar, S.D., & Wavhal, K.N. (1991). Effect of various growth regulators on physicochemical properties of ber cv. Umran. *Maharashtra Journal of Horticulture, 5*, 37–40.

Mastalerz, J.W. (1960). Gibberellic acid (GA) makes plant tall. *Spindly Sci. for the Farmer, 7*(3), 7.

Mehta, P.M., Shiva Raj, S., & Raju, P.S. (1986). Influence of fruit ripening retardants on succinate and malate dehydrogenases in papaya fruit with emphasis on preservation. *Indian Journal of Horticulture, 43*(3&4), 169–173.

Mohammadi, K., Khaligi, A., Moghadam, A.R.L., & Ardebili, O.Z. (2012). The effects of benzyl adenine, gibberellic acid and salicylic acid on quality of tulip cut flowers. *Int. Res. J. Applied Basic Sci. 4*(1), 152–154.

Mukhopadhyay, A. (1990). Responses of carnation to spray application of NAA and gibberellic acid. *Haryana J. Hort. Sci., 19*(3–4), 280–283.

Nagarajaiah, C., & Reddy, T.V. (1986). Quality of Queen Elizabeth cut roses as influenced by gibberellic acid. *Mysore J. Agric. Sci., 20*(4), 292–294.

Nanjan, K., & Muthuswamy, S. (1975). Growth and flowering responses of Edward rose (*Rose bourboriana*) to certain growth regulant sprays. *South Indian Hort., 23*, 94–99.

Olimpieri, I., Silicato, F., Caccia, R., Mariotti, L., Ceccarelli, N., Soressi, G.P., & Mazzucato, A. (2007). Tomato fruit set driven by pollination or by the parthenocarpic fruit allele are mediated by transcriptionally regulated gibberellin biosynthesis. *Planta, 226*, 877–888.

Osman, A.R., & Sewedan, E. (2014). Effect of Planting Density and Gibberellic Acid on Quantitative and Qualitative Characteristics of Solidago Canadensis "Tara" in Egypt. *Asian Journal of Crop Science, 6*(2), 89–100.

Osman, H.E., & Abu-Goukh, A.B.A. (2008). Effect of polyethylene film lining and gibberellic acid on quality and shelf life of banana fruit. *U K J Agric Sci., 16*, 242–261.

Ouzounidou, G., & Illias, I. (2005). Hormone-induced protection of sunflower photosynthetic apparatus against copper toxicity. *Biol Plant., 49*(2), 223–228.

Ozga, J.A., & Reinecke, D.M. (1999). Interaction of 4-chloroindole-3-acetic acid and gibberellins in early pea fruit development. *Plant Growth Regul., 27*, 33–38.

Ozkaya, O., Dundar, O., & Kuden, A. (2006). Effect of preharvest gibberellic acid treatments on postharvest quality of sweet cherry. *Journal of Food Agriculture and Environment, 4*, 189–191.

Padmapriya, S., & Chezhiyan, N. (2003). Effect of growth regulators on total phenol content and IAA oxidase activity in chrysanthemum (*Dendranthema grandiflora* Tzelev) cultivars. *Orissa J. Hort., 31*(1), 119–122.

Pandey, V. (1999). Effect of NAA and GA3 spray on fruit retention, growth, yield and quality of ber (*Zizyphus mauritiana* Lamk.) cv. Banarasi Karaka. *Orissa Journal of Horticulture*, *27*, 69–73.

Parmar, P.B., & Chundawat, B.S. (1989). Effect of various post-harvest treatments on the physiology of kesar mango. *Acta Hort.*, *231*, 679–684.

Parwal, R., Nagda, C.C., & Pundir, J.P.S. (1994). Influence of plant growth regulators on vegetative growth and flower earliness of damask rose. *South Indian Hort.*, *50*(1–3), 119–123.

Patel, B.M., & Patil, M.K. (1998). Effect of plant growth regulators on production of rose flower. Nutritional Symposium on Indian Floriculture in the New Millennium, Pune.

Patil, B., & Patil, V.J. (1979). Impacts of chemicals on ber (*Zizyphus mauritiana* Lamk.). *Pesticide*, *13*, 28–30.

Patil, V.J. (1981). Use of growth regulators to control pests and to increase yield. *Proceedings of Second world conference of chemical engineering*, Montreal, Canada.

Pessarakli, M. (2002). *Handbook of Plant and Crop Physiology. 2nd Ed.* New York: Marcel Dekker Inc.

Phuangchik, K. (1994). Influence of seasonal changes on pollination and effect of chemical on fruit set of 'Nam Dok Mai Tawai' mango (*Mangifera indica* L.). M.Sc. Thesis, Bangkok (original not seen).

Pogroszewska, E., Joniec, M., Rubinowska, K., & Najda, A. (2014). Effect of pre-harvest application of gibberellic acid on the contents of pigments in cut leaves of *Asarum europaeum* L. *Acta Agrobotanica*, *67*(2), 77–84.

Pramanik, D.K., & Bose, T.K. (1974). Studies on the effects of growth substances on fruit set and fruit drop in some minor fruits. *South Indian Horticulture*, *22*, 117–123.

Raifa, A.H., Hemmat, K.I., Khattab, H.M.E.B., & Mervat, S.S. (2005). Increasing the active constituents of sepals of roselle (*Hibiscus sabdariffa* L.) plant by applying gibberellic acid and benzyladenine. *Journal of Applied Sciences Research*, *1*(2), 137–146.

Rajesh, A.M. (2012). "Effect of Different Levels of Fertigation and Growth Regulators on Growth, Yield and Quality of Rose Cultivars under Polyhouse Condition." University of Agricultural Sciences, Bangalore, pp. 179.

Ram, R.B., Pandey, S., & Kumar, A. (2005). Effect of plant growth regulators (NAA and GA3) on fruit retention, physicochemical parameters and yield of ber (*Zizyphus mauritiana* Lamk.) cultivar Banarasi Karaka. *Biochemical and Cellular Archives*, *5*, 229–232.

Ramalingam, K. (2008). Effect of growth regulating substances on growth, yield and postharvest quality of cut Rose cv. Happy Hour, M.Sc. (Hort.) Thesis, Tamil Nadu Agricultural University, Coimbatore.

Ramesh, K., & Singh, K. (2003). Effect of growth regulator and shoot tip pinching on carnation. *J. Ornamental Hort.*, *6*(2), 134–136.

Randhawa, J.S., Dhillon, B.S., Sandhu, S.S., & Bhullar, J.S. (1980). Effect of postharvest application of GA, Cycocel and calcium chloride on the storage behavior of Leconte pear. *J. Res. Punj. Agri. Univ.*, *17*(4), 363–65.

Randhawa, J.S., Dhillon, B.S., Sandhu, S.S., & Bhullar, J.S. (1980). Effect of postharvest application of GA, Cycocel and calcium chloride on the storage behavior of Leconte pear. *J. Res. Punj. Agri. Univ.*, *17*(4), 363–65.

Rani, R., & Brahmachari, V.S. (2004). Effect of growth substances and calcium compounds on fruit retention, growth and yield of Amrapali mango. *Orissa Journal of Horticulture, 32*, 15–18.

Ranjan, A., Ray, R.N., & Prasad, K.K. (2005). Effect of post-harvest application of calcium salts and GA3 on storage life of mango (*Mangifera indica* L.) cv. Langra. *Journal of Applied Biology, 15*, 69–73.

Ranwala, A.P., & Miller, W.B. (2002). Effects of gibberellin treatments on flower and leaf quality of cut hybrid lilies. *Acta Hort. 570*, 205–210.

Rao, A.V.D.D. (2010). Standardization of production technology in garland chrysanthemum (*Chrysanthemum coronarium* L.), PhD (Hort.) Thesis, University of Agricultural Sciences, Dharwad.

Rathore, H.A., Tariq, M., Shehla, S., & Aijaz, H.S. (2009). Effect of storage on physicochemical composition and sensory properties of mango (*Mangifera indica* L.) variety Dosehari. *Pakistan Journal of Nutrition, 6*, 143–148.

Roberts, A.V., Blake, P.S., Lewis, R., Taylor, J.M., & Dunstan, D.I. (1999). The effect of gibberellins on flowering in roses. *J. Plant Growth Regulation, 18*(3), 113–119.

Sabehat, A., & Zieslin, N. (1995). Effect of GA3 on alterations in cell membranes and protein composition in petals of rose (Rosa x hybrida) flowers. *J. Plant Physiol., 144*, 513–517.

Sadanand, D.A., Ashok, A.D., & Rangaswamy, P. (2000). Effect of some growth regulating chemicals on growth and flowering of rose cv. First Red under greenhouse conditions. *J. Ornamental Horticulture,* New Series, *3*(1), 51–53.

Saeed, T., Hassan, I., Abbasi, N.A., & Jilani, G. (2014). Effect of gibberellic acid on the vase life and oxidative activities in senescing cut gladiolus flowers. *Plant Growth Regulation, 72*(1), 89–95.

Saffari, V.R., Ahmad, K., Hossein, L., Mesbah, B., & Obermaier, J.F. (2004). Effects of different plant growth regulators and time of pruning on yield components of *Rosa damascena* Mill. *International J. Agric. Bio., 6*(6), 1040–1042.

Sahare, H.A., & Singh, A. (2015). Effect of Pulsing on Post Postharvest Life and Quality of Cut Anthurium Flowers (*Anthurium andraeanum* L.) cv. Xavia. *Trends in Biosciences, 8*(2), 305–307.

Sainath (2009). Influence of spacing, fertilizer and growth regulators on growth, seed yield and quality in annual chrysanthemum (*Chrysanthemum coronarium* L.). M.Sc. (Agri.) Thesis, University of Agricultural Sciences, Dharwad.

Saini, H.S., Dhullon, B.S., Randhawa, J.S., & Singh, S. (1982). Effect of GA-3; application on the shelf life of sub-tropical peaches. *Indian Fd. Pckr., 36*(4), 33–35.

Sandu, S.S., Dhillon, B.S., & Singh, S. (1983). Post-harvest application at gibberellic acid and wrappers on the storage behavior at kinnow. *Indian Fd. Pckr, 37*(3), 65–71.

Sankhla, N., Agarwal, P., Mackay, W.A., Davis, T.D., Gehlot, H.S., Choudhary, R., & Joshi, S. (2006). Ber (*Zizyphus*) fruits: apple of arid regions of Indian Thar Desert. *Acta Horticulturae, 712*, 449–452.

Sarkar, S., & Ghosh, B. (2005). Effect of growth regulators on biochemical composition of mango cv Amrapali. *Environment and Ecology, 23*, 379–380.

Schwabe, W.W. (1973). The induction of parthenocarpy in 'cox's orange pippin.' *Acta Hort., 34*, 311–316.

Serrani, J.C., Sanjuán, R., Ruiz-Rivero, O., Fos, M., & Garcia-Martinez, J.L. (2007). Gibberellin regulation of fruit set and growth in tomato. *Plant Physiol.*, *145*, 246–257.

Seymour, G.B. (1993). Banana. In *Biochemistry of Fruit Ripening*. Springer, Netherlands.

Sharifani, M., Parizadeh, M., & Mashayekhi, K. (2005). The effects of pre-storage treatments on postharvest quality of cut anemone (*Anemone coronaria*) flowers. *Acta Hort.*, *682*, 701–708.

Sharma, M.K., & Tomar, Y.S. (2008). Studies of abscission and senescence of leaves by effect of growth promoters in *Rosa indica*. *Plant Archives*, *8*(1), 329–331.

Sharma, M.K., & Tomar, Y.S. (2009). Effect of exogenous application of regulator on biochemical content of leaf in *Rosa indica*. *Journal of Plant Development Sciences*, *1*(1/2), 73–74.

Sharma, M.K., Baljeet, S., Sanjay, K., & Tomer, Y.S. (2011). Effect of gibberelic acid (GA3) leaf content in intact and excised leaf of *Rosa indica*. *Journal of Plant Development Sciences*, *3*(1/2), 203–206.

Shaul, O., Elad, Y., & Zieslin, N. (1995). Suppression of Botrytis blight in cut rose flowers with gibberellic acid: effects of postharvest timing of the gibberellin treatment, conidial inoculation and cold storage period. *Postharvest Biology and Technology*, *6*(3–4), 331–339.

Shrivastava, D.K., & Jain, D.K. (2006). Effect of urea and GA3 on physiochemical properties of mango cv. 'Langra' during year. *Karnataka J. Agr. Sci.*, *19*, 754–756.

Siddiqui, M.W. (2015). Postharvest Biology and Technology of Horticultural Crops: Principles and Practices for Quality Maintenance. CRC Press, Boca Raton, Florida, USA. pp. 550.

Siddiqui, M.W. (2016). Eco-friendly technology for postharvest produce quality. Academic Press, Elsevier Science, USA. pp. 324

Siddiqui, M.W., Ayala-Zavala, J.F., & Hwang, C.A. (2016). Postharvest management approaches for maintaining quality of fresh produce. Springer, New York. pp. 222.

Siddiqui, M.W., Datta, P., Dhua, R.S., & Dey, A. (2014). Changes in biochemical composition of mango in response to pre-harvest gibberellic acid spray. *Agriculturae Conspectus Scientificus (ACS)*, *78*(4), 331–335.

Sindhu, S.S., & Singhrot, R.S. (1996). Effect of oil emulsion and chemicals on shelf life of Baramasi Lemon *(Citrus Limon Burn)*. *Haryana J. Hort. Set*, *25*(3), 67–73.

Singh, B.K. (1988). Post-harvest studies of Mango *(Mangifera indica* L.). PhD Thesis, RAU, Pusa, Bihar.

Singh, B.K., & Singh, T. (1993). Spoilage and economic life of Zardalu and Langra mangoes as influenced by certain post-harvest treatments. *Bihar J. Exp. Hort.*, *1*(1), 1–5.

Singh, K., & Randhawa, J.S. (2001). Effect of growth regulators and fungicides on fruit drop, yield and quality of fruit in ber cv. Umran. *Journal of Research PAU*, *38*, 181–184.

Singh, R.N. (1996). *Mango*. ICAR, New Delhi.

Srihari, D., & Rao, M.M. (1996). Induction of flowering in off phase mango trees by soil application of Paclobutrazol. *Acta Hort.*, *455*, 491–495.

Sudha, R., Amutha, R., Muthulaksmi, S., Baby Rani, W., Indira, K., & Mareeswari, P. (2007). Influence of pre- and postharvest chemical treatments on physical characteristics of sapota (*Achras sapota* L.) var. PKM 1. *Res. J. Agric. Biol. Sci.*, *3*, 450–452.

Sudhavani, V., & Sankar, C.R. (2002). Effect of pre-harvest sprays on shelf life and quality of Baneshan mango fruit under cold storage. *South Indian Hortic.*, *50*, 173–77.
Taiz, L., & Zeiger, E. (2006). *Plant Physiology, 4th Ed.*, Sinauer Associates: Sunderland, MA.
Teaotia, S.S., Pandey, I.C., & Mathur, R.S. (1961). Gibberellin induced parthenocarpy in guava (*Psidium guajava* L). *Current Science*, *30*(8), 312.
Turnbull, C.G.N. (1989). Identification and quantitative analysis of gibberellins in Citrus. *J. Plant Growth Regul.*, *8*, 273–282.
Umrao, V.K., Singh, R.P., & Singh, A.R. (2007). Effect of gibberellic acid and growing media on vegetative and floral attributes of gladiolus. *Indian Journal of Horticulture*, *64*, 73–76.
Van Buren, J. (1970). Fruit phenolics. *The Biochemistry of Fruits and Their Products*, *1*, 269–304.
Varga, A. (1966). The specificities of apple cultivars and of gibberellins in the induction of parthenocarpic fruits. *Koninkl. Ned. Akad. Wetenschap.* Proc. Ser. C., *69*, 641–644.
Verhoeff, K. (1980). The Infection Process and Post-Pathogen Interaction. In: Coley-Smith, J.R., Verhoeff, K., & Jarvis, W.R. (Eds.), The Biology of Botrytis. Academic Press, London, 153–180.
Verma, V.K., Sengal, O.P., & Dhiman, S.R. (2000). Effect of nitrogen and GA3 on carnation. *J. Ornamental Hort.*, New Series, *3*(1), 64.
Wakchaure, N.Y., Dhaduk, B.K., Patil, R.V., & Ingale, K.D. (2008). Effect of growth regulators on yield and quality of golden rod (*Solidago canadensis* L.). *J. Maharastra Agric. Univ.*, *33*, 110–111.
Wangbin, I., Wang, Y., Yue-hua, Z., Liu, J., & Xulin (2008). Effect of Molybdenum Foliar Sprays on Fruiting, Yield and Fruit Quality of Jujube. *Proceedings of 1st International Jujube Symposium.* pp. 55–56. Agricultural University of Hebei, Baoding, China.
Waters, W.E. (1966). The influence of postharvest handling techniques on vase-life of gladiolus flowers. *Proc. Fla. State Hort. Soc. 79*, 452–456.
Willemsen, K. (2000). *Growth Regulator Program for Stone Fruits*. Washington State University Tree Fruit Research & Extension Center.
Wisniewska-Grezeskiewicz, H., & Treder, J. (1989). Effect of gibberellic acid on the development of yield of 'Sonia' roses grown under plastic tunnel. *Acta Hort.*, *251*, 389–392.
Yadav, S., & Shukla, H.S. (2009). Effect of various concentrations of plant growth regulators and mineral nutrients on quality parameters and shelf-life of aonla (*Emblica officinalis* Gaertn.) fruit. *Indian J. Agric. Biochem.*, *22*, 51–56.
Zhang, G.X., Kang, L.X., Kao, Z.H., Zhu, S.L., & Gao, K.B. (1999). Effect of GA and CPPU on the quality of seedless loquat fruit. *J. Fruit Sci.*, *1*, 55–59.
Zhang, L., Li, L., Wu, J., Peng, J., & Wang, X. (2012). Cell expansion and microtubule behavior in ray floret petals of Gerbera hybrida: responses to light and gibberellic acid. *Photochem. Photobiol. Sci.*, *11*, 279–288.

CHAPTER 6

ADVANCES IN PACKAGING OF FRESH FRUITS AND VEGETABLES

ALEMWATI PONGENER[1] and B. V. C. MAHAJAN[2]

[1]ICAR-National Research Centre on Litchi, Mushahari, Muzaffarpur, 842002, Bihar, India, E-mail: alemwati@gmail.com

[2]Punjab Horticultural Postharvest Technology Centre, PAU, Ludhiana, 141004, Punjab, India

CONTENTS

6.1 Introduction 232
6.2 History of Fruit and Vegetable Packaging 233
6.3 Functions of Packaging 233
    6.3.1 Containment of Produce 234
    6.3.2 Protection of Content 234
    6.3.3 Convenience 234
    6.3.4 Communication 235
6.4 Packaging Requirements for Fruits and Vegetables 236
    6.4.1 Respiration 238
    6.4.2 Moisture Loss (Dehydration) 239
    6.4.3 Temperature 239
    6.4.4 Gas Composition 240
6.5 Kinds of Packaging Material 241
    6.5.1 Natural Materials 241
    6.5.2 Wood 242

6.5.3 Cardboard (Fiberboard) 242
6.5.4 Molded Plastics 243
6.5.5 Paper or Plastic Film 243
6.6 Modified Atmosphere Packaging 244
6.7 Intelligent Packaging and Smart Packaging 245
    6.7.1 Sensors 248
    6.7.2 Time-Temperature Indicators 250
    6.7.3 Gas Indicators 252
    6.7.4 Electronic Nose 252
    6.7.5 RFID and Wireless Sensing Technology 254
6.8 Future Directives 255
Keywords 257
References 257

6.1 INTRODUCTION

With increasing complexity of society, there is continuous demand for innovative and creative packaging from stakeholders (producers, processors, transporters, whole-sellers, retailers, and consumers) to guarantee food quality, safety, and traceability. Studies commissioned by the Food and Agricultural Organization of the United Nations estimates that about 45% of fresh fruits and vegetables constitute food loss in the postharvest chain (FAO, 2014). Despite globally recognized solutions in reducing food loss along the supply chain, postharvest loss remains a persistent challenge (Ahmad and Siddiqui, 2015). The reasons for these huge losses can be assigned to:

- lack of awareness about postharvest handling of fruits and vegetables;
- use of traditional packaging;
- abusive use of packages by labor;
- poor infrastructure for storage of horticulture produce;
- absence of cold chain for perishable produce.

Packaging is one of the most important steps in the complicated journey of fruits and vegetables from growers to consumers. The Packaging Institute International defines packaging as the enclosure of products, items or packages in a wrapped pouch, bag, box, cup, tray, can, tube, bottle, or other container form to perform one or more of the following functions: containment, protection, preservation, communication, utility and performance (Siddiqui, 2015). Therefore, packaging not only contains and protects during handling, storage, transportation, and distribution, but also serves as a symbol of value-addition and assurance of quality. It is also an important tool for marketing fresh produce as it provides for ease of handling and counted containers of uniform size. Standard-size packages reduce the need for repeated weighing. Several types of package are commonly used in the fresh produce industry, and include packages fabricated from paper and paper products (compressed cardboard, corrugated cardboard or fiberboard), wood and wood products, and plastics. The choice of each type depends on the utility, capacity to enhance value to produce, and cost factor.

6.2 HISTORY OF FRUIT AND VEGETABLE PACKAGING

Wooden crates and boxes have been the traditional mode of packaging. The history of modern packaging dates back to the mid-19th century with the concept of transforming flimsy sheets of paper into a rigid, stackable, and cushioning form of packaging. These were used for packaging of delicate goods such as bottles and glass lamp chimneys. First patents for making corrugated paper were granted in England in 1856. More than a decade later in 1871, patents were granted to Albert Jones in the USA for single-phase corrugated board. In 1874, Oliver Lang invented the corrugated board with liner sheet on both sides. Invention of corrugated box in 1890 is credited to the Scottish born Robert Gair. By the advent of the 20th century corrugated boxes began to replace wooden crates and boxes. Initially limited to packaging fragile goods like glass and pottery items, the corrugated boxes found extensive usage in enabling fruit and produce to be brought from farm to retail without bruise or damage. The introduction of corrugated boxes dramatically improved returns to the producers and opened up export markets.

6.3 FUNCTIONS OF PACKAGING

According to the Codex Alimentarius Commission food is packaged to preserve its quality and freshness, add appeal to consumers and to facilitate storage and distribution. Selecting packaging material for horticultural produce should take into account the functions the packaging is intended to perform. The functions of packaging are well defined and inter-related: containment, protection, convenience, and communication (Robertson, 2006).

6.3.1 CONTAINMENT OF PRODUCE

The most basic function of packaging is that of containment. Any produce should be contained within a package before it can be moved along the supply chain. Containment allows for convenience in handling and storage, reduces losses and makes large-scale transportation and marketing possible. Packaging helps to define portion sizes, and packaged food and beverage products normally are eaten in a single sitting and thus considered an 'individual serving' (Pomeranz and Miller, 2014). With increase in living standards internationally, portion sizes have also increased over the years. This has implications on consumption behavior of individuals. Containment function of packaging, therefore, has that important role of containing the right amount/size of produce or product that can lead to healthy eating habit.

6.3.2 PROTECTION OF CONTENT

Food packaging keeps food products in a limited volume, prevents it to leak or break-up, and protects it against possible contaminations and changes (Vanderroost et al., 2014). Packaging protects the contents from physical or mechanical injuries, cuts, tears, bruises, etc. Packaging acts as a barrier between the contents and the environment. It protects the contents from outside environmental effects like microorganisms, gases, odors, temperature abuse, dust, vibrations, compressions, and shocks. Interestingly, packaging also protects the outside environment from the contents.

6.3.3 CONVENIENCE

Packaging meets the demands of different stakeholders for convenience. Packaging provides convenience in storage, transportation, and distribution of horticultural produce. Food packaging allows for consumers to enjoy food they want, at their convenience. This holds greater meaning in the present era of increasing inter-dependence among nations for food and services. Over the years, modernization and industrialization have resulted in ever-increasing single-person-households and percentage of women in the workforce. This has resulted in drastic lifestyle changes with increasing demand for convenience food, which includes pre-prepared foods, fresh-cut, and minimally processed fruits and vegetables.

6.3.4 COMMUNICATION

Food packaging communicates important information about contained product and its nutritional content, together with guidelines about preparation. Packaging improves sales, and makes advertising and large-scale distribution possible. As mentioned earlier, packaging can act as a symbol of value addition and quality assurance. It communicates the quality of the content and the satisfaction a product offers. Modern scanning equipment at retail outlets rely on barcodes or UPC imprinted on packages. Packages communicate to the consumers about the nutritional value of contents and how best they can be made use of. Nutritional information on packages can help consumers make informed, healthful choices, thereby improving sales (Freedman and Connors, 2011).

Another important consideration before designing any package is the package environment. Packaging performs its function under three different environments: Physical, Ambient, and Human environment. Physical environment includes shocks, drops, falls, vibrations, compression, crushing, and all the physical damage that can happen to the product. Ambient environment is the one that surrounds the package. Factors such as gases, temperature, water and water vapor, light, microorganisms, insects, rodents, etc. make up the ambient condition that can cause damage to the product. The human environment takes into account the variability of intended consumers. Consideration of human environment in designing packages involves understanding specific consumer needs like convenience, portion size, nutritional information, content of the package etc. In order to make sure that the cost of packaging does not upset profits from any horticulture venture, a fine balance has to be struck between protecting the contents and the cost of the package. Currently, plastic crates and fiberboard boxes are commonly used by growers and traders. However, due to the cost involved, plastic crates are used for distribution in the local market. Plastic crates also offer the benefit of return to the owner after delivery. For transportation to distant markets cheaper disposable packages are preferred. If selected after careful and proper consideration fiberboard boxes measures up to most of these requirements. Polymers or plastics exhibit many desirable features like transparency, softness, heat-seal ability, and good strength to weight ratio with high tear and tensile strength. Efficient mechanical properties, good barrier to oxygen, and low cost make up the reasons for the extensive use of polymers for packaging.

6.4 PACKAGING REQUIREMENTS FOR FRUITS AND VEGETABLES

Fresh fruits and vegetables are very important components of human diet. They are rich sources of essential nutrients and vitamins, compounds that cannot be synthesized by the human body. Fresh produce has always been known and marketed as healthy. Consumption of fresh fruits and vegetables modifies the risk of many non-communicable diseases including cancer, stroke, high blood pressure, diabetes, and coronary heart disease (WHO, 2004). Maintaining freshness or harvest quality of produce has been one of the biggest challenges in the agri-food sector. An important event in the life of horticultural produce is the time of harvesting. Harvesting detaches the fruit or vegetable from the parent plant, and removes it from its source of food, water, and nutrients. Therefore, harvested produce has to rely on internal reserves to continue aerobic respiration, active metabolism, and maintain cellular integrity. In other words, fruits and vegetables are perishable. The rate of deterioration of any harvested produce is determined by how efficiently the internal reserves are used up. The quality of fruits and vegetables is determined by factors such as color, size, taste, flavor, texture etc. Quality is said to have deteriorated when one or more of these factors are said to be below the desirable level. Consumers want produce to be fresh, typified by the quality at harvest. Packaging plays an important role in protecting the fresh produce and delaying the process of deterioration. Individual fruits and vegetables differ in rate of respiratory metabolism (Table 6.1), transpiration or water loss, response to ethylene (Table 6.2), and storage environment (Tables 6.3 and 6.4). It is, therefore, important to understand the factors that determine the rate of produce deterioration.

Advances in Packaging of Fresh Fruits and Vegetables

TABLE 6.1 Classification of Vegetables According to Respiration Intensity

Class	Respiration Intensity at 10°C (mg CO_2 $kg^{-1}h^{-1}$)	Commodities
Very low	Below 10	Onions
Low	10–20	Cabbage, cucumber, melons, tomatoes, turnips
Moderate	20–40	Carrots, celery, gherkins, leeks, peppers, rhubarb
High	40–70	Asparagus (blanched), eggplant, fennel, lettuce, radishes
Very high	70–100	Beans, Brussels sprouts, mushrooms, savoy, cabbage, spinach
Extremely high	Above 100	Broccoli, peas, sweet corn

Adapted from Weichmann (1987).

TABLE 6.2 Classification of Fruits According to Their Maximum Ethylene Production Rate

Ethylene production rate $\mu L\ kg^{-1}\ h^{-1}$ at 20°C	Fruits
Very low (0.01–0.1)	Cherries, citrus, grapes, pomegranates, strawberries
Low (0.1–1.0)	Blueberries, kiwifruit, peppers, persimmon, pineapples, raspberries
Moderate (1.0–10.0)	Bananas, figs, honeydew melons, mango, tomatoes
High (10.0–100.0)	Apples, apricots, avocados, plums, cantaloupe, nectarines, papaya, peaches, pears
Very high (>100.0)	Cherimoya, mammy apples, passion fruit, sapota

Source: Kader (1980).

TABLE 6.3 Classification of Fruits and Vegetables According to Their Tolerance to Low O_2 Concentrations

Minimum O_2 concentration tolerated (%)	Commodities
0.5	Tree nuts, dried fruits and vegetables
1.0	Some cultivars of apples and pears, broccoli, mushrooms, garlic, onions, most cut or sliced (minimally processed) fruits and vegetables
2.0	Most cultivars of apples and pears, kiwifruit, apricots, cherries, nectarine, peaches, plums, strawberries, papaya, pineapple, olives, cantaloupe, sweet corn, green beans, celery, lettuce, cabbage, cauliflower, Brussel sprouts

TABLE 6.3 Continued

Minimum O$_2$ concentration tolerated (%)	Commodities
3.0	Avocados, persimmon, tomatoes, peppers, cucumber, artichoke
5.0	Citrus fruits, green peas, asparagus, potatoes, sweet potatoes

Source: Kader et al. (1989).

TABLE 6.4 Classification of Fruits and Vegetables According to Their Tolerance to Elevated CO$_2$ Concentrations

Minimum CO$_2$ concentration tolerated (%)	Commodities
2	Golden Delicious apples, Asian pears, European pears, apricots, grapes, olives, tomatoes, peppers (sweet), lettuce, endive, Chinese cabbage, celery, artichoke, sweet potatoes
5	Apples (most cultivars), peaches, nectarines, plums, oranges, avocados, bananas, mango, papaya, kiwifruit, cranberries, peas, peppers (chili), eggplant, cauliflower, cabbage, Brussels sprouts, radishes, carrots
10	Grapefruit, lemons, lime, persimmon, pineapple, cucumber, summer squash, snap beans, okra, asparagus, broccoli, parsley, leeks, green onions, dry onions, garlic, potatoes
15	Strawberries, raspberries, blackberries, blueberries, cherries, figs, cantaloupe, sweet corn, mushrooms, spinach, kale, Swiss chard

Source: Kader et al. (1989).

6.4.1 RESPIRATION

Respiration is a process in which energy-rich substrates are broken down into simpler molecules along with production of energy. Respiration rate is often a good indicator of storage life of horticultural produce, sharing an inverse correlation: lower the rate, longer the potential storage life, and vice versa (Siddiqui, 2016). The rate of respiration is specific to a particular harvested organ within species and variety, but may differ on maturity. Climacteric fruits exhibit a rise in respiration rate and ethylene evolution rate during ripening. Mechanical stress due to peeling, slicing, and de-stoning can accelerate the rate of respiration manifold. Such mechanical stresses can result in enzymatic changes in fresh produce such as browning and increase microbial spoilage (Watada and Qi, 1999).

6.4.2 MOISTURE LOSS (DEHYDRATION)

Moisture vapor is produced in association with aerobic respiration. Moisture loss occurs from the fruit surface through natural diffusion from a region of high concentration (fruit tissue) to a region of low concentration (atmosphere surrounding fruit). The rate of dehydration depends on factors such as temperature, surface area, air velocity around the fruit, and the extent of moisture saturation (relative humidity) within the vicinity of fruit. Fresh fruits and vegetables contain around 90% of water, and visually noticeable changes take place when more than 5% moisture loss occurs. Further loss of weight and moisture results in wilting, shriveling or wrinkling. Such produce not only lose their consumer appeal and marketability, but even a reduction in nutritional content is observed, especially those of water-soluble components (Siddiqui, 2016). Dehydrated produce are more susceptible to chilling injury and loses aroma and flavor. Moisture loss is detrimental to maintenance of membrane integrity due to loss in turgor pressure and results in oxidative damage and lipid peroxidation – hallmarks of senescence and cell death. Dehydration can also trigger acceleration in climacteric. Although undesirable in excess, some moderate dehydration may be beneficial to certain vegetables such as onions, garlic, or common mushrooms.

6.4.3 TEMPERATURE

The temperature at which most physiological processes go on normally in plants ranges from 0–40°C. Within this range, the rate of respiration in fresh fruits and vegetables increases 2 to 3-fold for every 10°C rise in temperature. Decreasing the storage temperature slows down enzymatic reactions and respiration according to Arrhenius relationship. Providing optimal ranges of temperature and relative humidity (RH) is the most important tool for maintaining quality and safety of intact and fresh-cut fruits and vegetables. Thus, increased precision in temperature and RH management is indispensable for quality maintenance from harvest to final consumption. Although refrigeration is the most appropriate method for preservation of fresh produce, not all fresh produce may benefit from refrigerated storage. Tropical fruits are prone to chilling and freezing injuries, and refrigerated storage may instead accelerate respiration and other deteriorative processes. Most products undergo irreversible damage at temperatures below −1°C. There is no substitute to maintaining the cool chain throughout the postharvest handling system for maintaining quality and safety of horticultural perishables.

6.4.4 GAS COMPOSITION

Gas composition surrounding the produce is another factor that determines postharvest quality and shelf life of fruits and vegetables. Modified or controlled atmosphere packaging/storage has been conceptualized based on this principle of response of fruits and vegetables to modifications in gaseous composition (Siddiqui et al., 2016). Lowering the oxygen level is effective in reducing respiration rate. Similarly, elevating the CO_2 level also reduces the respiration rate and can limit the production of ethylene. High CO_2 levels also act against aerobic microorganisms, although it may trigger development of anaerobic flora. However, lowering O_2 to levels below its tolerance limit may induce a shift to anaerobic respiration or anoxia leading to production of off-flavors and off-odors. Similarly, high CO_2 levels may cause tissue damage leading to an increase in the rate of respiration. Physiological disorders also result when fruits and vegetables are stored in an atmosphere with CO_2 levels more than the tolerance limit. Thus, maintaining the optimum gas composition, no matter how desirable, becomes a challenge, and depends on product respiration and sensitivity to CO_2. Packaging materials must, therefore, be properly chosen for each product. The permeability of some commonly-used polymeric films are summarized in Table 6.5.

All these factors make it clear that besides the traditional functions, modern packaging designs must also take into account the conditions that spoil or preserve fresh produce. It must perform several desirable functions such as prevention of mechanical damage, monitoring temperature, reduction in enzymatic reactions, respiration rate and tissue metabolism,

Advances in Packaging of Fresh Fruits and Vegetables

TABLE 6.5 Permeability Data of Some Polymeric Films, Air and Water

	Permeability x 10^{11} [mL(STP) cm cm^{-2} sec^{-1} (cm Hg)$^{-1}$]		Activation energy (kJ mol^{-1})		Permeability ratio (β) CO_2/O_2
	O_2	CO_2	O_2	CO_2	
Polyethylene (density 0.914)	30.0	131.6	42.4	38.9	4.39
Polypropylene	17.4	75.5	47.7	38.1	4.34
Poly(vinyl chloride)	0.47	1.64	55.6	56.9	3.49
Poly(vinylidene chloride)	0.055	0.31	66.5	51.4	5.64
Air	2.5 x 10^8	1.9 x 10^8	3.6	3.60	0.76
Water	9.0 x 10^2	2.1 x 10^4	15.8	15.8	23.33

Source: Kader et al. (1998).

reduction in excessive moisture loss from produce, and maintain critical gaseous composition.

6.5 KINDS OF PACKAGING MATERIAL

Packages of all kinds are used in different parts of the world for packaging and transportation of fresh fruits and vegetables.

6.5.1 NATURAL MATERIALS

In many developing markets, baskets and containers made of natural materials such as bamboo, straw, palm leaves, etc. constitute commonly used packages. Natural fibers such as jute, sisal, coconut coir, etc., are also used to prepare sacks or bags, either woven to a closed texture or as nets. These are mostly used for packaging produce that are less easily damaged, such as potato and onions. Natural materials are normally low in cost – both raw material and labor involved, and also offer the advantage of re-usability. However, disadvantages weigh more than the benefits in that,

- they lack rigidity and bend out of shape when stacked for long distances;

- they don't come in uniform shape and pose difficulty in loading;
- they often have sharp edges which usually cause cuts and puncture damages;
- they cause pressure damage when tightly filled.

6.5.2 WOOD

Wooden containers are still being commonly used in the form of reusable boxes or crates. They not only offer the benefits of strength and reusability, wooden boxes when made to a standard size stack well on trucks or storage rooms. Wooden boxes are now gradually being replaced by less expensive alternatives. The disadvantages of wooden containers are:

- they are not environment friendly as their usage leads to felling of trees;
- obtaining uniformity of weight is a problem;
- they are heavy and costly to transport;
- they may cause compression and vibration injuries if contents are over- or under-packed;
- they often have sharp edges, splinters, and nails, etc., which can easily cause damage to contents.

6.5.3 CARDBOARD (FIBERBOARD)

Nowadays, most containers are made from solid or corrugated cardboard. Containers closing with either fold over or telescopic (i.e., separate) tops are called boxes or cases, while shallower and open topped ones are called trays. Corrugated fiberboard boxes (CFB) are supplied in collapsed form and are usually set up by the user. This set up operation usually requires tapping, gluing, stapling, or fixing of interlocking tabs. Cardboard boxes are light and clean, and can readily and easily be printed upon with nutritional information, weight, amount, etc. Some CFB boxes may be collapsed and re-used. Besides, they are available in a wide range of sizes, designs, and strength. Because of its relatively low cost and versatility, it remains the dominant package container for fruits and vegetables. Most CFB boxes are made of three or more layers of paperboard manufactured by the Kraft process. The cardboard boxes may easily collapse when empty if multiple use is intended. They are seriously weakened if exposed to moisture and easily damaged by careless handling and stacking.

6.5.4 MOLDED PLASTICS

Molded plastics are strong, rigid, smooth, and are reusable. They are usually molded, to almost any specification, from high-density polythene and find wide application and usage in transportation of produce in many countries. Packages with a top and bottom that are heat-formed from one or two pieces of plastic are known as clamshells. Clamshells are gaining wide popularity for their versatility, low cost, and for providing excellent protection to the produce. They also present a very pleasing consumer package. Another advantage of molded plastic box is that they can easily stack when full of produce and nest when empty, thereby conserving storage space. However, they deteriorate rapidly on exposure to sunlight and require treatment with a UV inhibitor, thus, adding to the cost.

6.5.5 PAPER OR PLASTIC FILM

Paper or plastic film liners are usually used in packing boxes to reduce water loss from contents and prevent friction damage. Plastic film bags are widely used in fruit and vegetable industry. Owing to their low cost they are commonly used in consumer-convenient packs. They provide a clear view of the contents and allows for visual inspection of the content by the user. However, under conditions with no control over temperature plastic bags can lead to heavy build-up of water vapor and aggravates decay. Temperature can rise inside the polythene bags on exposure to sunlight and, along with water vapor, result in rapid deterioration. Therefore, consumer packs wrapped in plastic are not recommended under tropical environmental conditions except when/where refrigerated storage is available.

Choice of packaging material has been made to avoid unwanted interactions with product or content. Despite the contribution of traditional packaging materials towards early development of food distribution, growing complexity of human society with ever-increasing needs has brought insufficiency and insatiateness in modern packaging. Novel innovative packaging designs and methods are in constant demand among consumers and popular among researchers and industry alike. This is in large due to growing consumer demand for convenience food, increased regulatory requirements, market globalization, concern for food safety, and recent threat of food bioterrorism (Yam et al., 2005).

6.6 MODIFIED ATMOSPHERE PACKAGING

Modified atmosphere packaging may be described as the use of an atmosphere composition surrounding the product, which is different from that of normal air, usually consisting of reduced oxygen and increased carbon dioxide concentrations. Each product has a specific atmosphere composition that, in most cases combined with a low temperature, maximizes the product shelf-life. Atmosphere modification can be done in one of two ways: flushing the package with a gas having the required composition when the product is packed or relying on naturally occurring processes that tend to modify the composition over time. The first case is usually referred to as active packaging and is commonly used for meat products, bakery and dry foods, while the second is called passive packaging and it may be used for fresh and minimally processed fruits and vegetables (Siddiqui, 2016).

Use of polymeric films is very pronounced in packaging of fruits with a purpose to extend their storage life. Storing fruits in plastic films creates modified atmospheric conditions around the produce inside the package allowing lower degree of control of gases and can interplay with physiological processes of commodity resulting in reduced rate of respiration, transpiration and other metabolic processes of fruits (Zagory and Kader, 1988). The modified atmosphere conditions within the film packages can significantly reduce the rates of ripening and senescence primarily by reducing the synthesis and perception of ethylene (Burg and Burg, 1967; Abeles et al., 1992). Changes in respiration and starch, sugars, chlorophyll, and cell wall constituents during ripening and/or senescence can be delayed, and in some cases nearly arrested. Besides, modified atmospheric packaging (MAP) vastly improves moisture retention, which can have a greater influence on preserving quality. The composition of the atmosphere within a package results from the interaction of a number of factors that include the permeability characteristics of the package, the respiratory behavior of the produce, and the environment. Furthermore, packaging isolates the product from the external environment and helps to ensure conditions that, if not sterile, at least reduce exposure to pathogens and contaminants.

Polymeric films for creating MA conditions have been improvised through micro-perforation using laser beam, which allows for control of gas permeability of the film. Incorporation of anti-fogging, humidity buffering, and liquid water removal systems into polymeric film packaging alleviates the damage caused by moisture accumulation inside packages. Further improvements in MAP have been made, for example, through the use of temperature compensating films or breathable materials (BreatheWayTM) which assist in maintaining optimum atmosphere composition within limits even in situations of temperature abuse. Therefore, changes in the respiration rate caused by temperature fluctuation tend to be compensated for by changes in the film gases transmission rate, avoiding anoxic conditions. Micro-perforating the package film with laser creates patterns with precise round holes, and extends the shelf life by improving the atmosphere and optimizing the amount of O_2, CO_2, and moisture inside the package. Such a technology (e.g., Lasersharp®) also helps in reduction of condensation and bacterial growth while keeping the product fresh. Other means of obtaining optimum modified atmospheres include incorporation of oxygen and/or ethylene scavengers inside package. Commonly used oxygen scavengers are iron or ascorbic acid based, while ethylene scavengers make use of potassium permanganate activated carbon, or activated earth. Modified atmosphere packaging has been beneficially exploited for extending shelf life and maintaining freshness and quality of fruits and vegetables (Table 6.6). Summary of commercialized active packaging technology for enhancement of shelf life and maintenance of quality is presented in Table 6.7 (Figure 6.1).

TABLE 6.6 Modified Atmosphere Packaging to Maintain Quality and Extend Storage Life of Fruits and Vegetables

Crop	Film/coating	Altered physiology	Reference
Apple	LDPE	Extended shelf life	Geeson et al. (1994)
Avocado	Methyl cellulose	Reduced respiration rate, maintained color and firmness	Maftoonazad and Ramaswamy (2005)
Capsicum	Polypropylene	Reduced respiration rate and extended shelf life upto 49 days	Singh et al. (2014)
Carrot	Chitosan	Delayed microbial spoilage, and maintained color and texture	Leceta et al. (2015)
Cucumber	LDPE bags	Alleviated chilling injury, and reduced weight loss	Wang and Qi (1997)
Fig	Microperforated film	Minimized weight loss, decay, and extended shelf life	Villalobos et al. (2014)
Grape	Oriented PP	Prevented fruit decay and extended shelf life	Costa et al. (2011)
Litchi	PropaFresh™ PFAM	Maintained physiological and biochemical properties of fruit	Somboonlaew and Terry (2010)
Mango	Microperforated PE	Alleviated chilling injury	Pesis et al. (2000)
Papaya	Chitosan + Pectin	Extended shelf life and quality of fresh-cut papaya	Brazil et al. (2012)
Snap peas	Microperforated polypropylene bag	Maintained higher chlorophyll, vitamin C, SSC, and extended storage life	Elwan et al. (2015)
Sweet Cherry	LDPE	Fruits remained acceptable for 4–6 weeks	Meheriuk et al. (1995)

TABLE 6.7 Commercial Active Packaging Systems

Trade name	Manufacturer	Type
Ageless	Mitsubishi Gas Chemical Co. Ltd, Japan	Iron based O_2 scavenger
Freshmax	Multisorb Technologies, USA	Iron based O_2 scavenger
Bioca	Bioka Ltd., Finland	Enzyme based O_2 scavenger
Tenderpac®	SEALPAC, Germany	Moisture absorber
Agion®	Life Materials Technology Ltd., USA	Antimicrobial packing

Advances in Packaging of Fresh Fruits and Vegetables

TABLE 6.7 Continued

Trade name	Manufacturer	Type
Biomaster®	Addmaster Ltd., USA	Antimicrobial packing
Peakfresh	Peakfresh Products Ltd., Australia	Ethylene scavenger
Evert-Fresh	Evert-Fresh Corporations, USA	Ethylene scavenger

Adapted from Biji et al. (2015).

6.7 INTELLIGENT PACKAGING AND SMART PACKAGING

With fast-changing business requirements, strict quality standards for imported fruits and vegetables, along with technology development, the boundaries of traditional functions of packaging have expanded into development of dynamic packaging systems that are capable of carrying out intelligent functions (Ahmad and Siddiqui, 2015). Intelligent packaging is a system that provides reliable and correct information regarding the condition of the contained food product, the environment, or the integrity of packaging. According to the legal definition of EU (EC, 2009), intelligent packaging contains a component that enables the monitoring of the condition of packaged food or the environment surrounding the food during transport and storage. Therefore, it can be said that intelligent packaging is an extension of the communication function of packaging, and communicates to the user useful information based on ability to sense, record, or detect changes (desirable and/or undesirable) in the product or environment. These facilitate decision making to extend shelf-life, enhance safety,

FIGURE 6.1 Modified atmosphere packaging of peach, kinnow, and capsicum.

improve quality, provide information and warn about possible problems (Todorovic et al., 2014). Intelligent packaging should, however, not be confused with active packaging, which is an extension of the protection function of packaging. Active packaging is designed such that it contains a component that enables the release or absorption of substances into or from the packaged food or the environment surrounding the food (EC, 2009). Thus, active packaging refers to a system where the product, package, and the environment interact to extend shelf-life and improve the quality of the content, that otherwise cannot be achieved (Miltz et al., 1995). Smart packaging is the realization of synergistic association between intelligent packaging and active packaging; monitoring changes in the product and environment, and acting upon these changes to bring about desirable product quality and storability. Until now technologies that qualify for intelligent packaging include time-temperature indicators, gas indicators, sensors, indicators, nose systems, and radio frequency identification systems (Kerry et al., 2006).

6.7.1 SENSORS

As stated before, fruits and vegetables are living biological products and freshness of produce after harvest is affected by time, handling, environmental conditions, and metabolic processes. Fruits and vegetables also undergo various operations postharvest: pre-cooling, sorting, grading, washing, packaging, storage, transportation, and retail. Consumers expect produce to be fresh and safe. Quality sensing, therefore, becomes necessary at different stages of production or supply chain. Packaged produce in transit is prone to undesirable vagaries like physical injury, temperature abuse, microbial infection, etc. This is where sensors come into the picture helping users to detect and record changes in product, package, or environment. Also, these factors can have profound effect on freshness and quality of the produce (Siddiqui, 2015). There has been increasing research on sensor technology in the domain of fruit and vegetable. Quality attributes and disorders in fruits and vegetables studied through non-invasive computer vision technologies like nuclear magnetic resonance (NMR) relaxometry (MRR), NMR spectroscopy (MRS), ear infrared (NIR), mid infra-red (MIR), and magnetic resonance imaging (MRI)

Advances in Packaging of Fresh Fruits and Vegetables 249

TABLE 6.8 Use of NMR Relaxometry (MRR), NMR Spectroscopy (MRS), and MRI Techniques in Sensing Quality and Disorders in Fruit and Vegetable Crops

Crop	Maturity/ Sugar	Bruises/ Voids/ seeds	Tissue breakdown	Heat injury	Chill/ freeze injury	Infections
Apple	MRR/MRI	MRI	MRI/MRR			
Avocado	MRS/MRI					
Banana	MRR					
Cherimoya	MRR/MRI					
Cherry	MRR/MRS	MRI				
Cucumber					MRI	MRI
Durian	MRS/MRI		MRI			
Grape	MRS/MRR					MRI
Kiwifruit	MRR				MRI/ MRR	MRI
Mandarin	MRR	MRI				
Mango	MRS/MRR			MRI		MRI
Mangosteen	MRI					MRI
Melon	MRS	MRI	MRI			
Nectarine			MRI			MRR
Onion		MRI				
Orange	MRS/MRR	MRI			MRI	
Papaya				MRR		
Peach		MRI			MRR/ MRI	
Pear		MRI	MRR/MRI			
Persimon					MRI	
Pineapple	MRR/MRI					
Potato	MRR/MRI	MRI	MRI			MRI
Strawberry						MRI
Tomato	MRR/MRI					

Adapted from Ruiz-Altisent et al. (2010).

techniques are summarized in Tables 6.8 and 6.9. Physical injuries and wounds are of common occurrence during transportation of fruit and vegetable crops. Impact sensors are based on force versus time measurements

TABLE 6.9 Application of NIR-MIR Spectroscopy in Fruit and Vegetables

Sensor	Detection	Reference
NIR	Dry matter content of onions	Birth et al. (1985)
	Acidity and soluble solids content of apples	Lammertyn et al. (1998)
	Water content in mushroom	Roy et al. (1993)
	Brown heart in apples	Clark et al. (2003)
	Sensory attributes of apple	Mehinagic et al. (2004)
	Sugar and firmness of apple	Peng and Lu (2007)
	Softening in Nectarine	Zerbini et al. (2006)
MIR	Sugar and acids in tomato	Beullens et al. (2006)
	Changes in macromolecular constituents of hazelnut	Dogan et al. (2007)
	Olive oil acidity	Inon et al. (2003)

Adapted from Ruiz-Altisent et al. (2010).

and can be effectively used to ascertain the physical condition of produce (Chen and Ruiz-Altisent, 1996; Garcia-Ramos et al., 2003). Other than research prototypes, manufacturers have also entered this domain and impact sensors (Acoustic firmness sensor by Aweta, Intelligent firmness detector by Greefa, and Sinclair iQ firmness tester for example) have successfully been tested in fruits such as apples, avocadoes, citrus, kiwifruit, plums, nectarines, peaches, etc. (Howarth and Ioannides, 2002).

Biosensors are also increasingly being used in postharvest management of fruit and vegetables, with applications ranging from detection of pesticide residues through quality control in package atmospheres to detection of foodborne pathogens and toxins in fruits and vegetables. Low oxygen injury is common in postharvest systems like modified atmosphere packaging. The extent of low-O_2 injury can be ascertained by detection of ethanol using ethanol sensors (Velasco-Garcia and Mottram, 2003), the most advanced of which are bienzymatic sensors based on co-immobilization of alcohol oxidase with horseradish peroxidase (Azevedo et al., 2005). One of the greatest applications of biosensors is in detection of foodborne pathogens and toxins. Biosensors not only have the potential in toxicity detection but also act as indicators for freshness and spoilage of fruits and vegetables (Saaid et al., 2009; Velusamy et al., 2010). Summary on use of biosensors in supply chain of fresh fruits and vegetables is given in Table 6.10.

Advances in Packaging of Fresh Fruits and Vegetables 251

TABLE 6.10 Biosensors in Postharvest Management of Fruits and Vegetables

Crop	Biosensor	Detection	Reference
Fresh cut lettuce, cauliflower, broccoli, cabbage and carrot	Ethanol biosensor	Detection of ethanol in low oxygen injury in MAP	Smyth et al. (1999)
Guava	Bromophenol blue immobilized onto bacterial cellulose	Detects freshness of guava through sensing of package headspace pH	Kuswandi et al. (2013)
Kimchi	Irreversible chitosan-based ripeness indicator	Monitors package headspace CO_2 concentration	Meng et al. (2015)
Orange juice	Cytochrome c-based	Determination of antioxidant capacity	Cortina-Puig et al. (2009)
Fruits and vegetables	Laccase-based biosensor	Polyphenol determination	Gomes and Rebelo (2003)
Apple and Tomato	High-throughput enzymatic biosensor array system	Measurement of sugars and organic acids	Vermeir et al. (2005)

6.7.2 TIME-TEMPERATURE INDICATORS

Time-temperature indicators (TTIs) are small devices, typically self-adhesives, attached to the package, and provide a visual indication in response to temperature history within the package during the course of shipment to the end user.. The visual indication can be a result of enzymatic change, polymerization, melting point, diffusion, or pH change due to microbial growth. However, TTIs are because regardless of the mechanism of the indicator, the rate of visual change should be temperature-dependent. Also, TTIs can be of three basic types: full history TTIs that start integration immediately once activated, partial history TTIs that start integration on achieving a certain temperature threshold, and abuse indicator TTIs that show a visual change on achieving a certain temperature. Abuse indicator TTIs, therefore, do not indicate or respond to the time duration to which the product has been exposed. The mechanism of visually detectable indication include change in enzymatic reactions triggering a pH and color change, polymerization of a polymer that usually becomes darker at a temperature dependent rate, and melting or diffusion of substances that triggers a change in color in the indicator window. TTIs are now available commercially, for example, Monitor®, Checkpoint®, Fresh-Check®, OnVu™, eO®, TT Sensor™ (Poças et al., 2008). Time monitoring is also very important to judge the time that has elapsed since a package has been opened or activation of a label. Such an integration of time-temperature component would also reveal visual change in the indicator when a product has been exposed to temperature abuse for a specific duration, thus indicating suitability/unsuitability of the product for consumption.

6.7.3 GAS INDICATORS

Gas indicators are based on detection of volatile metabolites produced during food deterioration. Initially based on detection of oxygen and carbon dioxide for judging the freshness of food, the domain of gas indicators are now based on detection of metabolites like diacetyl, amines, ammonia, ethanol, and hydrogen sulfide. A fine example of commercial gas indicator is the RipeSense® sensor label developed by Jenkins group, New Zealand. RipeSense® is a ripeness indicator developed specifically for pears. The label consists of a sensor that detects the emitted natural aroma causing a change in color. The sensor changes in color from red through orange to yellow, indicating the ripeness level. Thus, consumers are able to make purchasing decisions by matching the color with their eating preferences.

6.7.4 ELECTRONIC NOSE

Gas measurement systems used with a particular methodology are sometimes referred to as 'Electronic noses' (Boeker, 2014). In other words, an electronic nose is any instrument that is capable of recognizing simple or complex odors. It comprises an array of electronic chemical sensors with partial specificity and an appropriate pattern recognition system (Gardner and Bartlett, 1994), and is based on the fact that aroma, odor, and volatiles are recognized through the sense of smell. A nose system mimics or exceeds the human sense of smell or taste by generating a unique response to each flavor or odor. The sensors evaluate the complex volatile mixture and collectively assemble the constituents via a transducer to form a digital pattern that is referred to as Electronic Aroma Signature Pattern (EASP). EASP is highly unique and specific to the particular gas mixture being analyzed (Baietto and Wilson, 2015). Aroma or odor is an important component of fruits and vegetables, and is a function of flavor and sensory quality that is so important from consumer's perspective. Some of the most important changes in fruit aroma take place during postharvest storage or shelf-life period. Aroma volatiles not only indicate the stage of ripeness of the fruit but also the physiological state of the fruit. Electronic nose technology incorporated into packaging can ascertain the state of ripeness, or if the produce inside has rotten or got spoiled. Electronic noses can, therefore, be used to

TABLE 6.11 Electronic Noses and Their Varied Applications in Postharvest Management of Horticultural Produce

Produce	Parameter detected	Reference
Apple	Quality of fruit during postharvest storage; Harvest date determination; detection of mealiness	Brezmes et al. (2001); Saevels et al. (2003); Li et al. (2007a)
Apple	Aroma profile during deteriorative shelf life	Li et al. (2007b)
Apple	Prediction of storage time	Hui et al. (2013)
Apricot	Ripening after harvest	Defilippi et al. (2009)
Banana	Monitoring ripeness	Sanaeifar et al. (2014)
Bell pepper	Quality assessment	Rosso et al. (2012)
Blueberry	Ripening stage and quality control	Simon et al. (1996)
Blueberry	Fruit disease detection	Li et al. (2010)
Grape	Postharvest treatments	Zoecklein et al. (2011)
Mandarin	Harvest date determination	Gomez et al. (2006)
Mango	Harvest date determination	Lebrun et al. (2008)
Mango	Determination of fruit ripeness	Salim et al. (2005)
Muskmelon	Determination of harvest date	Benady et al. (1995)
Orange	Detection of freeze injury	Tan et al. (2005)
Orange	Shelf life determination	Di Natale et al. (2001)

TABLE 6.11 Continued

Produce	Parameter detected	Reference
Pear	Prediction of acidity, soluble solids and pH	Zhang et al. (2008)
Peach	Volatile compounds and quality changes during cold storage	Rizzolo et al. (2013)
Persimon	Determination of harvest time and shelf life	Li et al. (2013)
Pineapple	Determination of shelf life	Torri et al. (2010)
Strawberry	Detection of fruit maturity	Du et al. (2010)
Tomato	Monitoring fruit with different storage time	Gomez et al. (2008)
Tomato	Determination of fruit maturity and shelf life	Wang and Zhou (2007)

develop consistent and reproducible non-destructive techniques to evaluate fruit quality inside packaged produce. Aroma analysis has been intensely researched for its potential as a non-destructive tool to evaluate produce quality (Table 6.11). Electronic noses offer many advantages over other methods including rapid and real-time detection of volatiles, lower costs, and ease of automation.

6.7.5 RFID AND WIRELESS SENSING TECHNOLOGY

Radiofrequency identification or RFID is a system for automated data acquisition based on tags that contain transponders emitting messages that are readable by RFID readers (Todorovic et al., 2014). RFID technology has found vast applications in a variety of fields such as development of biometric passports, social security cards, livestock identification, prescription drugs, food and beverages industry, access control, patient identification and hospital management, car immobilization, airline luggage management, traffic and toll control, archiving and aviation management, library and document management, logistics, retail and trade, and many more (Vlcek, 2006). An RFID tag attached to a fruit and vegetable package can be used to track its progress through the supply chain. RFID technology in conjunction with integrated sensors offers promise of a system designed to assess the status of content within a package without having to open it. Existing standards require that produce is clean, correctly labeled, and of marketable quality. In addition, tracking and tracing has become a critical part of trade practices. In order to comply with legislation and meet the standard requirements of food safety and quality, adoption of internal traceability system becomes an integral part (Bosona and Gebresenbet, 2013). RF technologies have been proposed to offer solutions to traceability in agro-food supply chain, but insufficiency exists in implementation of gapless traceability (Barchetti et al., 2009). Thus, RFID in conjunction with a continuous monitoring technology has evolved as solution to achieve gapless traceability (Mainetti et al., 2013). Wireless sensing technologies have opened up vast possibilities offering not just sensing but communication as well. This has been possible through the development of low-cost, low-power, sensor nodes, which enable environment sensing along with data processing. Sensor nodes are able to network with other sensor systems and exchange information and data with external users. Most of the wireless sensing technology applications makes use of IEEE 802.15.4/ZigBee protocol (IEEE 2003; van Tuijl et al., 2008). In this way, the environmental parameters like temperature, relative humidity, vibration, shock, light, gases composition, and microbial population can be sensed and detected within the package. Thus, a huge benefit of wireless sensing technology in packaging science is the feasibility of installation it brings within every package. Wireless sensing technology offers high potential in monitoring fruit and vegetable containers and storage facilities (Ruiz-Garcia et al., 2008). This technology assumes high significance in offering the possibility of monitoring quality, freshness, senescence, or spoilage of packaged fruit and vegetable during transport (Figure 6.2). Some applications of wireless sensing have been demonstrated for fruits and vegetables (Table 6.12).

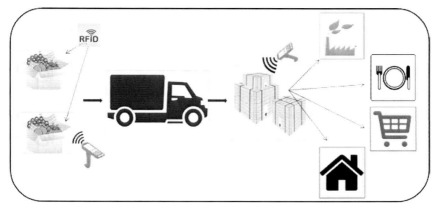

FIGURE 6.2 Schematic representation of intelligent packing through integration of RFID and sensors.

TABLE 6.12 Summary on Use of RFID and WSN in Fruit and Vegetable Packaging and Supply Chain

Crop/Application	RFID or WSN application	Reference
Apple	Monitoring fruit quality and conservation stage of fruit	Vergara et al. (2007)
Durian	WSN for monitoring maturity stage of fruit	Krairiksh et al. (2011)
Pineapple	Use of RFID for temperature mapping in supply chain	Amador et al. (2009)
Fresh ready-to-eat (RTE) vegetable products	Gapless traceability system using RF technologies and EPC global standard	Mainetti et al. (2013)
Fresh produce	Application of 2G-RFID system to Smart cold chain system (SCCS) to evolve Refined Smart Cold Chain System (RSCCS)	Chen et al. (2014)
Fruits and vegetables	'Intelligent Container' based on a cognitive sensor network that measures relevant parameters like temperature and humidity	Lang et al. (2011)
Lettuce	Cold chain transport using WSN technology with real-time monitoring of temperature, RH, door openings, truck stops, and psychometry	Ruiz-Garcia et al. (2010)
Perishable items	Integrated RFID-sensor network system for optimization of logistics systems operation	Mejjaoli and Babiceanu (2015)
Perishable food products	RFID tags with temperature sensor to model shelf life and produce quality during transportation	Jedermann et al. (2009)
Supply chain of perishable food	Real-time environment monitoring	Wang et al. (2015)

6.8 FUTURE DIRECTIVES

The export of fresh fruits and vegetables face many problems in terms of quality phytosanitary requirements, postharvest handling and packaging. Better storage and handling facilities at farm level, reduction in number of intermediaries in the chain, development of bulk handling systems including precooling and prepackaging can reduce loses. The shelf life of various fruits and vegetables can be increased by 5–15 days when pre-packed.

Introduction of consumer packs like polyethylene bags with ventilation holes, wrapping fruits and vegetables in stretch films and use of plastic punnets not only increases the shelf life of the produce but also adds value to the produce. Use of packages like plastic creates, leno bags, polypropylene boxes also helps in extending the shelf life of the fresh produce. Postharvest technologies including precooling, MAP, CAP, active packaging need to be adopted for quality preservation and extension of shelf life of various fruits and vegetables. If properly handled, temperature and packaging can act favorably in preserving fruits and vegetables. Many novel techniques and systems have become available for real-time measurement of produce within the package and its environment. Non-destructive techniques and technologies that allows for sensing and communication quality parameters without having to open the package are promising. However, successful commercialization of such technologies will depend on reliability of measurements and correlation with existing established techniques.

KEYWORDS

- Fruits and Vegetables
- Functions of Packaging
- Intelligent Packaging
- MAP
- Packaging Requirements
- Postharvest Quality
- Smart Packaging

REFERENCES

Abeles, F.B., Morgan, P.W., & Saltveit, M.E. (1992). Ethylene in Plant Biology. 2nd ed. Academic Press, San Diego CA.

Ahmad, M.S., & Siddiqui, M.W. (2015). Postharvest Quality Assurance of Fruits: Practical Approaches for Developing Countries. Springer, New York. pp. 265.

Amador, C., Emond, J.P., & Nunes, M.C.N. (2009). Application of RFID technologies in the temperature mapping of the pineapple supply chain. *Sensing and Instrumentation for Food Quality and Safety. 3*, 26–33.

Azevedo, A.M., Prazeres, D.M.F., Cabral, J.M.S., & Fonseca, L.P. (2005). Ethanol biosensors based on alcohol oxidase. *Biosensors and Bioelectronics. 21*, 235–247.

Baietto, M., & Wilson, A.D. (2015). Electronic nose applications for fruit identification, ripeness, and quality grading. *Sensors. 15*, 899–931.

Barchetti, U., Bucciero, A., De Blasi, M., Mainetti, L., & Patrono, L. (2009). Implementation and testing of an EPCglobal-aware discovery service for item level traceability. In: Proceedings of the International Conference on Ultra Modern Telecommunications and Workshops, pp. 1–8.

Benady, M., Simon, J.E., Charles, D.J., & Miles, G.E. (1995). Fruit ripeness determination by electronic sensing or aromatic volatiles. *Trans. ASAE. 38*, 251–257.

Beullens, K., Kirsanov, D., Irudayaraj, J., Rudnitskaya, A., Legin, A., Nicolai, B.M., & Lammertyn, J. (2006). The electronic tongue and ATR-FTIR for rapid detection of sugars and acids in tomatoes. *Sensors and Actuators B: Chemical 116*, 107–115.

Biji, K.B., Ravishankar, C.N., Mohan, C.O., & Srinivasa Gopal, T.K. (2015). Smart packaging systems for food applications: a review. *Journal of Food Science and Technology.* doi: 1-0.1007/s13197-015-1766-7.

Birth, G.S., Dull, G.G., Renfroe, W.T., & Kays, S.J. (1985). Nondestructive spectrophotometric determination of dry-matter in onions. *Journal of the American Society for Horticultural Science 110*, 297–303.

Boeker, P. (2014). On 'Electronic Nose' methodology. *Sensors and Actuators B: Chemical. 204*, 2–17.

Bosona, T., & Gebresenbet, G. (2013). Food traceability as an integral part of logistics management in food and agricultural supply chain. *Food Control 33*, 32–48.

Brasil, I.M., Gomes, C., Puerta-Gomez, A., Castell-Perez, M.E., & Moreira, R.G. (2012). Polysaccharide-based multi-layered antimicrobial edible coating enhances quality of fresh-cut papaya. *LWT-Food Sci Technol. 47*, 39–45.

Brezmes, J., Llobet, E., Vilanova, X., Orts, J., Saiz, G., & Correig, X. (2001). Correlation between electronic nose signals and fruit quality indicators on shelf-life measurements with pinklady apples. *Sensors and Actuators B: Chemical 80*, 41–50.

Burg, S.P., & Burg, E.A. (1967). Molecular requirements for the biological activity of ethylene. *Plant Physiology 42*, 114–52.

Chen, P., & Ruiz-Altisent, M. (1996). A low-mass impact sensor for high-speed firmness sensing of fruits. In: Proceedings International Conference of Agricultural Engineering, Madrid, Spain.

Chen, Y.Y., Wang, Y.J., & Jan, J.K. (2014). A novel deployment of smart cold chain system using 2G-RFID-Sys. *Journal of Food Engineering. 141*, 113–121.

Clark, C.J., McGlone, V.A., & Jordan, R.B. (2003). Detection of Brownheart in Braeburn apple by transmission NIR spectroscopy. *Postharvest Biology and Technology 28*, 87–96.

Cortina-Puig, M., Munoz-Berbel, X., Rouillon, R., Calas-Blanchard, C., & Marty, J.-L. (2009). Development of a cytochrome c-based screen-printed biosensor for the determination of the antioxidant capacity of orange juices. *Bioelectrochemistry. 76*, 76–80.

Costa, C., Lucera, A., Conte, A., Mastromatteo, M., Speranza, B., Antonacci, A., & Del Nobile, M.A. (2011). Effects of passive and active modified atmosphere packaging conditions on ready-to-eat table grape. *Journal of Food Engineering, 102*, 115–121.

Defilippi, B.G., San Juan, W., Valdes, H., Moya-Leon, M.A., Infante, R., & Campos-Vargas, R. (2009). The aroma development during storage of Castlebrite apricots as evaluated by gas chromatography, electronic nose, and sensory analysis. *Postharvest Biol. Technol. 51*, 212–219.

Di Natale, C., Macagnano, A., Martinelli, E., Paolesse, R., Proietti, E., & D'Amico, A. (2001). The evaluation of quality of post-harvest oranges and apples by means of an electronic nose. *Sens. Actuators B Chem. 78*, 26–31.

Dogan, A., Siyakus, G., & Severcan, F. (2007). FTIR spectroscopic characterization of irradiated hazelnut (*Corylusavellana* L.). *Food Chemistry 100*, 1106–1114.

Du, X., Bai, J., Plotto, A., Baldwin, E., Whitaker, V., & Rouseff, R. (2010). Electronic nose for detecting strawberry fruit maturity. *Proc. Fla. State Hort. Soc. 123*, 259–263.

EC (2009). *EU Guidance to the Commission Regulation on Active and Intelligent Materials and Articles Intended to Come into Contact with Food*. European Commission.

Elwan, M.W.M., Naser, I.N., El-Seifi, S.K., Hassan, M.A., & Ibrahim, R.E. (2015). Storability, shelf life and quality assurance of sugar snap peas (cv. Super sugar snap) using modified atmosphere packaging. *Postharvest Biology and Technology. 100*, 205–211.

Food and Agriculture Organization of the United Nations. (2014). Safe Food: Global Initiative on Food Loss and Waste Reduction. Rome, Italy. www.fao.org/safe-food

Freedman, M.R., & Connors, R. (2011). Point-of Purchase nutrition information influences food purchasing behaviors of college students: A pilot study. *J Am Diet Assoc. 111*, S42–S46.

Garcia-Ramos, F.J., Ortiz-Cãnavate, J., Ruiz-Altisent, M., Diez, J., Flores, L., Homer, I., & Chavez, J.M. (2003). Development and implementation of an on-line impact sensor for firmness sensing of fruits. *Journal of Food Engineering. 58, 53–57.*

Gardner, J.W., & Bartlett, P.N. (1994). A brief history of electronic noses. *Sens. Actuators B*(18–19), 211–220.

Geeson, J.D., Genge, P.M., & Sharples, R.O. (1994). The application of polymeric film lining systems for modified atmosphere box packaging of English apples. *Postharvest Biology and Technology. 4*, 35–48.

Gomes, S.A.S.S., & Rebelo, M.J.F. (2003). A new laccase biosensor for polyphenols determination. *Sensors. 3*, 166–175.

Gomez, A.H., Wang, J., Hu, G., & Pereira, A.G. (2006). Electronic nose technique potential monitoring mandarin maturity. *Sensors and Actuators B: Chemical 113*, 347–353.

Gomez, A.H., Wang, J., Hu, G., & Pereira, A.G. (2008). Monitoring storage shelf life of tomato using electronic nose technique. *Journal of Food Engineering 85*, 625–631.

Howarth, M.S., & Ioannides, Y. (2002). Sinclair IQ-firmness tester. In: Paper Read at Proc. Intnl. Conf. Agricultural Engineering, Budapest, Hungary.

Hui, G.H., Wu, Y.L., Ye, D.D., & Ding, W.W. (2013). Fuji apple storage time predictive method using electronic nose. *Food Anal. Method. 6*, 82–88.

IEEE (2003). Wireless Medium Access Control (MAC) and Physical Layer (PHY) Specifications. Specifications for Low-Rate Wireless Personal Area Networks (LR-WPANs).

IEEE Standard. 802.15.4. The Institute of Electrical and Electronics Engineers Inc. The Institute of Electrical and Electronics Engineers Inc., New York, USA.

Inon, F.A., Garrigues, J.M., Garrigues, S., Molina, A., & de la Guardia, M. (2003). Selection of calibration set samples in determination of olive oil acidity by partial least squares attenuated total reflectance-Fourier transform infrared spectroscopy. *Analytica Chimica Acta 489*, 59–75.

Jedermann, R., Ruiz-Garcia, L., & Lang, W. (2009). Spatial temperature profiling by semi-passive RFID loggers for perishable food transportation. *Computers and Electronics in Agriculture 65*, 145–154.

Kader, A.A. (1980). Prevention of ripening in fruits by use of controlled atmospheres. *Food Technology. 34*(3), 51–54.

Kader, A.A., Singh, R.P., & Mannappereruma, J.D. (1998). Technologies to extend the refrigerated shelf life of fresh fruits. In: Taub, I.A., & Singh, R.P. (Eds.), *Food Storage Stability*. CRC Press, Boca Raton, FL, Chapter 16.

Kader, A.A., Zagory, D., & Kerbel, E.L. (1989). Modified atmosphere packaging of fruits and vegetables. *Critical Reviews in Food Science and Nutrition. 28*, 1–29.

Kerry, J.P., O'Grady, M.N., & Hogan, S.A. (2006). Past, current, and potential utilization of active and intelligent packaging systems for meat and muscle-based products: a review. *Meat Science. 74*, 113–130.

Krairiksh, M., Varith, J., & Kanjanavapastit, A. (2011). Wireless sensor network for monitoring maturity stage of fruits. *Wireless Sensor Network. 3*, 318–321.

Kuswandi, B., Maryska, C., Jayus, Abdullah, A., & Heng, L.Y. (2013). Real time on-package freshness indicator for guavas packaging. *Journal of Food Measurement and Characterization 7*, 29–39.

Lammertyn, J., Nicolai, B., Ooms, K., De Smedt, V., & De Baerdemaeker, J. (1998). Nondestructive measurement of acidity, soluble solids, and firmness of Jonagold apples using NIR-spectroscopy. *Transactions of the ASAE 41*, 1089–1094.

Lang, W., Jedermann, R., Mrugala, D., Jabbari, A., Krieg-Brückner, B., & Schill, K. (2011). The "Intelligent Container"—a cognitive sensor network for transport management. *IEEE Sens. J 11*, 688–698.

Lebrun, M., Plotto, A., Goodner, K., Ducamp, M.-N., & Baldwin, E. (2008). Discrimination of mango fruit maturity by volatiles using the electronic nose and gas chromatography. *Postharvest Biology and Technology 48*, 122–131.

Leceta, I., Molinaro, S., Guerrero, P., Kerry, J.P., & de la Caba, K. (2015). Quality attributes of map packaged ready-to-eat baby carrots by using chitosan-based coatings. *Postharvest Biology and Technology 100*, 142–150.

Li, C., Heinemann, P.H., & Irudayaraj, J. (2007). Detection of apple deterioration using an electronic nose and zNose™. *Trans. ASABE. 50*, 1417–1425.

Li, C., Heinemann, P., & Sherry, R. (2007). Neural network and Bayesian network fusion models to fuse electronic nose and surface acoustic wave sensor data for apple defect detection. *Sensors and Actuators B: Chemical 125*, 301–310.

Li, C., Krewer, G.W., Ji, P., Scherm, H., & Kays, S.J. (2010). Gas sensor array for blueberry fruit disease detection and classification. *Postharvest Biology and Technology 55*, 144–149.

Li, J., Peng, Z., Xue, Y., Chen, S., & Zhang, P. (2013). Discrimination of maturity and storage life for 'Mopan' persimmon by electronic nose technique. *Acta Hortic. 996*, 385–390.

Maftoonazad, N., & Ramaswamy, H.S. (2005). Postharvest shelf-life extension of avocados using methyl cellulose-based coating. *LWT 8*, 617–624.

Mainetti, L., Patrono, L., Stefanizzi, M.L., & Vergallo, R. (2013). An innovative and low-cost gapless traceability system of fresh vegetable products using RF technologies and EPCglobal standard. *Computers and Electronics in Agriculture. 98*, 146–157.

Meheriuk, M., Girard, B., Moyls, L., Beveridge, H.J.T., McKenzie, D.-L., Harrison, J., Weintraub, S., & Hocking, R. (1995). Modified atmosphere packaging of 'Lapins' sweet cherry. *Food Research International. 28*, 239–244.

Mehinagic, E., Royer, G., Symoneaux, R., Bertrand, D., & Jourjon, F. (2004). Prediction of the sensory quality of apples by physical measurements. *Postharvest Biology and Technology 34*, 257–269.

Mejjaoli, S., & Babiceanu, R.F. (2015). RFID-wireless sensor networks integration: Decision models and optimization of logistics systems operations. *Journal of Manufacturing Systems. 35*, 234–245.

Miltz, J., Passy, N., & Mannheim, C. (1995). Trends and applications of active packaging systems. In: Jagerstad, M., & Ackerman, P. (Eds.) *Food and Packaging Materials – Chemical Interactions, 162*, 201–210. London, England: The Royal Soc. of Chemistry.

Peng, Y.K., & Lu, R.F. (2007). Prediction of apple fruit firmness and soluble solids content using characteristics of multispectral scattering images. *Journal of Food Engineering. 82*, 142–152.

Pesis, E., Aharoni, D., Aharon, Z., Ben-Arie, R., Aharoni, N., & Fuchs, Y. (2000). Modified atmosphere and modified humidity packaging alleviates chilling injury symptoms in mango fruit. *Postharvest Biology and Technology. 19*, 93–101.

Poças, M.F.F., Delgado, T.F., Oliveira, F.A.R. (2008). Smart packaging technologies for fruits and vegetables. In: Kerry, J., & Butler, P. (Eds.). Smart packaging technologies for fast moving consumer goods. John Wiley & Sons, Ltd., pp. 151–166.

Pomeranz, J.L., & Miller, D.P. (2014). Policies to promote healthy portion sizes for children. *Appetite*. doi. 10.1016/j.appet.2014.12.003

Rizzolo, A., Bianchi, G., Vanoli, M., Lurie, S., Spinelli, L., & Torricelli, A. (2013). Electronic nose to detect volatile compound profile and quality changes in "Spring Belle" peach (*Prunuspersica* L.) during cold storage in relation to fruit optical properties measured by time-resolved reflectance spectroscopy. *J. Agric. Food Chem. 61*, 1671–1685.

Robertson, G.L. (2006). Food Packaging Principles and Practices, 2nd edn. Boca Raton, Florida: CRC Press.

Rosso, F., Zoppellari, F., Sala, G., Malusa, E., Bardi, L., & Bergesio, B. (2012). A study to characterize quality and to identify geographical origin of local varieties of sweet pepper from Piedmont (Italy). *Acta Hortic. 936*, 401–409

Roy, S., Anantheswaran, R.C., Shenk, J.S., Westerhaus, M.O., & Beelman, R.B. (1993). Determination of moisture-content of mushrooms by VIS NIR spectroscopy. *Journal of the Science of Food and Agriculture. 63*, 355–360.

Ruiz-Altisent, M., Ruiz-Garcia, L., Moreda, G.P., Renfu, L., Hernandez-Sanchez, N., Correa, E.C., Diezma, B., Nicolai, B., & García-Ramos, J. (2010). Sensors for product characterization and quality of specialty crops-a review. *Computers and Electronics in Agriculture. 74*, 176–194.

Ruiz-Garcia, L., Barreiro, P., & Robla, J.I. (2008). Performance of ZigBee-based wireless sensor nodes for real-time monitoring of fruit logistics. *Journal of Food Engineering* 87, 405–415.

Ruiz-Garcia, L., Barreiro, P., Robla, J.I., & Lunadei, L. (2010). Testing ZigBee motes for monitoring refrigerated vegetable transportation under real conditions. *Sensors 10*, 4968–4982.

Saaid, M., Saad, B., Hashim, N.H., Mohamed Ali, A.S., & Saleh, M.I. (2009). Determination of biogenic amines in selected Malaysian food. *Food Chemistry. 113*, 1356–1362.

Saevels, S., Lammertyn, J., Berna, A.Z., Veraverbeke, E.A., Di Natale, C., & Nicolai, B.M. (2003). Electronic nose as a non-destructive tool to evaluate the optimal harvest date of apples. *Postharvest Biology and Technology 30*, 3–14.

Salim, S.N.M., Shakaff, A.Y.M., Ahmad, M.N., Adom, A.M., & Husin, Z. (2005). Development of electronic nose for fruits ripeness determination. 1st International Conference on Sensing Technology, Palmerston, New Zealand, pp. 515–518.

Sanaeifar, A., Mohtasebi, S.S., Ghasemi-Varnamkhasti, M., Ahmadi, H., & Lozano, J. (2014). Development and application of a new low cost electronic nose for the ripeness monitoring of banana using computational techniques (PCA, LDA, SIMCA, and SVM). *Czech J. Food Sci. 32*, 538–548.

Siddiqui, M.W. (2015). Postharvest Biology and Technology of Horticultural Crops: Principles and Practices for Quality Maintenance. CRC Press, Boca Raton, Florida, USA. pp. 550.

Siddiqui, M.W. (2016). Eco-friendly technology for postharvest produce quality. Academic Press, Elsevier Science, USA. pp. 324.

Siddiqui, M.W., Ayala-Zavala, J.F., & Hwang, C.A. (2016). Postharvest management approaches for maintaining quality of fresh produce. Springer, New York. pp. 222.

Simon, J.E., Herzroni, A., Bordelon, B., Miles, G.E., & Charles, D.J. (1996). Electronic sensing of aromatic volatiles for quality sorting of blueberries. *J. Food Sci. 61*, 967–970.

Singh, R., Giri, S.K., & Kotwaliwale, N. (2014). Shelf-life enhancement of green bell pepper (*Capsicum annuum* L.) under active modified atmosphere storage. *Food Packaging and Shelf Life. 1*, 101–112.

Smyth, A.B., Talasila, P.C., & Cameron, A.C. (1999). An ethanol biosensor can detect low-oxygen injury in modified atmosphere packages of fresh-cut produce. *Postharvest Biology and Technology. 15*, 127–134.

Somboonkaew, N., & Terry, L.A. (2010). Physiological and biochemical profiles of imported litchi fruit under modified atmosphere packaging. *Postharvest Biology and Technology 56*, 246–253.

Tan, E.S., Slaughter, D.C., & Thompson, J.F. (2005). Freeze damage detection in oranges using gas sensors. *Postharvest Biology and Technology 35*, 177–182.

Todorovic, V., Neag, M., & Lazarevic, M. (2014). On the usage of RFID tags for tracking and monitoring of shipped perishable goods. *Procedia Engineering. 69*, 1345–1349.

Torri, L., Sinelli, N., Limbo, S. (2010). Shelf life evaluation of fresh-cut pineapple by using an electronic nose. *Postharvest Biol. Technol. 56*, 239–245.

Vanderroost, M., Ragaert, P., Devlieghere, F., & De Meulenaer, B. (2014). Intelligent food packaging: The next generation. *Trends in Food Sci Technol. 39*, 47–62.

van Tuijl, B., van Os, E., & van Henten, E. (2008). Wireless sensor networks: state of the art and future perspective. In: DePascale, S., Mugnozza, G.S., Maggio, A., & Schettini, E. (Eds.), Proceedings of the International Symposium on High Technology for Greenhouse System Management, vols. 1 and 2.

Velasco-Garcia, M.N., & Mottram, T. (2003). Biosensor technology addressing agricultural problems. *Biosystems Engineering 84*, 1–12.

Velusamy, V., Arshak, K., Korostynska, O., Oliwa, K., & Adley, C. (2010). An overview of foodborne pathogen detection: in the perspective of biosensors. *Biotechnology Advances 28*, 232–254.

Vergara, A., Llobet, E., Ramirez, J.L., Ivanov, P., Fonseca, L., Zampolli, S., Scorzoni, A., Becker, T., Marco, S., & Wollenstein, J. (2007). An RFID reader with onboard sensing capability for monitoring fruit quality. *Sensors and Actuators B. Chemical. 127*, 143–149.

Vermeir, S. Nicolaï, B.M., & Lammertyn, J. (2005). High-throughput enzymatic taste biosensor. *Communications in Agricultural and Applied Biological Sciences. 70*, 289–292.

Villalobos, M.D.C., Serradilla, M.J., Martin, A., Ruiz-Moyano, S., Pereira, C., & Cordoba, M.D.G. (2014). Use of equilibrium modified atmosphere packaging for preservation of 'San Antonio' and 'Banane' breba crops (*Ficuscarica* L.). *Postharvest Biology and Technology. 98*, 14–22.

Vlcek, J. (2006). RFID-The new technology affecting logistics and the supply chain. ERPS Publication, pp. 1–23.

Wang, C.Y., & Qi, L. (1997). Modified atmosphere packaging alleviates chilling injury in cucumbers. *Postharvest Biology and Technology. 10*, 195–200.

Wang, J., & Zhou, Y. (2007). Electronic-nose technique: Potential for monitoring maturity and shelf life of tomatoes. *N.Z.J. Agric. Res. 50*, 1219–1228.

Watada, A.E., & Qi, L. (1999). Quality of fresh-cut produce. *Postharvest Biology and Technology. 15*, 201–205.

Weichmann, J. (1987). Low oxygen effects, In Postharvest Physiology of vegetables, Weichmann, J., Ed., Marcel Dekker, New York, Chapter 10.

World Health Organization. (2004). Joint FAO/WHO Workshop on Fruit and Vegetables for Health. (2004). "Fruit and vegetables for health: Report of a Joint FAO/WHO Workshop, 1–3 September, 2004," Kobe, Japan.

Yam, L., Takhistov, P., & Miltz, J. (2005). Intelligent packaging: concepts and applications. *Journal of Food Science. 70*, 1–10.

Zagory, D., & Kader, A.A. (1988). Modified atmosphere packaging of fresh produce. *Food Technology 42*(9), 70–77.

Zerbini, P.E., Vanoli, M., Grassi, M., Rizzolo, A., Fibiani, M., Cubeddu, R., Pifferi, A., Spinelli, L., & Torricelli, A. (2006). A model for the softening of nectarines based on sorting fruit at harvest by time-resolved reflectance spectroscopy. *Postharvest Biology and Technology 39*, 223–232.

Zhang, H., Wang, J., & Ye, S. (2008). Prediction of acidity, soluble solids and firmness of pear using electronic nose technique. *J. Food Eng. 86*, 370–378.

Zoecklein, B.W., Devarajan, Y.S., Mallikarjunan, K., & Gardner, D.M. (2011). Monitoring effects of ethanol spray on Cabernet franc and Merlot grapes and wine volatiles using electronic nose systems. *Am. J. Enol. Vitic. 62*, 351–358.

CHAPTER 7

FRESH-CUT PRODUCE: ADVANCES IN PRESERVING QUALITY AND ENSURING SAFETY

OVAIS SHAFIQ QADRI, BASHARAT YOUSUF, and ABHAYA KUMAR SRIVASTAVA

Department of Postharvest Engineering and Technology, Aligarh Muslim University, India, E-mail: osqonline@gmail.com, Tel.: +91-9419041070

CONTENTS

- 7.1 Introduction 266
- 7.2 Physiology (Effect of Cutting on Tissues) 268
  - 7.2.1 Physiological Effects of Cutting on Fruits and Vegetables 269
- 7.3 Microbiology.. 270
- 7.4 Quality and Safety of Fresh-Cut Fruits and Vegetables 275
- 7.5 Trends in Preserving Quality and Ensuring Microbiological Safety of Fresh Cut Produce 278
  - 7.5.1 Modified Atmosphere Packaging 278
  - 7.5.2 Edible Coatings 281
- Keywords 283
- References 283

7.1 INTRODUCTION

Fruit and vegetables comprise essential components of the human diet and are being linked to health and nutritional benefits associated with their consumption. Currently, benefits of fruit and vegetables through diet are now well known and are documented in the literature. Average daily-recommended intake of fruits and vegetables per capita, according to the World Health Organization (WHO, 2008) is more than 400 gram. Fresh-cut products offering convenience may play an important role to help meet the recommended daily intake or other essential dietary requirements (Siddiqui and Rahman, 2015).

There has been a continuous increase in consumer demand for convenient and minimally processed produce including fresh-cut fruits and vegetables. International Fresh-Cut Produce Association defines fresh-cut produce as "any fresh fruit or vegetable or combination thereof, physically altered from its original form, but remaining in a fresh state" (IFPA, 2001). Fresh-cut fruits and vegetables are minimally processed products prepared usually by trimming, peeling, washing, cutting the fruits or vegetables and packing them to offer convenience to consumers while maintaining the fresh like character of such products. Usually fresh-cut produce goes through preparation steps such as washing, peeling, cutting or slicing, packaging (Abadias et al., 2011). Nevertheless, there are no predefined operations for all types of fruits and vegetables, rather the operations may vary with respect to the particular type of product in consideration. In other words, the operations involved in the preparation of fresh-cut product are specific to fruit or vegetable being minimally processed. For instance, cabbage may only be shredded; carrots may be peeled and shredded, onions may be only peeled or peeled, sliced or diced; spinach may be washed and trimmed; pineapple may be cored and sliced; papaya may be halved and formed into slices or cubes and so on (Figure 7.1).

Fresh-cut fruits and vegetables are live tissues in a raw or fresh-like state, prepared without employing frequently used food-processing methods like freezing, canning, dehydrating, fermentation, acidification, or treatments with food additives. However they always require refrigerated storage.

Fresh-Cut Produce: Advances in Preserving Quality and Ensuring Safety   267

FIGURE 7.1 Fresh-cut Papaya prepared, packaged in polypropylene trays and stored under refrigerated conditions (4°C) at Food Processing Laboratory, Department of Postharvest Engineering and Technology, Aligarh Muslim University, India.

Due to their convenience, there is a high-demand for fresh-cut fruit and vegetables in the retail and food service industries. Also, there is a strong tendency among people toward consumption of minimally processed foods without addition of chemical preservatives. In the recent past, this industry has expanded to a multi-billion dollar sector and represents one of the fastest expanding segments of ready to use fresh foods. Foodservice establishments including restaurants, cafeterias and airlines and other outlets are increasingly relying on fresh-cut produce. Still, this sector has a plenty of room to grow.

Freshness and convenience are the primary characteristics of fresh-cut produce. However, food safety, nutritional and sensory quality during extended shelf life can in no way be ignored. Minimal processing leads to alteration in the integrity of the fruit tissue resulting in the increase in respiration and consequently accelerates degradation and shortens shelf life. Fresh-cut vegetables deteriorate faster than intact produce. The degradation due to increased respiration includes biochemical deteriorations such as enzymatic browning, development of off flavor or breakdown of texture. In addition minimal processing operations will result in damage to the fruit and vegetable tissue by cellular disruption and de-compartmentalization.

7.2 PHYSIOLOGY (EFFECT OF CUTTING ON TISSUES)

Fruits and vegetables are biological entities that are living even after their detachment from the parent plant. The physiological systems of fruits and vegetables are maintained postharvest as the metabolic reactions continue (Table 7.1). The processes like respiration and transpiration continue after harvest but due to detachment of produce from the plant, neither respirable substrate loss nor moisture loss is compensated, making the produce entirely dependent on its own reserves of food and moisture. This reduction in food reserves and moisture of produce after harvest initiates deterioration of quality, giving the perishable character to these commodities. The causes of deterioration may be internal (including respiration rate, ethylene production and action, mechanical injuries, rates of compositional changes, sprouting and rooting, physiological disorders) or external (temperature, relative humidity, atmospheric composition and sanitation procedures) or a combination of both.

Fresh cut fruits and vegetables are subjected to different minimal processing procedures depending upon the type of commodity, which may include cutting, trimming, shredding, stoning and peeling, resulting in their physical injury. This physical damage of the fresh cut produce tissues further enhances the deteriorative changes, increasing the perishability of such products manifold.

TABLE 7.1 Metabolic Changes in Postharvest Fruits and Vegetables

Degradative	Synthetic
Destruction of chloroplast	Synthesis of carotenoids and anthocyanins
Breakdown of chlorophyll	Synthesis of flavor volatiles
Starch hydrolysis	Synthesis at starch
Organic acid catabolism	Synthesis of lignin
Oxidation of substrate	Preservation of selective membranes
Inactivation of phenolic compounds	Interconversion of sugars
Hydrolysis of pectin	Protein synthesis
Breakdown of biological membranes	Gene transcription
Cell wall softening	Formation of ethylene biosynthesis pathway

Adapted from Baile and Young (1981).

7.2.1 PHYSIOLOGICAL EFFECTS OF CUTTING ON FRUITS AND VEGETABLES

The quality changes encountered in fresh cut fruits and vegetables are basically a combined effect of different physiological processes, which may be induced or enhanced as result of injury. The extent and severity of these physiological changes depend on different factors including cultivar, post-harvest crop management, physiological maturity, pre- and post-cutting treatments, atmospheric composition. (Toivonen and DeEll, 2002).

The wounding of fruits and vegetables as a result of minimal processing has been associated with increase in ethylene production in the close vicinity of the wound area. Ethylene may have potential effects on plant tissues depending on the type and physiology of tissue. Increase in ethylene production after cutting has been reported for many fruits and vegetables including galia, kiwifruit, tomato, papaya, and strawberry. Few studies have also reported contradictory results depicting no change in ethylene content upon slicing of banana (Watada et al., 1990) and decrease in ethylene production in cantaloupe upon cutting (Luna Guzman et al., 1999).

Respiration of fruits and vegetables has been correlated to their shelf life, higher respiration corresponding to lower shelf life (Kader, 1987). Wounding of fruits and vegetables may enhance the respiration rate which will eventually decrease the shelf life which may be attributed to the assumption that a structural change of mitochondria and increase in number of mitochondria is induced by wounding justifies this statement (Siddiqui, 2015, 2016). However, the increase in respiration rate of fresh cut fruits and vegetables shows a lot of variation. Watada et al. (1990) reported an increase in respiration rate of kiwifruit on cutting but no such increase was observed in banana.

Fruits and vegetables are mostly enclosed within a covering called peel and at microscopic level; also there exist many membranes like cell wall, cell membrane, organelle membrane. The function of these coverings is to maintain the integrity of the produce. These membranes get damaged during minimal processing which may trigger series of processes detrimental to the quality of the product. Browning, developments of off-flavors (Brecht, 1995), free radical production (Thompson et al., 1987), membrane lipid breakdown (Galliard 1979), enhanced ethylene production (Sheng et al., 2000) are some of the reported effects of membrane deterioration due to wounding. These changes are however, specific to certain commodities and may not occur in every fruit and vegetable on injury. Theologis and Laties (1980) reported some fruits and vegetables like carrot, avocado, and banana do not show wound-induced membrane lipid breakdown.

Accumulation of secondary metabolites is another physiological effect that has been associated to wounded tissues. Wounded has been shown to increase the activity of enzyme phenylalanine ammonia lyase which initiated the phenolic accumulation. Increase in unpleasant sulfur compounds has been attributed to membrane deterioration, which exposes several sulfur containing substrates to enzymes like cysteine sulphoxidelyase and results in their oxidation.

Water loss mainly as a result of transpiration is continuous process in fruits and vegetables and minimizing this loss is directly related to quality, otherwise resulting in greater susceptibility to wilting and shriveling. The peeling and cutting of fruits and vegetables results in reduction or elimination of outer periderm or cuticle, which provide resistance against transpirational movement of water vapors, resulting in increased rate of water loss. Agar et al. (1999) demonstrated that slicing of kiwifruit increased rate of water loss and peeling further enhanced the rate.

Increased susceptibility to microbial spoilage is one of the most important concerns in fresh cut fruits and vegetables and exposing the fresh cut produce to normal conditions may provide an open invitation to microorganisms. Watada et al. (1996) reported that broken cells and their adjacent tissues are the areas of largest microbial population in stored fresh cut produce, which may be attributed to the fact that damaged tissues, and broken cells provide easy access to nutrients, ensuring a protective environment for growth of wide range of microorganisms.

7.3 MICROBIOLOGY

Increased awareness of healthy eating habits lead to overall increased produce consumption in terms of fresh fruits and vegetables. Most of these fruits and vegetables provide feasible conditions for the survival, growth and proliferation of various types of microorganisms. Microbial decay is the major source of spoilage of fresh-cut produce.

In general, hazards associated with food may be categorized as physical, chemical or microbiological. In fresh-cut industry, microbiological safety is the main issue of concern (Table 7.2). Microorganisms are natural contaminants associated with all the agricultural produce. These

TABLE 7.2 Microbial Populations Present on Fresh-Cut Fruits and Vegetables

Fresh Produce	Microbial Groups	Microbial population (Log cfu g^{-1})	References
Broccoli	Total Mesophilic Count	4.7	Jacques and Morris (1995)
	Coliform Count	2.1	
	Yeast and Mold	3.25	
Cantaloupe pieces	Total Mesophilic Count	1.05	Lamikanra et al. (2000)
	Yeast and Mold	6.34	Tournas et al., (2006)
Carrot	Total Mesophilic Count	5.5	Jenni et al. (2013)
	C. jejuni	6.5 to 6.9	Karenlampi and Hanninen (2004)
	E. coli O157:H7	6.1	Lacroix and Lafortune (2004)
Shredded carrots	Total Mesophilic Count	2.9	Chervin and Boisseau (1994)
	Lactic Acid bacteria	1.1	
Minimally processed broadleaf endive	Total Mesophilic Count	3.83–4.82	Carlin et al. (1996)
Cut chicory endive	Total Mesophilic Count	4.00	Bennick et al. (1998)
Chicory endive (shredded)	Total Mesophilic Count	5.2	Jacxsens et al. (1999)
	Lactic Acid bacteria	2.63	
	Yeast and Mold	3.0	

TABLE 7.2 continued

Fresh Produce	Microbial Groups	Microbial population (Log cfu g^{-1})	References
Cut Cabbage	Total aerobic bacteria	5.2	Koide e al. (2009)
	Yeast and Mold	3.9	
	Escherichia coli O157:H7	4.10–4.90	Lee et al. (2014)
Fresh-cut celery	Total bacterial Count	5.08	Zhang et al. (2005)
Lettuce	Total Mesophilic Count	6.39–7.69	Gras et al. (1994)
	Coliform Count	4.14–5.29	
Chopped lettuce	Total Mesophilic Count	4.85	Odumeru et al. (1997)
Processed lettuce	Total Mesophilic Count	2.5–6.2	Francis and O'Beirne (1998)
Shredded lettuce	Total plate Count	4.28	Delaquis et al. (1999)
	Lactic Acid bacteria	<1	
	Yeast and Mold	2.07	
Lettuce salad	Total Mesophilic Count	7.23–7.61	Jayasekara (1999)
Chopped bell peppers	Total Mesophilic Count	3.5	Izumi (1999)
Mixed fruit pieces	Yeast and Mold	6.34	Tournas et al. (2006)
Watermelon chunks	Yeast and Mold	6.26	Tournas et al. (2006)
Mixed vegetables	Total Mesophilic Count	8	Manzano et al. (1995)
	Lactic Acid bacteria	5.5–6.3	
	Yeast and Mold	4–4.2	
Mixed salad	Total Mesophilic Count	1.84–2.99	Martinez-Tome et al. (2000)
	Coliform Count	0.7–1.90	
Fresh-cut mushrooms	Total Mesophilic Count	8.3	Sapers and Simmons (1998)

TABLE 7.2 continued

Fresh Produce	Microbial Groups	Microbial population (Log cfu g^{-1})	References
Packaged garden salad (iceberg lettuce, carrot, red cabbage)	Total Mesophilic Count	5.3–8.9	Hagenmaier and Baker (1998)
	Yeast and Mold	0.9–3.85 yeasts	
		<0.3–2.2 molds	
Prepackaged ready-to-serve salad	Total Mesophilic Count	5.5–8.3	Lack et al. (1996)
	Yeast and Mold	<3–6.75	
Potato strips	Total Mesophilic Count	2.00	Gunes et al. (1997)
Potato salad	Total Mesophilic Count	5.41–4.98	Jayasekara (1999)
Japanese radish shreds	Total Mesophilic Count	3.9	Izumi (1999)
Raw vegetables	Total Mesophilic Count	5.7	Kaneko et al. (1999)
	Coliform Count	2.3	
Ready-to-use mixed salad	Total Mesophilic Count	7.18	Vescovo et al. (1995)
	Coliform Count	6.60	
	Lactic Acid bacteria	5.3	
Salad mix	Total Mesophilic Count	5.35	Odumeru et al. (1997)
Trimmed spinach leaves	Total Mesophilic Count	4.00	Izumi (1999)
Fresh-cut pawpaw	Bacterial count	1.8–2.5	Daniyan & Ajibo (2011)
Melon	Bacterial count	3.5–9	Daniyan & Ajibo (2011)
Fresh-cut Pineapple	Bacterial count	1.6–8.4	Daniyan & Ajibo (2011), Mantilla et al. (2013)
	Yeast and Mold	3.86	Mantilla et al. (2013)
Yellow onions (diced & sliced)	Aerobic plate count	4.19	Juneja et al. (2002)

TABLE 7.2 continued

Fresh Produce	Microbial Groups	Microbial population (Log cfu g⁻¹)	References
Fresh-cut apple slices	Total aerobic counts 4.3		Rupasinghe et al. (2006)
Tomato (diced)	Aerobic plate count 4.78		Juneja et al. (2002)
Cut strawberries	Yeast and Mold 4.36		Tournas et al. (2006)
Fresh-cut watermelon	Aerobic plate count 3.1		Sipahi et al. (2013)
	Yeast and Mold 1.4		
	Coliforms 1.5		

microorganisms decrease the economic value of fresh-cut products and pose a threat to public health. Contamination of fresh produce can occur at any stage from farm until its final consumption.

There could be many sources of surface contamination of fruits and vegetables by microorganisms, such as soil, water, animals, birds, and insects during the growing stage (Siddiqui, 2015). Further contamination may occur during different processes like harvesting, washing, cutting, packaging, and shipping could. Fruits and vegetables frequently are exposed to soil, insects, animals, or humans during growing or harvesting. Poor agronomic practices, use of contaminated water irrigation of crops, use of improperly composted manure and lack of training among field workers on good personal hygiene may be considered as some of the reasons for contamination. Furthermore, inefficient sanitary control during various postharvest operations may also elevate the microbial load on fruit and vegetable products (Siddiqui et al., 2016).

Next crucial stage is preparation of fresh-cut produce, which involves a number of preparatory/processing steps. In fresh-cut produce production chain, there are several processing steps and in each of these steps many points for potential microbial contamination may exist. During these preparatory steps, the fruit loses its natural protection (outer skin or peel) and become more susceptible to microbial growth and subsequent degradation of the product (Martin-Belloso et al., 2006). Furthermore, cut surfaces can provide favorable conditions for growth of both foodborne pathogens and spoilage microorganisms (Trias et al., 2008). Besides this, the high water activity and approximately neutral (vegetables) or low acidic (many fruits) tissue pH, facilitate rapid microbial growth (Parish et al., 2003). Cutting will present the risk that rapidly reproducing spoilage microorganisms will establish within open wound sites. Juice leakage from the damaged/cut tissue favors the growth of microorganisms including various bacteria and yeasts. Another critical factor affecting microbiology is cross contamination, which may occur during any of these preparatory steps.

Literature available on the occurrence of microorganisms in minimally processed fruit and vegetable products emphasizes mostly on total bacterial populations and microbial groups, such as *coliforms*, fecal *coliforms*, *pectinolytic* species and yeast and mold counts. Microflora associated with most vegetables is dominated by gram-negative bacteria, while as dominant microflora associated with raw fruits mostly includes yeasts and molds. (Tournas, 2005). Strains of pathogenic bacteria, such as *Listeria monocytogenes*, *Salmonella* species, *Shigella* species, *Aeromonashydrophila*, *Yersinia enterocolitica* and *Staphylococcus aureus*, as well as some *Escherichia coli* usually found in contaminated vegetable and fruit products (Beuchat, 2002; Breidt & Fleming, 1997).

Usually, the natural microflora of raw fruits and vegetables is non-pathogenic and may be present at the time of consumption (Ahvenainen, 1996). This non-pathogenic category of microflora has an important role to play. A large population of non-pathogenic bacteria is in-fact a barrier to reduce the risk of food borne illness from fresh-cut products. They do not necessarily or directly prevent the growth of pathogens but give an indication of temperature abuse and age of the produce by causing spoilage in terms of visible deteriorative changes. If there are no such visible changes on spoilage, the product safety may sometimes be compromised, either intentionally or unintentionally.

7.4 QUALITY AND SAFETY OF FRESH-CUT FRUITS AND VEGETABLES

Quality of fresh-cut fruit and vegetable products is a combination of different parameters such as appearance, texture, flavor, and nutritional value (Kader, 2002). Quality determines the value of product to the consumer and the relative importance of each quality parameter varies with the commodity concerned. At the time of purchase, a person may judge quality of fresh-cut fruit and vegetables primarily on the basis of appearance and freshness. However, on subsequent purchases he will also consider texture, nutritional quality and safety of fresh-cut products. Quality of fresh-cut fruits and vegetables is of key importance because any compromise in the quality may lead the consumers to doubt its safety and there may be potential risks to the public health. Since fresh-cut produce is extremely sensitive, it is a huge challenge to maintain quality efficiently for a longer period. The shelf life of these products is mainly limited by microbial spoilage, desiccation, discoloration or browning, softening or texture breakdown and development of off flavors and off odors.

Quality of fresh-cut fruits and vegetables is influenced by various factors. It is primarily and largely dependent on the selection of raw material for preparation of such products. The quality of raw material (fresh fruits and vegetables) is in turn affected by factors like agricultural practices, soil fertilizers, climate and harvesting conditions. Therefore, these factors directly or indirectly affect the final quality of fresh-cut fruits and vegetables (Ahvenainen, 1996). Additionally, the quality of fresh-cut products may also be affected by the cultivar selected. For instance, a cultivar of a fruit may develop browning rapidly or extensively than the other cultivars of the same fruit. All this directs for the careful selection of raw material for fresh-cut processing.

Postharvest and processing are among prime factors determining the quality of fresh-cut products. Processing operations or preparatory steps cause mechanical injury to the tissue. Removal of peel and loss of tissue integrity makes the product more susceptible to quality loss. Exposure of tissue to air and release of endogenous enzymes leads to detrimental changes in quality.

According to Hodges and Toivonen (2008) quality of Fresh-cut is affected by both internal and external factors. Internal factors represent metabolic characterizations which affect fresh-cut processing and storage such as morphological, physiological, and biochemical defense mechanisms, genotype, stress-induced senescence programs, and processing maturity. External factors are environmental situations, which inhibit or exacerbate the manifestation of the internal factors such as storage temperature, humidity, cutting-knife sharpness, and chemical treatments.

Safety must be of primary concern in any food including the fresh-cut products. Fresh-cut fruits and vegetables are considered to be a source of food borne outbreaks in many parts of the world. Fresh-cut produce has emerged as a potential vehicle for deadly foodborne pathogens (Beuchat, 1998). Disease outbreaks that could spurt as a result of microbial growth during the extended shelf life of fresh-cut produce are posing a challenge to food processors in the global commercialization of these products (Alzamora & Guerrero, 2003).

Fresh-cut produce does not receive any 'lethal' treatment or "kill step" that kills all pathogens prior to consumption. Natural microflora composed of many spices, present on fresh-cut fruits and vegetables, has been reported to compete with or exhibit antagonistic activity towards the pathogens (Liao & Fett, 2001; Ukuku et al., 2004). The shelf-life of fresh-cut produce is reduced and is generally, less than that of intact fruits and vegetables because of the wounding of tissues during their preparation. The limited shelf life is due to both microbial and physiological deterioration. Considering the difficulties linked with the processing and preservation of living tissues, the preparation of sound fresh-cut products presents a challenge for the food industry. Nevertheless, development of safe production and disinfection methods is of critical importance to ensure the safety and quality of fresh-cut fruits and vegetables (Abadias et al., 2011).

Since the contamination with microorganisms can occur at any stage from farm to final consumer, both good agricultural practices (GAP's) and good manufacturing practices (GMP's) are important to ensure safety of fresh-cut products. Good agricultural practices help to ensure sanitation of produce in field where as good manufacturing practices limits product contamination during different processing operations in the processing plant.

Safety of fresh-cut produce is a serious concern for consumers due to:

- there is no kill step in the preparation of fresh-cut produce, which will ensure safety.
- fresh-cut products are consumed raw unlike cooked products, which are thermally processed.

- wounded surfaces pose additional risks of contamination and growth of microorganisms and subsequent degradation.
- fresh-cut products contain no preservatives which will guarantee their safety.

Texture, color, aroma and sweetness are critical factors limiting the shelf life of fresh-cut products. Fresh-cut processing promotes faster deterioration of fruit and vegetable tissues in comparison with their intact counterparts.

7.5 TRENDS IN PRESERVING QUALITY AND ENSURING MICROBIOLOGICAL SAFETY OF FRESH CUT PRODUCE

The fragile nature of fresh cut fruits and vegetables in addition to their increasing popularity has resulted in the surge of research in this field development of new technologies to at least maintain, if not improve, the quality of fresh cut products for longer duration. Few approaches to preserve the quality of fresh cut produce presently under extensive research include modified atmosphere packaging (Bai et al., 2001), surface treatments through application of coatings (Bai and Baldwan, 2002), irradiation (Xanthopoulos et al., 2012) and others.

7.5.1 MODIFIED ATMOSPHERE PACKAGING

Modified atmosphere packaging is a method of packaging aimed at increasing the shelf life foods with minimal loss of quality. In modified atmospheric packaging, the composition of air surrounding the packaged food is changed in order to slow down the deterioration of the product. Modified atmosphere packaging is a potential preservative technique which can be effectively used for various foods which, when packed, influence the mixture of gases inside the package (Sandhya, 2010). Interaction between rate of respiration of the product and transfer of gases through the packaging material help in achieving the modified atmosphere condition within the package (Caleb et al., 2012). Fruits and vegetables are metabolically active products which continue to respire till senescence and can change the composition of surrounding atmosphere if packed hermetically. The rate of respiration for most of the fresh cut fruits and vegetables is more than the intact ones, making modified atmosphere packaging a technique of choice for preserving quality and increasing shelf life of such products (Kader and Watkins, 2000).

Modified atmosphere packaging is passive when the product is sealed with natural air inside and the change of gaseous composition to the desired levels within the package is relied upon the product respiration. Establishment of equilibrium gaseous state within the package takes time in passive modified atmosphere package depending upon respiration rate and film permeability characteristics. In active modified atmosphere packaging, gas flushing or gas replacement or use of gas scavenging agents helps in achieving the desired gas composition within the package. Excessive accumulation of gases (CO_2 or O_2), which can occur in passive modified atmosphere packaging, can be detrimental to the quality of fresh cut produce and their accumulation can be minimized with the help of CO_2 and O_2 scavengers.

Many factors that influence modified atmosphere packaging include film thickness and surface area, product weight, free space within the pack and temperature (Caleb et al., 2012). Different types of packaging materials used for modified atmosphere packaging are flexible packages, rigid containers, engineered oxygen transmission rate (OTR) polymers, microperforated materials, or a combination of these. Selection of a packaging material for a specific product depends on various factors like product type, product quantity, market application, package dimensions, stiffness, graphics, marketing, cost, environmental impact, reusability (Toivonen et al., 2009). The physical properties, chemical properties and gas transmission rate are specific for different packaging materials and for packaging of fresh cut fruits and vegetables, gas transmission rate especially O_2 transmission rate and CO_2 transmission rate are important attributes.

Modified atmosphere packaging has been reported to increase the shelf life of fresh cut products (Table 7.3). The prolonging of shelf life of any fresh cut product can be achieved by reduction in respiration rate, decrease in ethylene biosynthesis and action, preventing microbial growth and contamination, and delaying of senescence. Reduction of

TABLE 7.3 Modified Atmosphere Storage of Different Fresh Cut Products

Fresh cut Product	MAP Conditions	Days of Storage	References
Antichoke	5 kPaO$_2$, 15 kPaCO$_2$, 5°C	4	Ghidelli et al. (2015)
Apple	0 kPaO$_2$, 4°C	21	Soliva-Fortuiny et al. (2001)
	1 kPaO$_2$, 30–35 kPaCO$_2$, 4°C	28	Aguayo et al. (2010)
Butterhead lettuce	80 kPaO$_2$, 10–20 kPaCO$_2$, 1°C	5	Escalona et al. (2006)
Carrot	5% O$_2$, 5% CO$_2$, 2°C	13	Alasalvar et al. (2005)
Jackfruit	3% O$_2$, 5% CO$_2$, 6°C	35	Saxena et al. (2008)
Kiwifruit	90% N$_2$O, 5% O$_2$, 5% CO$_2$, 4°C	12	Rocculi et al. (2005)
Lotus roots	8.7% O$_2$, 6.9% CO$_2$, 4°C	8	Xing et al. (2010)
Mango	21 kPa O$_2$, 0.03 kPaCO$_2$, 4°C	7	Beaulieu and Lea (2003)
Melon	70% O$_2$, 30% N$_2$, 5°C	10–14	Oms-Oliu et al. (2008)
Mushroom	10–20% O$_2$, 2.5% CO$_2$, 4°C	12	Simon et al. (2005)
Pear	2.5% O$_2$, 7% CO$_2$, 4°C	14	Oms-Oliu et al. (2008)
Pepper	5 kPaO$_2$, 5 kPaCO$_2$, 5°C	7–10	Rodoni et al. (2015)
Pineapple	8% O$_2$, 10% CO$_2$, 0°C	14	Marrero and Kader (2005)
Pomegranate arils	6.5% O$_2$, 11.4% CO$_2$, 5°C	10	Palma et al. (2009)
Tomato	3 kPa O$_2$, 4 kPa CO$_2$, 0°C	14	Aguayo et al. (2004)

O$_2$ and increase in CO$_2$ levels under modified atmosphere conditions effectively reduce the respiration rate and control ethylene biosynthesis. As far as microbial contamination of fresh cut products is concerned, the effect of modified atmosphere packaging is diverse and depend upon microbial species, fresh cut produce type and storage condition (Wang et al., 2004). The conditions commonly applied in modified atmosphere packaging for fresh cut produce do not have biocidal effects on microorganisms but it can affect the rate of growth of different species. An increase in shelf life of fresh cut products using modified atmosphere packaging can only be achieved if it has been appropriately designed and can successfully reduce enzymatic browning, respiration rate, moisture loss, and some microbial growth.

7.5.2 EDIBLE COATINGS

Edible coatings are the surface treatments, using edible materials, given to a food product that coats the outer surface of the product and provides a barrier to moisture, oxygen, and solute movement for the food (Dhall, 2013). Being edible in nature these coatings are derived from natural sources and thus are environmental friendly. As per the basic material used in the formulation, the coatings may be classified into three broad categories; polysaccharide, protein or lipid based. In order to achieve maximum desired properties in a single coating a combination of these basic coating might also be used which are known as composite coatings. The important polysaccharides that can be used in formulation of coatings include starches, dextrin, pectin, cellulose and its derivatives, chitosan, alginate, carrageenan, gellan and so on. Similarly proteins like gluten, collagen, zein, casein, and whey protein, and lipids like carnauba wax, beeswax waxes, acylglycerols or fatty acids also possess the desired properties and can be developed into effective coatings. Use of edible coatings on fresh cut produce will only be successful if the coating material possesses certain properties including efficient water vapor barrier capacity, stability under high relative humidity, efficient oxygen and carbon dioxide barrier capacity, good mechanical properties, easy adhesion with product, physico-chemical and microbial stability and reasonable cost. The application of edible coating on fresh cut produce can serve many purposes including reduce water loss, delay ripening of climacteric fruits, delay color changes, improve appearance, reduce aroma loss, reduce exchange of humidity between fruit pieces, act as carriers of antioxidants, texture enhancers, volatile precursors, nutraceuticals, reduce quality losses and extend the shelf life of these products. The use of edible coating isolates the coated product from the environment, giving an effect similar to that of modified atmosphere packaging (Olivas and Barbosa-Canovas, 2005). Many studies have reported encouraging results on use of edible coatings in fresh cut produce and few of them are depicted in Table 7.4.

TABLE 7.4 Edible Coatings Applied to Different Fresh Cut Products

Type of coating	Fresh cut product	References
Polysaccharide based	Jackfruit bulbs	Saxena et al. (2011)
	Apple	Banasaz et al. (2013); Pan et al. (2013); Chauhan et al. (2011)
	Kiwifruit	Benitez et al. (2013)
	Garlic cloves	Geraldine et al. (2008)
	Cantaloupe	Krasaekoopt & Mabumrung (2008); Martinon et al. (2014)
	Melon	Oms-Oliu et al. (2008); Raybaudi-Massilia et al. (2008)
	Watermelon	Sipahi et al. (2013)
	Papaya	Gonzalez-Aguilar et al. (2009); Brasil et al. (2012); Tapia et al. (2008)
	Broccoli	Moreira et al. (2011)
	Mango	Nongtaodum & Jangchud (2009); Chien et al. (2007); Chiumarelli et al. (2011)
	Pear	Ochoa-velasco & Guerrero-Beltran (2014); Xiao et al. (2011); Mohamed et al. (2013)
	Carrot	Vargas et al. (2009), Costa et al. (2012)
	Banana	Bico et al. (2009)
	Pineapple	Azarakhsh et al. (2012); Bierhals et al. (2011);
Protein based	Apple	Ghavidel et al. (2013)
	Papaya	Cortez-Vega et al. (2014)
Composite	Pineapples	Mantilla et al. (2013)
	Papaya	Brasil et al. (2012)
	Apple	Perez-Gagoa et al. (2005); Perez-Gagoa et al. (2006)
	Cantaloupe	Martinon et al. (2014)
Lipid based	Apple	Khan et al. (2014)

KEYWORDS

- Fresh Cut Microbiology
- Fresh-Cut Produce
- Microbiological Safety
- Physiological Response
- Quality Preservation
- Shelf Life

REFERENCES

Abadias, M., Alegre, I., Usall, J., Torres, R., & Vinas, I. (2011). Evaluation of alternative sanitizers to chlorine disinfection for reducing food borne pathogens in fresh-cut apple. *Postharvest Biol. Technol. 59*, 289–297.

Agar, I.T., Massantini, R., Hess-Pierce, B., & Kader, A.A. (1999). "Postharvest CO_2 and ethylene production and quality maintenance of fresh-cut kiwifruit slices." *J. Food Sci. 64*, 433–440.

Aguayo, E., Escalona, V., & Artes, F. (2004). Quality of fresh-cut tomato as affected by type of cut, packaging, temperature and storage time. *European Food Research and Technology, 219*(5), 492–499.

Aguayo, E., Requejo-Jackman, C., Stanley, R., & Woolf, A. (2010). Effects of calcium ascorbate treatments and storage atmosphere on antioxidant activity and quality of fresh-cut apple slices, *Postharvest Biology and Technology, 57*(1), 52–60.

Ahvenainen, R. (1996). New approaches in improving the shelf life of minimally processed fruit and vegetables. *Trends in Food Science and Technology, 7*, 179–187.

Alasalvar, C., Al-Farsi, M., Quantick, P.C., Shahidi, F., & Wiktorowicz, R. (2005). Effect of chill storage and modified atmosphere packaging (MAP) on antioxidant activity, anthocyanins, carotenoids, phenolics and sensory quality of ready-to-eat shredded orange and purple carrots. *Food Chem. 89*, 69.

Alzamora, S.M., & Guerrero, S. (2003). Plant antimicrobials combined with conventional preservatives for fruit products. In: Roller, S. (Ed.), *Natural Antimicrobials for the Minimal Processing of Foods* (pp. 235–249). Boca Raton, FL: CRC Press LLC.

Azarakhsh, N., Osman, A., Ghazali, H.M., Tan, C.P., & Mohd Adzahan, N. (2012). Optimization of alginate and gellan-based edible coating formulations for fresh-cut pineapples. *International Food Research Journal, 19*(1), 279–285.

Bai, J.H., Saftner, R.A., Watada, A.E., & Lee, Y.S. (2001). Modified atmosphere maintains quality of fresh-cut cantaloupe. *Journal of Food Science, 66*(8), 1207–1211.

Banasaz, S., Hojatoleslami, M., Razavi, S.H., Hosseini, E., & Shariaty, M, A. (2013). The Effect of Psyllium seed gum as an edible coating and in comparison to Chitosan on the textural properties and color changes of Red Delicious Apple. *International Journal of Farming and Allied Sciences, 2*(18) 651–657.

Beaulieu, J.C., & Lea, J.M. (2003). Volatile and quality changes in fresh-cut mangoes prepared from firm-ripe and soft-ripe fruit, stored in clamshell containers and passive MAP. *Postharvest Biology and Technology, 30*(1), 15–28.

Bennick, M.H.J., Vorstman, W., Smid, E.J., & Gorris, L.G.M. (1998). The influence of oxygen and carbon dioxide on the growth of prevalent *Enterobacteriaceae* and *Pseudomonas* species isolated from fresh and controlled-atmosphere-stored vegetables. *Food Microbiology, 15*(5), 459–469.

Beuchat, L.R. (1998). Surface Decontamination of Fruits and Vegetables Eaten Raw: A Review. Food Safety, WHO.

Beuchat, L.R. (2002). Ecological factors influencing survival and growth of human pathogens on raw fruits and vegetables. *Microbial Infections, 4*, 413–423.

Biale, J.B., & Young, R.E. (1981). Respiration and ripening in fruits—retrospect and prospect. In: Friend, J., & Rhodes, M.J.C. (Eds.), *Recent Advances in the Biochemistry of Fruits and Vegetables.* Academic Press, New York, NY.

Bico, S.L.S., Raposo, M.F.J., Morais, R.M.S.C., & Morais, A.M.M.B. (2009). Combined effects of chemical dip and/or carrageenan coating and/or controlled atmosphere on quality of fresh-cut banana. *Food Control, 20*, 508–514.

Bierhals, V.S., Chiumarelli, M., & Hubinger, M.D. (2011). Effect of cassava starch coating on quality and shelf life of fresh-cut pineapple (Ananas Comosus L. Merrilcv "P'erola"). *J. Food Sci. 76*, 62–72.

Brasil, I.M., Gomes, C., Puerta-Gomez, A., Castell-Perez, M.E., & Moreira. R.G. (2012). Polysaccharide-based multilayered antimicrobial edible coating enhances quality of fresh-cut papaya. *Food Science and Technology, 47*, 39–45.

Brecht, J.K. (1995). "Physiology of lightly processed fruits and vegetables." *HortScience 30*, 18–22.

Breidt, F., & Fleming, H.P. (1997). Using lactic acid bacteria to improve the safety of minimally processed fruits and vegetables. *Food Technology, 51*, 44–51.

Caleb, O.J., Opara, U.L., & Witthuhn, C.R. (2012). Modified atmosphere packaging of pomegranate fruit and arils: a review. *Food Bioprocess Technol. 5*(1), 15–30.

Chauhan, O.P., Raju, P.S., Singh, A., & Bawa, A.S. (2011). Shellac and aloe-gel-based surface coatings for maintaining keeping quality of apple slices. *Food Chemistry, 126*, 961–966.

Chien, P.J., Sheu, F., & Yang. F.H. (2007). Effects of edible chitosan coating on quality and shelf life of sliced mango fruit. *Journal of Food Engineering, 78*, 225–229.

Chiumarelli, M., Ferrari, C.C., Sarantópoulos, C.I.G.L., & Hubinger, M.D. (2011). Fresh cut 'Tommy Atkins' mango pre-treated with citric acid and coated with cassava (Manihotesculenta Crantz) starch or sodium alginate. *Innovative Food Science and Emerging Technologies, 12*, 381–387.

Cortez-Vega, W.R., Pizato, S., Andreghetto de Souza, J.T., & Prentice, C. (2014). Using edible coatings from Whitemouth croaker (Micropogoniasfurnieri) protein isolate and organo-clay nanocomposite for improve the conservation properties of fresh-cut 'Formosa' papaya. *Innovative Food Science and Emerging Technologies, 22*, 197–202.

Costa, C., Conte, A., Buonocore, G.G., Lavorgna, M., & Del Nobile, M.A. (2012). Calcium alginate coating loaded with silver-montmorillonite nanoparticles to prolong the shelf-life of fresh-cut carrots. *Food Research International, 48*, 164–169.

Carlin, F., Nguyen-The, C., Da Silva, A.A., & Cochet, C. (1996). Effects of carbon dioxide on the fate of *Listeria monocytogenes*, aerobic bacteria and on the development of spoilage in minimally processed fresh endive. *International Journal of Food Microbiology, 32*, 159–172.

Chervin, C., & Boisseau, P. (1994). Quality maintenance of ready-to-eat shredded carrots by gamma irradiation. *Journal of Food Science, 59*(2), 359–365.

Daniyan, S.Y., & Ajibo, C.Q. (2011) Microbiological examination of sliced fruits sold in Minna metropolis. *International Research Journal of Pharmacy, 2*, 124–129.

Delaquis, P.J., Stewart, S., Toivonen, P.M.A., & Moyls, A.L. (1999). Effect of warm, chlorinated water on the microbial flora of shredded iceberg lettuce. *Food Research International, 32*, 7–14.

Dhall, R.K. (2013). Advances in Edible Coatings for Fresh Fruits and Vegetables: A Review. *Critical Reviews in Food Science and Nutrition, 53*(5), 435–450.

Escalona, V.H., Verlinden, B.E., Geysen, S., & Nicolai, B.M. (2006). Changes in respiration of fresh-cut butter head lettuce under controlled atmospheres using low and super atmospheric oxygen conditions with different carbon dioxide levels. *Postharvest Biology and Technology, 39*(1), 48–55.

Francis, G.A., Thomas, C., & O'Beirne, D. (1999). The microbiological safety of minimally processed vegetables. *International Journal of Food Science and Technology, 34*(1), 1–22.

Gras, M.H., Druet-Michaud, C., & Cerf, O. (1994). La florebactérienne des feuilles de saladefraiche. *Sciences des Aliments, 14*(2), 173–188.

Gunes, G., Splittstoesser, D.F., & Lee, C.Y. (1997). Microbial quality of fresh potatoes: effect of minimal processing. *Journal of Food Protection, 60*(7), 863–866.

Galliard, T. (1979). "The enzymatic degradation of membrane lipids in higher plants." In: Advances in the Biochemistry and Physiology of Plant Lipids. In: Appelqvist, L. A., & Liljenberg, C. (eds.). Elsevier, North Holland Biochemical Press, Amsterdam, pp. 121–132.

Geraldine, R.M., Soares, N.F.F., Botrel, D.A., & Goncalves, L.A. (2008). Characterization and effect of edible coatings on minimally processed garlic quality. *Carbohydrate Polymers, 72*, 403–409.

Ghavidel, R.A., Davoodi, M.G., Asl, A.F.A., Tanoori, T., & Sheykholeslami, Z. (2013). Effect of selected edible coatings to extend shelf-life of fresh-cut apples. *International Journal of Agriculture and Crop Sciences, 6*(16), 1171–1178.

Ghidelli, C., Mateos, M., Rojas-Argudo, C., & Pérez-Gago, M.B. (2015). Novel approaches to control browning of fresh-cut artichoke: Effect of a soy protein-based coating and modified atmosphere packaging, *Postharvest Biology and Technology*, vol. 99.

Gonzalez-Aguilar, G., Valenzuela-Soto, E., Lizardi-Mendoza, J., Goycoolea, F., Martínez-Téllez, M., Villegas-Ochoa, M., et al. (2009). Effect of chitosan coating in preventing deterioration and preserving the quality of fresh-cut papaya 'Maradol.' *Journal of the Science of Food and Agriculture, 89*, 15–23.

Hodges, D.M., & Toivonen, P.M.A. (2008). Quality of fresh-cut fruits and vegetables as affected by exposure to abiotic stress. *Postharvest Biology and Technology, 48*, 155–162.

International Fresh-cut Produce Association (IFPA) (2001). "Fresh-cut Produce: Get the Facts!" www.fresh-cuts.org.

Hagenmeier, R.D., & Baker, R.A. (1998). A survey of the microbial population and ethanol content of bagged salad. *Journal of Food Protection, 61*(3), 357–359.

Izumi, H. (1999). Electrolyzed water as a disinfectant for fresh cut vegetables. *Journal of Food Science, 64*(3), 536–537.

Jacques, M.A., & Morris, C.E. (1995). Bacterial population dynamics and decay on leaves of different ages of ready-to-use broad leaved endive. *International Journal of Food Science and Technology, 30*, 221–236.

Jacxsens, L., Devlieghere, F., Falcato, P., & Debevere, J. (1999). Behavior of *Listeria monocytogenes* and *Aeromonas spp.* on fresh-cut produce packaged under equilibrium modified atmosphere. *Journal of Food Protection, 62*(10), 1128–1135.

Jayasekara, N.Y. (1999). Ecological, physiological and biotechnological properties of pseudomonads isolated from mineral waters and salads." PhD Thesis, The University of New South Wales, Sydney, Australia, pp. 4–58.

Jenni, M., Marja, L., Risto, K., Hanna-Riitta, K, & Maarit, M. (2013). Microbiological Quality of Fresh-Cut Carrots and Process Waters. *Journal of Food Protection, 7*, 1240–1244.

Juneja, V.K., Novak, J.S., & Sapers, G.M. (2002). Microbial Safety of Minimally Processed Foods. CRC Press.

Kaneko, K.I., Hayashidani, H., Ohtomo, Y., Kosuge, J., Kato, M., Takahashi, K., Shiraki, Y., & Ogawa, M. (1999). Bacterial contamination of ready-to-eat foods and fresh products in retail shops and food factories. *Journal of Food Protection, 62*(6), 644–649.

Karenlampi, R., & Hanninen, M.L. (2004). Survival of *Campylobacter jejuni* on various fresh produce. *Int. J. Food Microbiol 97*, 187–195.

Koide, S., Takeda, J., Shi, J., Shono, H., & Atungulu, G.G. (2009). Disinfection efficacy of slightly acidic electrolyzed water on fresh cut cabbage. *Food Control, 20*, 294–297.

Kader, A.A. (1987). "Respiration and gas exchange of vegetables." In: Postharvest Physiology of Vegetables. In: J. Weichmann, (ed.), Marcel Dekker, Inc., New York, pp. 25–43.

Kader, A.A. (2002). Quality parameters of fresh-cut fruit and vegetable products. In: Lamikanra, O. (Ed.), Fresh-Cut Fruits and Vegetables. Science, Technology and Market. Boca Raton, FL: CRC Press.

Khan, M.K.I., Cakmak, H., Tavman, S., Schutyser, M., & Schroen, K. (2013). Anti-browning and barrier properties of edible coatings prepared with electrospraying. *Innovative Food Science and Emerging Technologies, 25*, 9–13.

Krasaekoopt, W., & Mabumrung, J. (2008). Microbiological evaluation of edible coated fresh-cut cantaloupe. *Kasetsart Journal: Natural Science, 42*, 552–557.

Liao, C.H., & Fett, W.F. (2001). Analysis of native microflora and selection of strains antagonistic to human pathogens on fresh produce. *J. Food Prot. 64*, 1110–1115.

Luna-Guzmán, I., Cantwell, M., & Barrett, D.M. (1999). "Fresh-cut cantaloupe: effects of CaCl2 dips and heat treatments on firmness and metabolic activity." *Postharvest Biol. Technol. 17*, 201–213.

Lack, W.K., Becker, B., & Holzapfel, W.H. (1996). Hygienic quality of pre-packed ready-to-serve salads in 1995. *Archiv. Fuer-Lebensmittel Hygiene. 47*(6), 129–152.
Lacroix, M., & Lafortune, R. (2004). Combined effects of gamma-irradiation and modified atmosphere packaging on bacteria resistance in grated carrots (Daucuscarota). Rad Phys Chem 71,77–80.
Lamikanra, O., Chen, J.C., Banks, D., & Hunter, P.A. (2000). Biochemical and microbial changes during the storage of minimally processed cantaloupe. *Journal of Agriculture and Food Chemistry, 48*, 5955–5961.
Lee, H.H., Hong, S.I., & Kim, D. (2014). Microbial reduction efficacy of various disinfection treatments on fresh-cut cabbage. *Food Science and Nutrition, 2*(5), 585–590.
Mantilla, N., Castell-Perez, M.E., Gomes, C. & Moreira, R.G. (2013). Multilayered antimicrobial edible coating and its effect on quality and shelf-life of fresh-cut pineapple (Ananascomosus). *Food Science and Technology, 51*, 37–43.
Manzano, M., Citterio, B., Maifreni, M., Paganessi, M., & Comi, G. (1995). Microbial and sensory quality of vegetables for soup packaged in different atmospheres. *Journal of the Science of Food and Agriculture, 67*(4), 521–529.
Martinez-Tome, M., Vera, A.M., & Murcia, M.A. (2000). Improving the control of food production in catering establishments with particular reference to the safety of salads. *Food Control, 11*, 437–445.
Marrero, A., & Kader, A.A. (2006). Optimal temperature and modified atmosphere for keeping quality of fresh-cut pineapples. *Postharvest Biology and Technology, 39*(2), 163–168.
Martin-Belloso, O., Soliva-Fortuny, R., & Oms-Oliu, G. (2006). Fresh-cut fruits. In: Hui, Y.H. (Ed.), *Handbook of Fruits and Fruit Processing*. Blackwell Publishing, Oxford, pp. 129–144.
Martinon, M.E., Moreira, R.G., Castell-Perez, M.E., & Gomes, C. (2014). Development of a multilayered antimicrobial edible coating for shelf life extension of fresh-cut cantaloupe (*Cucumismelo* L.) stored at 4°C. *Food Science and Technology, 56*, 341–350
Mohamed, A.Y.I., Aboul-Anean, H.E., & Hassan, A.M. (2013). Utilization of edible coating in extending the shelf life of minimally processed prickly Pear. *Journal of Applied Sciences Research, 9*(2), 1202–1208.
Moreira, M.R., Roura, S.I., & Ponce. A. (2011). Effectiveness of chitosan edible coatings to improve microbiological and sensory quality of fresh cut broccoli. *Food Science and Technology, 44*, 2335–2341.
Nongtaodum, S., & Jangchud, A. (2009). Effects of edible chitosan coating on quality of fresh-cut mangoes (fa-lun) during storage. *Kasetsart Journal: Natural Science, 43*, 282–289.
Ochoa-Velasco, C.E., & Guerrero-Beltran, J.A. (2014). Postharvest quality of peeled prickly pear fruit treated with acetic acid and chitosan. *Postharvest Biology and Technology, 92*, 139–145.
Olivas, G.I., & Barbosa-Canovas, G.V. (2005). Edible coatings of fresh-cut fruits. *Critical Reviews in Food Science and Nutrition 45*, 657–670.
Oms-Oliu, G., Raybaudi-Massilia Martínez, R.M., Soliva-Fortuny, R., & Martín-Belloso, O. (2008). Effect of super atmospheric and low oxygen modified atmospheres on shelf-life extension of fresh-cut melon. *Food Control. 19*(2), 191–199.

Odumeru, J.A., Mitchell, S.J., Alves, D.M., Lynch, J.A., Yee, A.J., Wang, S.L., Styliadis, S., & Farber, J.M. (1997). Assessment of the microbiological quality of ready-to-use vegetables for health-care food services. *Journal of Food Protection, 60*(8), 954–960.

Oms-Oliu, G., Soliva-Fortuny, R., & Martín-Belloso, O. (2008). Physiological and microbiological changes in fresh-cut pears stored in high oxygen active packages compared with low oxygen active and passive modified atmosphere packaging. *Postharvest Biol. Tec. 48*(2), 295–301.

Palma, A., Schirra, M., D' Aquino, S., La Malfa, S., & Continella, G. (2009). Chemical properties changes in pomegranate seeds packaged in polypropylene trays. In: A.I. Özgüven (ed.) Proceedings of the 1st IS on pomegranate, vol. 818. Acta Horticulturae, ISHS., pp. 1–4

Pan, S.Y., Chen, C.H., & Lai, L.S. (2013). Effect of Tapioca Starch/Decolorized Hsian-Tsao Leaf Gum-Based active Coatings on the Qualities of Fresh-Cut Apples. *Food Bioprocess Technology, 6*, 2059–2069.

Parish, M.E., Beuchat, L.R., Suslow, T.V., Harris, L.J., Garrett, E.H., Farber, J.N., & Busta, F.F. (2003). Methods to reduce/eliminate pathogens from fresh cut produce. Comprehensive Review Food Sci. *Food Safety, 2*, 16–173.

Perez-Gago, M.B., Serra, M., & Rio, M.A. (2006). Color change of fresh-cut apples coated with whey protein concentrate-based edible coatings. *Postharvest Biology and Technology, 39*, 84–92.

Perez-Gagoa, M.B., Serrab, M., Alonsoa, M., Mateosb, M., & del Rio, M.A. (2005). Effect of whey protein- and hydroxypropyl methylcellulose-based edible composite coatings on color change of fresh-cut apples. Postharvest Biology and Technology, 36,77–85.

Raybaudi-Massilia, R.M., Mosqueda-Melgar, J., & Martin-Belloso, O. (2008). Edible alginate-based coating as carrier of antimicrobials to improve shelf-life and safety of fresh-cut melon. *International Journal of Food Microbiology, 121*, 313–327.

Rocculi, P., Romani, S., & Rosa, M.D. (2005). Effect of MAP with argon and nitrous oxide on quality maintenance of minimally processed kiwifruit. *Postharvest Biology and Technology, 35*(3), 319–328.

Rodoni, L., Vicente, A., Azevedo, S., Concellón, A., & Cunha, L.M. (2015). Quality retention of fresh-cut pepper as affected by atmosphere gas composition and ripening stage, LWT. *Food Science and Technology, 60*(1), 109–114.

Rupasinghe, H.P.V., Boulter-Bitzer, J., Ahn, T., & Odumeru, J.A. (2006). Vanillin inhibits pathogenic and spoilage microorganisms in vitro and aerobic microbial growth in fresh-cut apples. *Food Research International, 39*, 575–580.

Sapers, G.M., & Simmons, G.F. (1998). Hydrogen peroxide disinfection of minimally processed fruits and vegetables. *Food Technology, 52*(2), 48–52.

Saxena, A., Bawa, A.S., & Raju, P.S. (2008). Use of modified atmosphere packaging to extend shelf-life of minimally processed jack fruit (*Artocarpusheterophyllus* L.) bulbs. *Journal of Food Engineering, 87*(4), 455–466.

Saxena, A., Saxena, T.M., Raju, P.S., & Bawa, A.S. (2011). Effect of controlled atmosphere storage and chitosan coating on quality of fresh-cut jackfruit bulbs. *Food Bioprocess Technology*, 1–8.

Sandhya. (2010). Modified atmosphere packaging of fresh produce: current status and future needs. *Food Science and Technology, 43*, 381–392.

Siddiqui, M.W. (2015). Postharvest Biology and Technology of Horticultural Crops: Principles and Practices for Quality Maintenance. CRC Press, Boca Raton, Florida, USA. pp. 550.

Siddiqui, M.W. (2016). Eco-friendly technology for postharvest produce quality. Academic Press, Elsevier Science, USA. pp. 324

Siddiqui, M.W., Ayala-Zavala, J.F., & Hwang, C.A. (2016). Postharvest management approaches for maintaining quality of fresh produce. Springer, New York. pp. 222.

Siddiqui, M.W., & Rahman, M.S. (2015). Minimally Processed Foods Technologies for Safety, Quality, and Convenience. Springer, USA. pp. 1–306.

Simon, A., Gonzalez-Fandos, E., & Tobar, V. (2005). The sensory and microbiological quality of fresh sliced mushroom (*Agaricus bisporus* L.) packaged in modified atmospheres. *International Journal of Food Science and Technology, 40*(9), 943.

Sheng, J., Luo, Y., & Wainwright, H. (2000). Studies on lipoxygenase and the formation of ethylene in tomato. *J. Hort. Sci. Biotechnol. 75*, 69–71.

Sipahi, R.E., Castell-Perez, M.E., Moreira, R.G., Gomes, C., & Castillo, A. (2013). Improved multilayered antimicrobial alginate-based edible coating extends the shelf life of fresh-cut watermelon (Citrulluslanatus). *Food Science and Technology, 51*, 9–15.

Soliva-Fortuny, R.C., Grigelmo, M.N., Odriozola-Serrano, I., Gorinstein, S., & Martin-Belloso, O., (2001). Browning evaluation of ready-to-eat apples as affected by modified atmosphere packaging. *J. Agric. Food Chem.* 49, 3685–3690.

Theologis, A., & Laties, G.G. (1980). "Membrane lipid breakdown in relation to the wound induced and cyanide resistant respiration in tissue slices." *Plant Physiol. 66*, 890–896.

Thompson, J.E., Legge, R.L., & Barber, R.F. (1987). "The Role of Free Radicals in Senescence and Wounding." *New Phytol. 105*, 317–344.

Toivonen, P.M.A., Brandenburg, J.S., & Luo, Y. (2009). Modified atmosphere packaging for fresh-cut produce. In: Yahia, E. (Ed.), Modified and Controlled Atmospheres for Storage, Transportation and Packaging of Horticultural Commodities. Taylor and Francis Group, Boca Raton, FL, pp. 463–489.

Toivonen, P.A., & DeEll, J. (2002). Physiology of Fresh-cut Fruits and Vegetables. In: Fresh-Cut Fruits and Vegetables. CRC Press.

Tournas, V.H. (2005). Moulds and yeasts in fresh and minimally processed vegetables, and sprouts. *International Journal of Food Microbiology, 99*(1), 71–77.

Tournas, V.H., Heeres, J., & Burgess, L. (2006). Moulds and yeasts in fruit salads and fruit juices. *Food Microbiology, 23*, 684–688.

Trias, Rosalia, Bañeras, Lluís, Badosa, Esther, & Montesinos, Emilio (2008). Bioprotection of Golden Delicious apples and Iceberg lettuce against foodborne bacterial pathogens by lactic acid bacteria. *International Journal of Food Microbiology, 123*, 50–60.

Ukuku, D.O., Fett, W.F., & Sapers, G.M., (2004). Inhibition of Listeria monocytogenes by native microflora of whole cantaloupe. *J. Food Saf. 24*, 129–146.

Vescovo, M., Orsi, C., Scolari, G., & Torriano, S. (1995). Inhibitory effect of selected lactic acid bacteria on microflora associated with ready-to-use vegetables. *Letters in Applied Microbiology, 21*, 121–125.

Vargas, M., Chiralt, A., Albors, A., & Gonzalez-Martínez, C. (2009). Effect of chitosan-based edible coatings applied by vacuum impregnation on quality preservation of fresh-cut carrot. *Postharvest Biology and Technology, 51*, 263–271.

Wang, H., H. Feng, & Luo, Y. (2004). Microbial reduction and storage quality of fresh-cut cilantro washed with acidic electrolyzed water and aqueous ozone. *Food Res. 37*, 949–956.

Watada, A.E., Abe, K., & Yamauchi, N. (1990). Physiological activities of partially processed fruits and vegetables. *Food Technol. 116&118*, 120–122.

Watada, A.E., Ko, N.P., & Minott, D.A. (1996). "Factors affecting quality of fresh-cut horticultural products." *Postharvest Biol. Technol. 9*, 115–125.

World Health Organization (2008). WHO European Action Plan for Food and Nutrition policy. http://www.euro.who.int/data/assets/pdf_file/0017/74402/E91153.pdf

Xanthopoulos, G., Koronaki, E.D., & Boudouvis, A.G. (2012). Mass transport analysis in perforation-mediated modified atmosphere packaging of strawberries. *Journal Food Engineering, 111*, 326–335.

Xiao, Z., Luo, Y., Luo, Y., & Wang, Q. (2011). Combined effects of sodium chlorite dip treatment and chitosan coatings on the quality of fresh-cut d'Anjou pears. *Postharvest Biology and Technology, 62*, 319–326.

Xing, Y., Li, X.H., Xu, Q.L., Jiang, Y.H., Yun, J., & Li, W.L. (2010). Effects of chitosan-based coating and modified atmosphere packaging (MAP) on browning and shelf life of fresh-cut lotus root (Nelumbonucifera Gaerth). *Innovative Food Science and Emerging Technologies 11*, 684–689.

Zhang, L., Lu, Z., Yu, Z., & Gao, X. (2005). Preservation of fresh-cut celery by treatment of ozonized water. *Food Control 16*(3), 279–283.

CHAPTER 8

POSTHARVEST PATHOLOGY, DETERIORATION AND SPOILAGE OF HORTICULTURAL PRODUCE

S. M. YAHAYA

Department of Biology, Kano University of Science and Technology, Wudil, P.M.B. 3244, Nigeria, E-mail: sanimyahya@yahoo.com

CONTENTS

8.1 Introduction 292
    8.1.1 Importance of Postharvest Diseases 293
8.2 The Postharvest Pathogens 294
    8.2.2 How Pathogens Attack Plants? 294
    8.2.3 Postharvest Spoilage causing Pathogens 295
        8.2.3.1 Yeast 298
        8.2.3.2 Mold 299
        8.2.3.3 Bacteria 301
8.3 Factors Affecting Food Spoilage and Shelf Life 302
    8.3.1 Vegetables 303
    8.3.2 Fruits and Juices 303
8.4 Control of Postharvest Spoilage Pathogens 304
    8.4.1 Management of Mycotoxins Contamination 305
        8.4.1.1 Preharvest Control 305
        8.4.1.2 Postharvest Control 306
8.5 Conclusion 307

Keywords 308
References 308

8.1 INTRODUCTION

Pathogenic micro-organisms account for substantial losses of grains, fruits, and vegetables at both pre- and postharvest stages of crop production. Narayanasamy (2006) emphasized that "the responsibilities of the plant pathologist do not end with the harvest of satisfactory yields of plant products and that harvesting marks the termination of one phase of plant protection and the beginning of another." This clearly indicates that the second phase of plant protection—of seeds, fruits, vegetables, and other economic plant parts from the time of harvest until they reach the consumer—is equally important. Postharvest pathology, earlier termed "market pathology," deals with the science of, and practices for the protection of harvested produce during harvesting, packaging, transporting, processing, storing and distribution (Doyle, 2007).

Previous researches have reported post-harvest losses occurring between harvest and subsequent utilization of the agricultural produce by the consumers as enormous, yet accurate record of losses is not available (Yahaya and Alao, 2008; Siddiqui, 2015, 2016). The agricultural produce are generally divided into: (i) durable crops which include grain legume, cereal grains, and oilseeds; (ii) perishable crops which consist of succulent storage organs which include fleshy fruits, rhizomes, vegetables and tubers. The harvested produce are mainly dormant plant organs which has physiological functions entirely different from the tissues of the mother plant. Generally the susceptibility of the stored agricultural produce to deterioration caused by microorganism increases at ripening/senescence stage (Siddiqui et al., 2016). Therefore, storage could be regarded as an abnormal state for organs of a living plant, which requires overcrowding of large volume of agricultural produce in intimate contact in a limited space. These unfavorable conditions will predispose the produce to various types of diseases, which may occur due to pathogenic or physiological causes (Doyle, 2007).

8.1.1 IMPORTANCE OF POSTHARVEST DISEASES

Generally there is a wide spread spoilage of agricultural produce perishable and durables due to the activities of post-harvest diseases caused by pathogens. Higher losses of agricultural produce are recorded in developing countries due to in adequate storage facilities and handling methods causing high level of injury or wound at harvest and transit. The durable agricultural produce are generally stored in a dry state with level of moisture not exceeding 12%, however, the perishable produce have high level of moisture which may reach 50% or more at storage time. Therefore, the harvested produce might have already been infected during time of transit and storage. It is therefore, estimated that, in the tropics, about 25% of harvested perishables crops are lost at the time interval between harvest and consumption. In general losses of durable produce such as oilseeds, cereals and pulses are about 10% on a worldwide basis (Waller, 2002). These losses include both quantitative and qualitative (Table 8.1).

Under favorable condition, post-harvest losses due to pathogenic attack may be higher than economic gains achieved by improvements of primary production. Many researches on post-harvest diseases are essentially aimed at preventing economic loss from spoilage of harvested effects of mycotoxins produce by pathogenic fungi contaminating both perishables and durables. Mycotoxins are generally known to be carcinogenic, responsible for many serious ailments in humans and animals (Narayanasamy, 2002) (Table 8.2).

TABLE 8.1 Assessment of Losses Caused by Postharvest Diseases in Nigeria

Crop	Loss(%)
Tomato	20–60
Pepper	10–35
Cabbage	37
Carrot	20–25
Onion	15–20
Citrus	20–25
Okro	20

*Source Alao (2000).

TABLE 8.2 Some Estimates of Production and Losses of Selected Commodities in Developing Country (Philippines, 2008)

Crop	Production (m. tons)	Production Value ($)	Percentage Loss	Loss Value ($)
Fruits	2,763,443	403,909,220	28.1	113,498,490
Vegetables	1, 640, 541	248, 564, 310	42. 2	104,894,130
Total	4,403,984	652,473,530	—	218,392,620

*Source: FAO Corporate document Repository (2009).

Studies by Alao (2000) reported that the long chain and complex marketing system of agricultural produce between the producers and consumers make it very difficult to determine accurate level of losses of many crops in Nigeria. He therefore, concluded that Nigeria justified it inclusion in the list of poor countries because food security rating is very low and not guaranteed in Nigeria. Report by FAO (2009) showed that agricultural produce in Nigerian were not been able to meet world standard due to poor post-harvest handling. Therefore, this necessitated the need to gather eases, conditions which favor diseases development, and other methods of developing more effective systems of disease's control, so as to reduce cost and increases availability.

8.2 THE POSTHARVEST PATHOGENS

8.2.2 HOW PATHOGENS ATTACK PLANTS?

Pathogens attack plants because during their evolutionary development, they acquired the ability to live off the substances manufactured by the host plants, and some of the pathogens depend on these substances for survival (Agrios, 1999). Many substances are contained in the protoplast of the cells; however, if pathogens are to gain access to them, they must first penetrate the outer barriers formed by the cuticle and/or cell walls. Even after the outer cell wall has been penetrated, further invasion of the plant by the pathogen necessitates the penetration of more cell walls. Furthermore, the plant cell contents are not always found in forms immediately utilizable by pathogens and must be broken down to units that the pathogen can absorb and assimilate. Moreover, the plant, reacting to the presence and activities of the pathogen; if the pathogen is to survive and to continue living off the plant, it must be able to overcome such obstacles.

Therefore, for a pathogen to infect a plant it must be able to make its into and through the plant, obtain nutrients from the plant, and neutralize the defense reactions through secretions of chemical substances that affect certain or metabolic mechanisms of their hosts. Penetration and inversion, however, seem to be aided by, or in some cases be entirely the results of, the mechanical force exerted by certain pathogens on the cell walls of the plant.

8.2.3 POSTHARVEST SPOILAGE CAUSING PATHOGENS

There are thousands of genera and species of pathogenic microorganism. Several hundred are associated, in one-way or another, with food products (Agrios, 1999). The pathogens that are principally involved in food deterioration are yeast, mold and bacteria. Mold and bacteria causing postharvest diseases usually can attack healthy, living tissue, which they disintegrate and cause to rot. Often, however, other mold and bacteria follow them and live saprophytically on the tissues already killed and macerated by the former. Many of the postharvest diseases of fruits and vegetables, grains and legumes are the results of infections by pathogens in the field. Symptoms from "field infections" may be too inconspicuous to be noticed at harvest. In fleshy fruits and vegetables, field infections continue to develop after harvest, whereas in grains and legumes, they cease to develop soon after harvest. In fleshy fruits and vegetable's new infections may be caused in storage by the same or other pathogens, whereas in grains and legume's storage infections are usually caused by pathogens other than those causing field infections (Doyle, 2007) (Table 8.3).

Postharvest diseases caused by yeast, mold and bacteria are favored greatly by high moisture and high temperatures. However, except where these microorganisms are especially cultivated by selective inoculation or by controlled conditions to favor their growth over that of less desirable

TABLE 8.3 Some Important Food Crops and their Principal Diseases

Crop	Fungal	Viral	Nematode	Bacterial
Tomato *Lycopersicon esculentum*	*Botrytis cinerea* *Aspergillus* spp *Penicillium* spp			
Pepper *capsicum annum*	*Botrytis cinerea* *Aspergillus* spp *Alternaria alternate*			
Lettuce *Lactuca sativa*	*Botrytis cinerea* *Aspergillus* spp			
Pea *pisum sativa*		Mosaic virus, *Ascochyta* spp		
Carrot *Daucus carrota*	*Alternaria dauci*			
Beans *Phaseolus* spp				Xanthomonas campestris, Psedomonas syringe
Millet				
Common millet (Panicum miliaceum)				Downy mildew: Sclerospora graminicola
Finger millet (Eleusine coracana)	*Blast: Pyricularia setariae* *Leaf blight: Cochliobolus nodulosus*			
Maize (*Zea mays*)	Northern corn leaf blight: Helminthosporium turcicum (*Setosphaeria turcica*)	Chlorotic dwarf: maize chlorotic dwarf machlovirus	Stewart's wilt: *Erwinia stewartii*	Downy mildew: Sclerospora spp. and others
	Southern corn leaf blight: *H. maydis* (*Cochliobolus heterostrophus*)	Streak: maize streak Gemini virus	Corn stunt disease: Spiroplasma kunkelii	

TABLE 8.3   Continued

Crop	Fungal	Viral	Nematode	Bacterial
	Rust: *Puccinia* spp.	Yellow dwarf: barley yellow dwarf luteo virus		
	Smut: *Ustilago zeae*			
	Stalk and ear rots: *Gibberella zeae*, *Diplodia* spp. and others			
Sweet potato (*Ipomoea batatas*)	Scab: *Sphaceloma batatas* (Elsino batatas)	Feathery mottle: sweet potato feathery mottle potyvirus	Root-knot nematode: *Meloidogyne* spp.	Soil rot: *Streptomyces ipomoea*
	Fusarium wilt: *Fusarium oxysporum*			Little leaf: sweet potato little leaf phytoplasma
	Black rot: *Ceratocystis fimbriata*			
	Java black rot: *Botryodiplodia theobromae*			

*Source: Narayanasamy, et al. (2006).

types, microorganism multiplication on or in foods is a major cause of food deterioration. The microorganisms will attack virtually all food constituents. Some will ferment sugars and hydrolyze starches and cellulose. Others will hydrolyze fats and produce rancidity. Still others will digest proteins and produce putrid and ammonia-like odors. Some will form acid and make food sour. Others will produce gas and make food foamy. Some will form pigments, and a few will produce toxins and give rise to foodborne illnesses. When food is contaminated under natural conditions, several types of organisms will be present together. Such mixed organisms contribute to a complex of simultaneous or sequential changes, which may include acid, gas, putrefaction, and discoloration (Doyle, 2007).

8.2.3.1 Yeast

Yeast is a subset of a large group of organisms called fungi that also includes molds and mushrooms. They are generally single –celled organisms that are adapted for life in specialized, usually liquid, environments, and, unlike some molds and mushroom's yeast, do not produce toxic secondary metabolites. Yeast can grow with or without oxygen (facultative) and are well-known for their beneficial fermentations that produce bread and alcoholic drinks. They often colonize foods with a high sugar or salt content. Fruits and juices with a lowpH are another target (Kurtzman, 2006). There are four main groups of spoilage yeast:

(a) **Zygosaccharomyces** and related genera tolerate high sugar and high salt concentrations and are the usual spoilage organisms in foods such as honey, dried fruit, jams and soy sauce. They usually grow slowly, producing off-odors and flavors and carbon dioxide that may cause food containers to swell and burst. *Debaryomyces hansenii* can grow at salt concentrations as high as 24%, accounting for its frequent isolation from salt brines used for cured meats, cheese, and olives. This group also includes the most important spoilage organisms in salad dressings (Campos et al., 2003).

(b) **Saccharomyces** spp. are best known for their role in production of bread and wine but some strains also spoil wines and other beverages by producing gassiness, turbidity and off-flavors associated with hydrogen sulfide and acetic acid. Some species grow on fruits, including yogurt-containing fruit and some are resistant to heat processing (Malfeito-Ferreira, 2003).

(c) **Candida and related genera** are heterogeneous group of yeasts, some of which also cause human infections. They are involved in spoilage of fruits, some vegetables and dairy products (Casey, 2003).

(d) **Dekkera/Brettanomyces** are principally involved in spoilage of fermented foods, including beverages and some dairy products. They can produce volatile phenolic compounds responsible for off-flavor (Hogg et al., 2003).

8.2.3.2 Mold

Molds are filamentous fungi that do not produce large fruiting bodies. Molds are very important for recycling dead plant and animals remains in nature but also attacks a wide variety of foods. Mold is larger than bacteria and yeast and more complex in structure (Chang and Kang, 2004). Majority of postharvest diseases causing mold belong to the class Ascomycetes. They are well adapted for growth on and through solid substances, generally produce airborne spores, and require oxygen for their metabolic processes. Most molds grow at a pH range of 3 to 8 and some can grow at very low water activity levels (0.7–0.8) on dried foods. Spores can tolerate hash environmental conditions but most are sensitive to heat treatment. An exception is *Byssochlammys*, whose spores have a D value of 1–12 min at 90°C. Different mold species have different optimal growth temperatures, with some able to grow in refrigerators. They have a diverse secondary metabolism. Some spoilage molds are carcinogenic while others are not (Chang and Kang, 2004). Spoilage molds can be categorized into four groups:

(a) *Zygomycetes* are considered relatively primitive fungi but are widespread in nature, growing rapidly on simple carbon source in soil and plant debris, and their spores are commonly present in indoor air. Generally, they required high-water activities for growth and are notorious for causing rots in a variety of stored fruits and vegetables, including strawberries and sweet potatoes. Some common bread molds also are Zygomycetes. Some Zygomycetes are also utilized for production of fermented soy products, enzymes, and organic chemicals. The most common spoilage species are *Mucor* and *Rhizopus*. Zygomycetes are not known for producing mycotoxins but there are some reports of toxic compounds produced by a few species (Doyle, 2007).

(b) *Penicillium* and related genera are present in soils and plant debries both tropical and Antarctic conditions but tend to dominate spoilage in temperate regions. They are distinguished by their reproductive structures that produce chains of conidia. Although they can be useful to humans in producing antibiotics and blue cheese, many species are important spoilage organisms, and some produce potent mycotoxins (patulin, ochratoxin, citreoviridin, penitrem). *Penicillium* spp. Cause visible rots on citrus, pear, and apple fruits and causes enormous losses in these crops. They also spoil other fruits and vegetables, including cereals. Some species can attack refrigerated and processed foods such as jams and margarine.

(c) *Aspergillus* and related molds generally grow faster and more resistant to high temperatures and low water activity than *Penicillium* spp. and tend to dominate spoilage in warmer climates. Many Aspergilli produce mycotoxins: aflatoxins, ochratoxin, territrems, cyclopiazonic acid. Aspergilli spoil a wide variety of food and non-food items (paper, leather, etc.) but are probably best known for spoilage of perishables, grains, dried beans, peanuts, tree nuts, and some spices.

(d) *Botrytis cinerea* is the causal agent of the postharvest disease called gray mold. Grey mold is one of the most common and destructive diseases of green house and outdoor crops. *B. cinerea* can infect over 230 crops (Govrin and Levine, 2000; Zhao et al., 2009) and is estimated to cause greater economic loss of ornamentals and vegetables than any other disease (Samir and Amnon, 2007). *B. cinerea* appear as blossom blights, fruit rots as well as damping off, stem cankers or rots, leaf spots and tuber, corn bulb and root rots. Grey mold can be a serious problem during both short and long term storage and subsequent shipment of most types of horticultural produce, (Agrios 1997). Williamson et al. (2007) consider *B. cinerea* as the most widely distributed disease of vegetables, ornamentals fruits and field crops throughout the world.

Other molds, belonging to several genera, have been isolated from spoiled food. These generally are not major causes of spoilage but can be a problem for some foods. *Fusarium* spp. Causes plant diseases and produce several important mycotoxins but are not important spoilage organisms. However, their mycotoxins may be present in harvested grains and pose a health risk (Agrios, 1999).

8.2.3.3 Bacteria

Bacteria are unicellular microorganisms of many forms, although three principal shapes of the individual cells predominate. These are the spherical shape represented by several forms of cocci, the rod shape of the bacilli, and spiral forms possessed by the spirilla. Some bacteria produce spores, which are remarkably resistant to heat, chemicals and other more adverse conditions. Bacterial spores are far more resistant than yeast or mold spores. And more resistant to most processing conditions than natural food enzymes. All bacteria associated with foods are small. Most are of the order of one to a few microns in cell length and somewhat smaller than this in diameter. (A micron is one-thousandth of a millimeter (0.001 mm) or about 0.00004 inches). Postharvest spoilage causing bacteria can be grouped into the followings:

(a) **Spore-forming bacteria** are usually associated with spoilage of heat-treated foods because their spores can survive high processing temperatures. These Gram-positive bacteria may be strict anaerobes or facultative (capable of growth with or without oxygen). Some spore-formers are thermophilic, preferring growth at high temperatures (as high as 53°C). Some anaerobic thermophiles produces hydrogen sulfide (Desulfotomaculum) and others produce hydrogen and carbon dioxide (Thermoanaerobacterium) during growth on canned/hermetically sealed foods kept at high temperatures, Other thermophiles, (*Bacillus* and *Geobacillus* spp.) cause a flat sour spoilage of high or low pH canned foods with little or no gas production (Pepe et al., 2003). Mesophilic anaerobes, growing at ambient temperature causes, several types of spoilage of vegetables (*Bacillus* spp.): putrefaction of canned products, early blowing of species, and butyric acid production in canned vegetables and fruits (*Clostridium* spp.) and 'medicinal' flavors in canned low-acid foods (*Alicylobacillus*) (Chang and Kang, 2004). Psychrotolerant sporeformers produce gas and sickly odors in chilled meats and brine-cured hams (*Clostridium* spp.) while others produce off-odors and gas in vacuum-packed, chilled foods and milk (*Bacillus* spp.).

(b) **Lactic acid bacteria** (LAB) are group of Gram-positive bacteria, including species of Lacto-bacillus, *Psdiococcus*, *Leuconostoc* and *Oenococcus,* some of which are useful in producing fermented foods such as yogurt and pickles. However, under low oxygen, low temperature, and acidic conditions, these bacteria become the predominant spoilage organisms on a variety of foods. Undesirable changes caused by LAB include gas formation in cheeses (blowing), pickles (bloater damage), and canned or package vegetables. LAB may also produce large amounts of an exopolysaccharide that causes ropy spoilage in some beverages (Doyle, 2007).

(c) *Pseudomonas* and related genera are aerobic, Gram-negative soil bacteria, some of which can degrade a wide variety of unusual compounds. They generally require a high-water activity for growth (0.95 or higher) and are inhibited by pH values less than 5.4. Some species grow at refrigeration temperatures (Psychrophilic) while others are adapted for growth at warmer, ambient temperatures. Four species of Pseudomonas (*P. fluorescens, P. fragi, P. lundensis*, and *P. viridiflava*), *Shewanella putrefaciens,* and *xanthomonas campestris* are the main food spoilage organisms in this group. Soft rots of plant-derived foods occur when pectins that hold adjacent plant cells together are degraded by pectic lyse enzymes secreted by *Campestris, P. fluorescens* and *P. viridiflava.* These two species of Pseudomonas comprises up to 40% of the naturally-occurring bacteria on the surface of fruits and vegetables and cause nearly half of post-harvest rot of fresh produce stored at cools temperatures (Fonnesbech Vogel et al., 2005).

(d) *Entrobacteriaceae* are Gram-negative, facultative anaerobic bacteria that include a number of spoilage organisms. These bacteria are wide spread in nature in soil, on plant surfaces. Erwinia carotovora is one of the most important bacteria causing soft rot of vegetables in the field or stored at ambient temperatures (Rasch et al., 2005).

8.3 FACTORS AFFECTING FOOD SPOILAGE AND SHELF LIFE

Foods by their nature are rich in carbohydrates, proteins and lipids that microbes as well as humans find very nutritious. Living plants have structures and chemical defenses to prevent microbial colonization, but once they are harvested, dead or in a dormant state, these systems deteriorate and become less effective. Many different microbes may potentially be able to use the nutrients in a food, but some species have a competitive advantage under certain conditions (Pitt and Hocking, 2007). Different food categories present different challenges for inhibition of spoilage pathogens.

8.3.1 VEGETABLES

Vegetables are a very good source of nutrients for spoilage pathogens because of their near neutral pH and high-water activity. Although vegetables are exposed to a multitude of soil microbes, not all of these can attack plants and some spoilage microbes are not common in soil, for example, lactic acid bacteria. Most spoilage losses are not due to microorganisms that cause plant diseases but rather to bacteria and mold that take advantage of mechanical and chilling damage to plant surfaces. Some microbes are found in only a few types of vegetables while others are widespread. *Botrytis cinerea* and *Erwinia carotovora* are the most common spoilage pathogenic fungi and bacterium respectively and has been detected in virtually every kind of vegetables (Tournas, 2005).

Bacterial spoilage first causes softening of tissues as pectins are degraded and the whole vegetables may eventually degenerate into a slimy mass. Starches and sugars are metabolized next, and unpleasant odors and flavors develop along with lactic acid and ethanol. Besides *carotovora*, several *Pseudomonas* spp. and lactic acid bacteria are important spoilage bacteria.

Molds belonging to several genera, including *Rhizopus, Alternaria*, and *Botrytis,* causes a number of vegetables rots described by their color, texture, or acidic products. The higher moisture content of vegetables as compared to grains allows different fungi to proliferate, but some species of *Aspergillus* attack onions (Chang and Kang, 2004).

8.3.2 FRUITS AND JUICES

Intact, healthy fruits have many microbes on their surfaces but can usually inhibit their growth until after harvest. Ripening weakens cell walls and decreases the amounts of antifungal chemicals in fruits, and physical damage during harvesting causes breaks in outer protective layers of fruits that spoilage organisms can exploit. Molds are tolerant to acidic conditions and low water activity and are involved in spoilage of citrus fruits, apple, pears, and other fruits. *Penicillium, Botrytis*, and *Rhizopus* are frequently isolated from spoiled fruits (Calho et al., 2007). Yeast and some bacteria, including *Erwinia* and *Xanthomonas,* can also spoil some fruits and these may particularly be a problem for fresh cut packaged fruits (Ngarmsak et al., 2006).

Fruits juices generally have relatively high levels of sugar and a low pH and this favors growth of yeast, molds and some acid-tolerant bacteria. Spoilage may be manifested as surface pellicles or fibrous mats or molds, cloudiness, and off-flavors. *Saccharomyces and Zygosaccharomyces* are resistant to thermal processing and are found in some spoiled juices (Fitzgerald et al., 2004). *Aliccyclobacillus* spp., an acidophilic and thermophilic spore-forming bacteria, has emerged as an important spoilage microbe, causing a smoky taint and other off-flavors in pasteurized juices (Chang and Kang, 2004).

8.4 CONTROL OF POSTHARVEST SPOILAGE PATHOGENS

Microbial pathogens, depending on their pathogenic potential (virulence), the level of susceptibility/resistance of crop cultivar, and the existence of favorable environmental conditions, may cause losses of marketable produce to a varying extent. In order to counter the adverse effects of microbial pathogens, one or more strategies for disease management that are compatible with each other have to be evaluated. Different diseases management strategies that have to provide more effective disease control than that possible with a single approach (Narayanasamy, 2002). Many economically important crop diseases have been managed by integrating various strategies, such as crop sanitation, certification, crop rotation (crop sequence), adoption of suitable planting/sowing date, and use of resistant cultivars, with nominal use of fungicides or other chemicals. Acknowledge of pathogen biology and ecology, source of inoculum, and epidemiological factors and storage conditions favoring disease incidence and spread may be useful in selecting and integrating suitable strategies for the effective management of the postharvest diseases of both durable and perishable commodities (Narayanasamy, 2002).

8.4.1 MANAGEMENT OF MYCOTOXINS CONTAMINATION

Seed microflora may comprise both beneficial and harmful microorganisms. However, the adverse effects caused by seed-borne fungal and bacterial pathogens outweigh the usefulness of beneficial microbes. In addition to failure of seed germination, reduction in seedling vigor, and seed deterioration caused by pathogens, production of mycotoxins by some fungal pathogens and contamination of foods and feeds have been the cause of concern, due to the possible health hazards. As mycotoxins contamination is unavoidable and unpredictable, it poses a unique challenge to food safety (FAO, 2009). The quality and safety of feeds is of vital importance to ensure that markets are not compromised by the distribution and sale of poor quality or unsafe food. Development of integrated system of management is considered to be the most effective approach to meet the risk associated with mycotoxins contamination. The Hazard Analysis and Critical Control (HACCP) approach identifies, evaluates and controls hazards that are significant for food safety. It is a structured, systematic approach for control of food safety throughout the commodity system, from the 'seed to spoon' (Wareing, 1999). Mycotoxins production and contaminations may be contained by taking measures during pre- and postharvest stage. The control parameters include various factors, such as time of harvesting, temperature and moisture during storage, selection of agricultural products prior to processing, decontamination conditions, addition of chemicals and final products storage (Lopez-Garcia et al., 2004).

8.4.1.1 Preharvest Control

The first step in ensuring a postharvest control and safe final products is prevention through preharvest control. Some seeds are contaminated with mycotoxins in the field. If the infections occur in the field, as in the case of wheat, barley, and corn, the fungal pathogens (*Fusarium* spp.) will continued to develop during postharvest stages and storage. Mycotoxins, such as fumonisin B1, are invariably produced preharvest. Aflatoxins may be produced both pre- and postharvest. Drying the seed to a safe water activity level is one to 14% for maize and 9.5% for groundnuts at 20°C it is possible to reduce the growth of *Aspergillus flavus* (Wareing, 1999).

Insect infestation of seed results in greater level of damaged kernels, favoring higher incidence of *A. flavus* and *A. parasiticus*. Hence, control of insect infestation may help prevent proliferation of *Aspergillus* spp. and aflatoxins production. Proper disposal of infected crop residues that may form the source of inoculum for infection of the next crop and adoption of a proper crop sequence (rotation) have been suggested to reduce infection by fungi producing mycotoxins. A maize-soybean rotation may result in a reduction in the incidence of *Fusarium* spp compared to monoculture of maize. Soil fertility and drought stress appear to have some influence on the level of preharvest aflatoxins contamination of maize. Drought followed by high moisture conditions has been found to be favorable for the development of *Fusarium moniliforme* and fumonism production. Development of cultivars resistant to toxigenic fungal pathogens may be the ideal approach for the management of mycotoxin contaminations. Some investigations have indicated the possibility of producing wheat and corn cultivars with resistance to the pathogens producing mycotoxins.

8.4.1.2 Postharvest Control

Factors such as timeliness, clean-up, and drying to maintain safe levels are important during harvesting. As crops left on the field for longer periods shows higher levels of mycotoxins contamination, it is essential to harvest the crops at the right time, followed by adequate drying, for mycotoxins decontamination, biological methods have been explored (Siddiqui, 2016). The possibility of degrading aflatoxins using certain fungi, which produce peroxidases was reported by Lopez-Garcia and Park (1998). Among the several chemicals evaluated for their ability to inactivate and reduce the hazard of mycotoxins, ammonification has been shown to be an effective process. Aflatoxin's contamination in maize, peanuts, and cotton could be significantly reduced by the ammonification process (Lopez-Garcia and Park, 1998).

Due to the unpredictable and heterogeneous nature of mycotoxins production and contamination, it may not be possible to achieve 100% destruction of all mycotoxins in all food systems. However, it is considered that the use of HACCP-based hurdle system, in which contamination is monitored and controlled throughout production and postproduction operations, may be effective. The development of suitable integrated mycotoxins management systems may control at points from the field to the consumer (Lopez-Garcia et al., 2004).

8.5 CONCLUSION

The importance of postharvest microbial pathogens yeast, fungi, and bacteria with potential to inflict substantial quantitative and qualitative losses of harvested produce has been recognized. Various strategies have been found to be effective at varying levels under range of conditions, which interact with each other and in many cases control is achieved through the less expensive, but very effective biological and cultural control methods. Chemical control offers the best choice but unfortunately this cannot be used at the time when the produce is ready for consumption, since this is the most susceptible stage of the host to the pathogen. In addition, chemicals are expensive, and excessive use results in pathogen resistance rendering them less effective, and they can be detrimental to the environment. Therefore, host resistance remains the most cost effective method for managing postharvest spoilage and deterioration of food. However, for this to be achieved, it is necessary to have a good understanding of the host defense mechanisms against the pathogen. Also better understanding of the epidemiology of the most common postharvest pathogen is needed to effectively tackle important stages in their life cycles.

KEYWORDS

- Food Spoilage
- Horticultural Produce
- Postharvest Pathology
- Quality preservation
- Safety measures
- Shelf Life
- Spoilage Pathogens

REFERENCES

Agrios, G.H. (1999). Plant Pathology. Academic Press, London.

Ahmad, M.S., & Siddiqui, M.W. (2015). Postharvest Quality Assurance of Fruits: Practical Approaches for Developing Countries. Springer, New York. pp. 265.

Alao, S.E.L. (2000). The Importance of post-harvest loss prevention. Paper presented at Graduation Ceremony of School of Torage Technology. Nigerian Stored Product and Research Institute Kano, pp. 1–10.

Calvo, J., Caivente, V., De Orellano, M.E., Benuzzi, D., & Tosetti, D. (2007). Biological Control of postharvest spoilage caused by *Penicillium* expansum and *Botrytis* cinerea in apple by using the bacterium Rahnella aquatilis. *Int J. Food Microbial 113*, 251–257.

Casalinuovo, I.A., Di Pierro, D., Coletta, M., & Di Francesco, P. (2006). Application of Electronic noses for diseases diagnosis and food spoilage detection. *Sensors 6*, 1428–1439.

Casey, G., & Dobson, D. (2003). Molecular detection of *Candida krusei* contamination in fruit juice using the citrate synthase gene csl and a potential role for this gene in the adaptive response to acetic. *J. Appl. Microbiol. 95*, 13–22.

Castro, M. Garro, O. Gerschenson, L.N., & Campos, C.A. (2003). Interaction between potassium sorbate, oil and Tween 20, its effect on the growth and inhibition of *Z. bailii* in model salad dressings. *J. Food Safety 23*, 1–55.

Chang, S.S., & Kang, D.H. (2004). *Alicyclobacillus* spp. in the fruit juice industry: history, characteristics, and current isolation/detection procedures. *Crit. Rev. Micro. Biol. 30*, 55–74.

Couto, J.A. Neves, F. Campos, F., & Hogg, T. (2005). Thermal inactivation of the Wine spoilage yeasts Dekkera/Brettanomyces. *Int. J. Food Microbiol. 104*, 337–344.

Doyle, M.E. (2007). Microbial Food spoilage-losses and control strategies. A Review of the Literature. Food Research Institute, University of Wisconsin-Madison. Wi53706.

Fitzgerald, D.J.. Straford, M. Gasson, M.J., & Narbad, A. (2003). *Anal. Inhibition Food Microbiol. 86*, 113–122.

Fonnesbech Vogel, B., Venkateswaran, K., Satomi, M., & Gram, L. (2005) Identification of Shewanella baltica as the most important H2S-producing species during iced storage of Danish marine fish. *Appl. Environ. Microbiol.* 71, 6689–6697.

Food and Agricultural Organization (2009). Corporate Document Repository Report, pp. 1–10.

Govrin, E.M., & Levine, A. (2000). The hypersensitive response facilitates plant infection by Necrotrophic pathogen Botrytis cinerea. *Current Biology.* 10(13), 751–757.

Kurtzman, C.P. (2006). Detection, identification, and enumeration methods for spoilage yeast, In: Blackburn, C. de W. (Ed.), *Food Spoilage Microorganisms*. CRC Press, LLC., Boca Raton FL, p. 28–54.

Lopez–Garcia, R., & Park, D.L. (1998). Effectiveness of postharvest procedures in management of mycotoxins hazards. Mycotoxins in Agriculture and Food Safety. In: Bhatnagar, D., & Sinha, S. (eds.), Marcel Dekker, Inc., New York, pp. 407–433.

Lopez-Garcia, R., Park, D.L., & Phillips, T.D. (2004). Integrated mycotoxins management system. Internate Resource: File: 11A Mycotoxins management systems foods, nutrition and agriculture. Available atwww.fao.org/document/show_cdr.asp.

Loureiro, V., & Malfeito-Ferreira, M. (2003). Spoilage yeast in the wine industry. *Int. J. Food Prot.* 69, 2729–2737.

Narayanasamy, P. (2002). Microbial Plant Pathogens and Crop Disease Management. Science Publishers, Inc., Enfield, New Hampshire, USA.

Narayanasamy, P. (2006). Postharvest Pathogens and Diseases Management. John Wiley and Sons.

Ngarmsak, M., Delaquis, P., Toivonen, P., Ngarmsak, T., Ooraikul, B., & Mazza, G. (2006). Antimicrobial activity of vanillin against spoilage microorganisms in stored fresh-cut mangoes. *J. Food Prot.* 69, 1724–1727.

Pepe, O., Blaiotta, G., Moschetti, G., Grco, T., & Villani, F. (2003). Rope-producing strains of bacillus spp. from wheat bread and strategy for their control by lactic acid bacteria. *Appl. Environ. Microbiol.* 69, 2321–2329.

Pitt, J.I., & Hocking, A.D. (1997). Fungi and Food Spoilage. Blackie Academic and Professional, New York.

Rasch, M., Anderson, J.B., Nelson, K.F., Flodgaard, L.R., Christensen, H., Givskov, M., & Gram, M. (2005). Involvement of bacterial quorum-sensing signals in spoilage of bean sprouts. *Appl. Environ. Microbiol.* 71, 3321–3330.

Samir, D., & Amnon, L. (2007). Postharvest Botrytis infection: Etiology, Development and management. In: Botrytis: Biology Pathology and Control, Elad, Y. (Eds.) 349–367. Springer Dordrecht, The Netherlands.

Siddiqui, M.W. (2015). Postharvest Biology and Technology of Horticultural Crops: Principles and Practices for Quality Maintenance. CRC Press, Boca Raton, Florida, USA. pp. 550.

Siddiqui, M.W. (2016). Eco-Friendly Technology for Postharvest Produce Quality. Academic Press, Elsevier Science, USA. pp. 324.

Siddiqui, M.W., Ayala-Zavala, J.F., & Hwang, C.A. (2016). Postharvest Management Approaches for Maintaining Quality of Fresh Produce. Springer, New York, pp. 222.

Tournas, V.H. (2005). Spoilage of vegetable crops by bacteria and fungi and related health hazards. *Crit. Rev. Microbiol.* 31, 33–44.

Waller, J.M. (2002). Postharvest diseases. In: Plant Pathologist Pocketbook. Waller, J.M., Leanne, M., & Waller, S.J. (Eds.). pp. 33–54. CAB International, UK.
Wareing, P. (1999). The application of Hazard Analysis Critical Control Point (HACCP) approach to the control of mycotoxins in foods and feeds. *Postharvest Convention,* Cranfield University, Bedford, UK.
Williamson, B., Tudzynski, B., Tudzynski, P., & VanKan, J.A.K. (2007). *Botrytis cinerea:* the cause of graymold diseases. *Molecular Plant Pathology, 8,* 561–580.
Yahaya, S.M., & Alao, S.E. (2008). An assessment of postharvest losses of tomato *Lycopersicon esculentum* and *pepper capsicum* annum in selected irrigation areas of Kano State, Nigeria. *Inter. Journal of Research in Bioscience,* 2006, pp. 53–56.
Zhao, M., Zhao, J., Wei, S., & You Jiu, T. (2009). Boty-II, a novel LTR retrotransposon in *Botrytis cinerea* BO5. 10 Revealed by Genomic Sequence. *Electronic Journal of Biotechnology, 12*(3), 2–3.

CHAPTER 9

NATURAL ANTIMICROBIALS IN POSTHARVEST STORAGE AND MINIMAL PROCESSING OF FRUITS AND VEGETABLES

MUNIR ABBA DANDAGO

Department of Food Science and Technology, Faculty of Agriculture and Agricultural Technology, Kano University of Science and Technology, Wudil, Kano State, Nigeria

CONTENTS

9.1 Introduction 312
9.2 New Alternative Methods 312
9.3 Postharvest Applications of Natural Antimicrobials in
    Fruits and Vegetables 313
9.4 Postharvest Applications of Herbs, Spices and
    Essential Oils in Fruits and Vegetables 316
9.5 Postharvest Applications of Natural Coatings in Fruits and
    Vegetables 318
9.6 Postharvest Applications of Antogonistic
    Micro-Organisms in Fruits and Vegetables 319
9.7 Conclusion and Recommendations 319
Keywords 320
References 320

9.1 INTRODUCTION

Fresh fruit and Vegetables are perishable and susceptible to postharvest diseases, which limit the storage period and marketing life, Moreover, postharvest decay results in substantial economic losses around the world (Zhang et al., 2011). The high rate of microbial attack in fruits and vegetables after harvest is due to the loss in their natural resistance, high water and nutrient contents. Postharvest losses many reach up to 50% and more in the field depending on the crop, harvest method, length of storage, marketing condition, etc. (Appleton and Nigro, 2003).

Fungal diseases are the major cause of losses in postharvest industry and to consumers. Infections occurring either before harvest, between flowering and maturity, or during harvest, handling and storage is currently controlled with synthetic chemical fungicides (Appleton and Nigro, 2003).

Synthetic fungicides treatments have long being the main method of controlling postharvest diseases. However, the use of synthetic fungicides has many limitations and disadvantages (Siddiqui, 2016). These include progressively restrictive legislations, social rejections due to toxicological problems affecting humans and environment and the development of resistance (Appleton and Nigro, 2003). Owing to the toxicity, synthetic chemicals are no longer recommended for prevention of microbial spoilage in food products. According to Suslow (2000) the rapid rise in the demand for organically produced fruits and vegetables all over the world is increasing the demand for natural pesticides and therefore making synthetic chemicals irrelevant (Ahmad and Siddiqui, 2015). There is also increasing international concern over the indiscriminate use of synthetic fungicides on crops because of the possible harmful effects on human health and the emergence of pathogen resistance to fungicides. Therefore, new alternatives for controlling postharvest diseases, which have good efficiency, low resistance, and little or no toxicity to nontarget organisms are in urgent demand (Zhang et al., 2011).

9.2 NEW ALTERNATIVE METHODS

Consumers are nowadays more concerned about both the fresh and processed food they eat and so there is increasing interest to replace the synthetic preservatives with natural, effective, non toxic compounds (Marija, 2009; Siddiqui et al., 2016).

According to Sanzani and Ippolito (2011) the development of resistance together with increasing concern about possible adverse effects on human health and environment caused by synthetic fungicides have contributed to arouse interest in the development of alternative means of controlling plant pathogens capable of integrating if not totally replacing synthetic fungicides. Substantial progress has been made in finding alternatives to synthetic fungicides for the control of postharvest diseases of fruits and vegetables (Ippolito et al., 2004; Paluo et al., 2008; Sanzani et al., 2009; Scheneria et al., 2007; Sharmer et al., 2009; Zhang et al., 2009).

In the last few years, one of the priorities of Food Industry's researches has been to look for a healthier, safer and more sustainable alternative treatments and compounds to substitute the traditional chemical products used for Food preservation particularly due to increasing consumer demands (Gonzalez et al., 2010). So there is a clear need to develop new alternative methods of controlling postharvest diseases and the emerging technologies for the control of postharvest diseases are essentially threefold; application of antagonistic microorganisms, application of natural antimicrobials and the application sanitizing products (Mari et al., 2003). The natural antimicrobials and sanitizing products in the first place are plant extracts and essential oils of herbs and spices. As natural foodstuff, herbs and spices appeal to all who question the safety of synthetic food additives and demand high quality products that are safe and stable (Marija, 2009).

According to Sanzani and Ippolito (2011) many alternative control agents provided limited success under field conditions. This was attributed to uncontrollable environmental conditions. However, the likelihood of success greatly increases during postharvest phase due to better control of the environment.

9.3 POSTHARVEST APPLICATIONS OF NATURAL ANTIMICROBIALS IN FRUITS AND VEGETABLES

The natural antimicrobial compounds are compounds derived from natural sources that are believed to have the ability to kill or inhabit the growth of microorganisms, much has been written about the potential for natural antimicrobials to replace or reduce reliance synthetic food preservatives. Antimicrobial agents are substances or mixtures of substances that are used in the food industry to destroy or suppress the growth of harmful microorganisms on food and other surfaces (Hirnersen et al., 2010). Studies have shown that antimicrobials can be applied either by dipping, or spraying on the surface to reduce the bacterial growth on surfaces of produce (Beuchat, 1998).

Plants produce a large number of secondary metabolites with antimicrobial effects on postharvest pathogens. Detailed studies have been conducted on aromatic compounds, essential oils, volatile substances and isothiocyanites with encouraging results (Mari et al., 2003). Some of chemicals with antimicrobial activity are natural constituents of the plant while others are produced in response to physical injury, which allows an enzyme contact with its substrate, and some are produced in response to microbial inversion (Adams, 2003). In addition to their protective role in the living plant, their potential as antimicrobials in foods has long been a subject of review in the area (Shelf, 1983; Beuchat, 1994; Nychas, 1995; Smid and Gorros, 1999; Nychese et al., 2003; Ippolito and Nigro, 2003; Alzamora and Guerrero, 2003).

In recent years, interests in natural substances have increased and numerous studies on the biocidal activity of a wide range of secondary metabolites have been reputed (Man et al., 2003). Certain volatile components produced by fruits like apples and pear during ripening show antifungal activity. For example, acetaldehyde has been found to be effective in postharvest control of *P. expansum* in apples and pear (Man et al., 2003).

A study by Chanda et al. (2010) investigated the antimicrobial activity of seven fruits and vegetables peels against eleven microorganisms. Result showed that the peel of mango had the best and promising antimicrobial activity.

Bukar and Magashi (2008) determined the antimicrobial/sanitizing effect of aqueous extract of *Parkia biglobosa* as washing solution for tomatoes, peppers and oranges. Results have shown the potential use of the extract of *P. biglobosa* as an antimicrobial washing solution to reduce the count and/or eliminate potential spoilage and pathogenic organisms on the surface of fruit and vegetables prior to storage.

Green mold caused is *Penecillium digitatum* is a common and serious postharvest disease of fruits particularly citrus in the Mediterranean climates. Applications of pomegranate peel extract (PPE) as postharvest dip provides an eco friendly and safe method for control of *P. digitatum* and the method could serve as an efficient postharvest treatment in packing houses (Tayel et al., 2009).

Recently, there has also been increasing interest in the use of Aloe Vera gel in the food industry as a functional ingredient in foods, drinks, beverages and creams. Nevertheless, most of the so called aloe Vera products may contain very small amounts of active compounds because of the processing techniques used to obtain the A. Vera gel affects the bioactive compounds (Valuede, 2005).

The antifungal activity of Aloe Vera pulp has been documented including several postharvest applications to fruit pathogens such as *pencillium digitatum, P. expansum, B. cinerea, Alternaria alternata* (Jasso de Rodreguez et al., 2005; Sacks et al., 1999). The antifungal activity of Aloe Vera was based on the suppression of germination and inhibition of mycelial growth and this could be attributed to the presence of more than one active compound with antifungal activity.

In addition, Aloe vera gel has been proven to reduce the growth of 17 bacteria, although the compounds responsible for the antibacterial activity were not elucidated even though compounds such as saponins and anthraquinones are known to have antibiotic activity (Reynolds and Dweck, 1999).

In a study by Valverde et al. (2005) on the use of Aloe vera gel to maintain quality of table grapes; uncoated dusters of table grapes shared rapid deterioration with an estimated shelf life of 7 days at 1°C plus 4 days at 20°C based on the fast weight loss, Color change accelerated suffering and ripening, browning and cough incidence of berry decay. On contrary, Aloe Vera treated dusters showed significant delay in postharvest quality losses as measured by those quality parameters. Storability was extended to 35 days at 1°C and the edible coating was able to reduce initial microbial load for mesophilic aerobic bacteria, yeast, and molds which increased the storage period (Valverde et al., 2005).

In a similar study, Aloe Vera gel was used to coat table grapes and it extended the shelf life to 35 days at 1°C. The gel worked as a barrier of

O_2 and CO_2 thereby creating a modified atmosphere and acted as a moisture barrier reducing weight loss, browning, softening, growth of yeasts and molds. The material reportedly contains antimicrobial compounds and thus prevents decay. Aloe Vera contains malic acid, acetylated carbohydrate that demonstrated anti-inflammatory activity.

Themgavela et al. (2004) studied the used of Jatropha leaf extract for control of postharvest disease of Banana. Results showed that *J. carcass* leaf extract was able to control the anthracnose disease in three Banana varieties. According to Rahman et al. (2011), *Jatropha curcas* fruit and seeds could be used as antifungal components to control major postharvest diseases of fresh horticultural produce in vitro. In the study *J. carcass* seed and pulp extract were found to be more effective than the whole fruit extract in the control of anthracnose disease of papaya fruits. The study concluded that *J. carcass* seeds and pulp extract have the potential to be used as natural fungicide against fungal phyto-pathogens to replace synthetic fungicides in agricultural applications.

9.4 POSTHARVEST APPLICATIONS OF HERBS, SPICES AND ESSENTIAL OILS IN FRUITS AND VEGETABLES

Common food preservation such as nitrites sodium benzoate, sodium metabiosulfate have a long history of safe use (Gould and Russell, 2003). However, there are occasional reports of allergic reactions in sensitive individuals and the formulation of potential carcinogenic compounds such as nitrosamines from nitrites, which have raised concerns about potential detrimental effects of preservatives on health (Roller, 2003).

Herbs and spices have been added to foods since ancient times not only as flavoring agent but also as food preservatives. Spices occupy a prominent place in traditional culinary practices and are indispensable part of daily diet of millions of people all over the world. They are essentially flavoring agents used in small quantities and reputed to have both beneficial effects and antimicrobial properties.

Nowadays, herbs and spices are added to foods for their antimicrobial activities and medical effects in addition to their flavor and fragrance qualities (Marija et al., 2009).

Essential oils have been shown to possess antibacterial, antifungal, antiviral insecticidal and antioxidant properties (Dobre et al., 2011; Bourt, 2004). Currently, about 300 essential oils out of about 3000 are commercially important especially in pharmaceuticals Agriculture, food, cosmetics and perfume industries (Bourt, 2004; Dalamare et al., 2007; Dobre et al., 2011).

Cinnamon as an antimicrobial agent has been used in apple Juice (Yaste and Fung, 2004; Muthuswamy et al., 2008); ethanolic extract of Cinnamon bark (1–2%) could reduce the *E. coli* in vitro. Citrus peel extracts have also been applied successfully in postharvest treatments of fruits and vegetables (Fisher and Phillips, 2008).

According to Marija (2009) numerous studies have been published on the antimicrobial activities of plant extradites against different microbes including food borne pathogens (Banchat, 1994; Smith-Palmer et al., 1998; Harakudo et al., 2004).

Lopez-Malo et al. (2006) reviewed some antimicrobial components that have been identified in spices and herbs such as Eugenol from cloves, thymol from thyme, vanillin from vanilla allocin from garlic, cinnamic aldehyde from cinnamon etc.

Rasooli (2007) has comprehensively reviewed the use of various essential oils from plant materials and different food commodities. It has been reported that spices owe their antimicrobial properties mostly due to the presence of alkaloids, phenols, glycosides, steroids, essential oils, coumarins and Tannins (Ebana et al., 1991).

In another study, Gotterez et al. (2009) reported that effectiveness of oregano in decontamination of carrots was comparable to that of chlorine. When carrots were treated with oregano the initial total viable counts was significantly lower than water treated samples. Since carrots treated with oregano were acceptable in terms of sensory qualities so the application; oregano could offer a natural alternative for the washing and preservation of minimally processed fruits and vegetables.

Chemicals, such as Methyl Jasmonite (MJ) a natural compound detected from Jasmine and other plant species is known to extend the shelf life of whole or fresh cut Mango, Guava and strawberry (Roller and Seedhar, 2002). Carvacerol and Cinnamic acid to delay microbial spoilage of fresh cut melon and kiwi fruit (Corbo et al., 2010) Essential oils from coriander mint, vanlla, persely, citrus peels and isothiocynates obtained from cruciferous vegetables were also tested (Carbo et al., 2010).

9.5 POSTHARVEST APPLICATIONS OF NATURAL COATINGS IN FRUITS AND VEGETABLES

An edible coating according to Ganzales-angular et al. (2010) is a thin layer of edible material (hydrocolloid or lipid) applied on the surface of a food product for the purpose of generating a semi-permeable barrier to gases, water vapor and volatile compounds. Edible coatings were fund to be able to extend the shelf life of fresh cut products by decreasing respiration and color.

Coatings most commonly used as edible coatings include chitosan, starch, cellulose, alginate, zein, gluten, whey, carnauba wax, beeswax, etc. Coatings are considered one of the most promising technologies not only in the postharvest of fruit and vegetables but also in pre-harvest applications (Siddiqui, 2015). Coatings based on edible and biodegradable hydrocolloids, interparating natural compounds with anti-microbial actually to control fungal growth in fresh fruits and Bacterial growths in minimally processed product is nowadays a promising technology in food preservation. Natural compounds with antimicrobial activity include a wide number of products from plants, sea organisms, insects and microorganisms (Gonzalez et al., 2010).

Edible coatings made of natural materials are also used to coat fresh fruits and vegetables as well as fresh cuts. The advantages of using edible coatings on fruit and vegetables are reduction of packaging waste because they are considered biodegradable and also for the development of new products. Evidence for example fresh cucumbers was coated with wax, cellulose, lipids starch, zein, alginate and mucilage (Valverde et al., 2005).

Additives of natural origin such as Lysozome, Nisin, organic acids, and essential oils are added to the coatings formulation to help in the preservation of the quality of fresh cut, productions and in particular, the functionality of edible coating can be extended by incorporating antimicrobial compounds (Olivas et al., 2005; Ayala et al., 2008).

Different antimicrobials used as dipping and for filling solutions and treatments with edible coatings were found to be able to extend shelf life of fresh cut fruits (Allende et al., 2006; Chien et al., 2007; Lanciotti et al., 2004; Gonzale–Anguilar et al., 2010; Rico et al., 2007; D Amato et al., 2010; Companiello et al., 2008).

Many researches have shown that the use of antimicrobials can reduce or eliminate specific microorganisms but it may also produce favorable conditions for others. Thus combining essential oils could lead to useful efficacy against both spoilage and pathogenic organisms (Marija et al., 2009).

9.6 POSTHARVEST APPLICATIONS OF ANTOGONISTIC MICRO-ORGANISMS IN FRUITS AND VEGETABLES

During the past decade Antogonistic microorganisms effective against postharvest diseases have gained considerable attention and achieved practical applications. Antogonistic microorganisms are the type of organisms capable of suppressing the growth of other organisms. Control of postharvest diseases through the use of Antogonistic microorganisms is also a new frontier that needs to be exploited to the fullest. The mechanism of biocontrol among antagonists appears to be competition for nutrients and space but other mechanism may also involve the production of anti fungal metabolites, direct parasitism and induced resistance sometimes associated with reduction of pathogen enzyme activity (Mari et al., 2003).

Ippolito and Nigro (2003) stated that the success of large-scale studies on microbial antagonists has generated interest of researchers and agrochemical companies in the area of prevention and/or control of postharvest diseases using antagonistic microorganisms.

Currently there are four antagonistic microorganisms to control postharvest diseases of fresh fruits and vegetables commercially available. These are *C. oleophila* (Aspire™), *C. albidas* (Yieldplus™) and two strains of *Pseudomonas syringae* (Biosave™).

9.7 CONCLUSION AND RECOMMENDATIONS

Natural antimicrobials and antagonistic microorganisms are viable alternatives to synthetics chemicals in postharvest storage as well as minimal processing of fruits and vegetables. They have the potential to completely or partially replace the synthetics when fully harnessed.

It is recommended that developing countries where the high postharvest losses are high should intensify their researches on the use of natural antimicrobials common within a region and to be specific on a particular fruit or vegetable. Results of researches should also be disseminated to farmers and other stakeholders in the commodity chain through adequate extension services.

KEYWORDS

- Antogonistic Micro-Organisms
- Fruits and Vegetables
- Minimal Processing
- Natural Antimicrobials
- Natural Coatings
- Postharvest Storage

REFERENCES

Adams, M. (2003). Nisin in multifactorial food preservation. In: Roller, S. (Ed). *Natural Antimicrobials for Minimal Processing of Foods.* Wood Head Publishing Limited, Cambridge, pp. 10–33.

Ahmad, M.S., & Siddiqui, M.W. (2015). Postharvest Quality Assurance of Fruits: Practical Approaches for Developing Countries. Springer, New York, pp. 265.

Alzamara, S.M., & Guerrero, S. (2003). Plant antimicrobials combined with conventional preservatives for Fruit products. In: Roller, S. (Ed). *Natural Antimicrobials for Minimal Processing of Foods.* Wood Head Publishing Limited, pp. 235–249.

Anato, D., Sinigaglia, M., & Corbo, M.R. (2012). Use of chitosan, honey and pineapple juice as filling ligands for increasing the microbiological shelf life of a fruit based salad. *International Journal of Food Science and Technology.* 45, 1033–1041.

Ayala-Zavala, J.F., Del-Toro, S.L., Alvarez Parilla, E., & Gonzalez-Angular, G.A. (2008). High relative humidity in package of fresh cut fruits and vegetables: Advantages or disadvantages considering microbiological problems and antimicrobial delivering systems. *Journal of Food Science, 73(R),* 41–47.

Beuchat, I.R. (1994). Antimicrobial properties of spices and their essential oils. In: Dillon, V.M., & Board, R.G. (Eds.). *Natural Antimicrobial Systems and Food Preservation.* CAB International Wallingford, pp. 167–179.

Beuchat; L.R. (1998). Surface Determination of fruit and vegetables eaten raw: A review. *Food Safety Issue,* World Health Organization Geneva WHO/FSF/FOS/982.

Bukar, A., & Magashi, A.M. (2007). Preliminary investigation on the use of Aqueous extracts as Antimicrobial washing solutions on Tomatoes, peppers and oranges. *International Journal of Pure and Applied Sciences. 2*(1), 22–26.

Burt, S. (2004). Essential oils: Their antimicrobial properties and potential applications in Food. *Journal of Applied Microbiology. 94*(1), 223–253.

Chien, P.J., Shen, F., & Yang, F.H. (2007). Effects of Edible chitosan coating on the quality and shelf life of sliced mango fruit. *Journal of Food Engineering. 78,* 225–229.

Corbo, M.R., Lanciotti, R., Gardiani, F., Sinigaglia, M., & Guerzoni, M.F. (2000). Effects of Hexanal, Trans-z-Hexanal and storage temperature on shelf life of fresh sliced Apples. *Journal of Agricultural and Food Chemistry, 48,* 2401–2408.

Dalemare, A.P.L., Pistrollo, I.T.M., Liane, A., & Serafini, Echeverrigary, S. (2007). Antibacterial activity of the essential oils of *Salvia officinalis* L., & *Salvia triloba* cultivated in south Brazil. *Food Chemistry, 100*(1), 603–608.

Dobre, A.A., Gagiu, V., & Nuculita, P. (2011). Preliminary studies on the Antimicrobial activity of Essential oils against food Borne Bacteria and Toxigenic Fungi AUDJG *Food Technology, 35*(2), 16–26.

Goherez Jorge et al. (2009). Impact of essential oils on microbiological, organoleptic and quality markets of minimally processed vegetables. *Innovative Food Science and Technologies. 10*(2), 135–296.

Gonzales-Angular, G.A., Ayala-Zuvala, J.F., Olivas, G.I., de la Rosu, L.A., & Alvarez-Parilla, E. (2010). Preserving quality of fresh cut products using safe Technologies. *Journal fur Verbracherschutz, 5*(1), 65–72.

Gutierez, J., Barry-Ryan, C., & Burke, P. (2008). The antimicrobial efficiency of plant essential oil combinations and interaction. *Food Microbiology, 124*(1), 91–97.

Hirneisen, K.A., Black, E.P., Coscarino, J.L. Fino, V.R., Hoover, D.G., & Kniel, K.E. (2010). Viral Inactivation in Foods: A Review of Traditional and Novel Processing Technologies. *Comprehensive Reviews in Food Science and Food Safety. 9*(1), 3–20.

Ippolito, A., & Nigro, F. (2003). Natural antimicrobials in postharvest storage of fresh fruits and vegetables. In Roller, S. (Ed). *Natural Antimicrobials for the Minimal Processing of Foods.* Wood Head Publishing Limited, Cambridge, pp. 201–224.

Ippolito, A., Nigro, F., & Schena, L. (2004). Control of postharvest diseases of fresh fruits and vegetables by preharvest application of antagonistic microorganisms. In: Dris, R., Niskanen, R., & Jain, S.M. (Eds.). *Crop Management and Postharvest Handling of Horticultural Products.* Vol. 4, pp. 1–4. Science Publishers, Enfield, USA.

Jasso de Rodriguez, D., Harnandez-Catillo, D., Rodriguez Garcia, R., & Angulo, J.L. (2005). Antifungal activity in vitro of Aloe vera pulp and liquid fraction against plant pathogenic fungi. *Industrial Crop Production, 21,* 81–87.

Lanciotti, R., Gianotti, A., Patrignani, F., & Gurdini, F. (2004). Use of Natural aroma compounds to improve shelf life and safety of minimally processed fruits. *Trends in Food Science and Technology, 15,* 201–208.

Mari, M., Bertolini, P., & Pratellen, G.C. (2003). Non conventional methods for the control of Postharvest Pear Diseases. *J. Appl. Microbiol., 94*(1), 761–766.
Marija, M., & Nement, N.T. (2009). Antimicrobial effects of spies and Herbs essential oils. *APTEFF 40*(1), 195–209.
Nychas, G. (1995). Natural antimicrobials from plants. In: Gould, G.W. (Ed). *New Methods of Food Preservation.* Blakie Academic, London, pp. 58–89.
Olivas, G.I., & Barbose-Canovas, G.V. (2005). Edible Coatings for Fresh cut Fruits. *Critical Reviews in Food Science and Nutrition, 45*, 657–663.
Palou, L. Smilanick, J.L., & Droby, S. (2008). Alternatives to conventional fungicides for the control of Citrus Postharvest Green and Blue mold. *Stewart Postharvest Review 4*(2), 1–6.
Rahman, M., Ahmed, S.H., Mohamed, S.T.M., & Rahman, M.Z. (2011). Extraction of *Jatropha curcas* fruits for antifungal activity against anthracnose of papaya. *African Journal of Biotechnology, 10*(48), 9796–9799.
Rasooli, I. (2007). Food preservation: A bio preservative approach. *Food 1*(2), 111–134.
Reynolds, T., & Dweck, A.C. (1999). Aloe Vera gel: A review update. *Journal of Ethnopharmacology. 4*(1), 1745–1755.
Rico, D., Martin-Diana, A.B., Barat, J.M., & Barry, C. (2007). Extending and measuring the quality of fresh cut fruits and vegetables: a review. *Trend Food Science and Technology, 18*, 373–386.
Roller, S. (2003). Introductions. In: Roller, S. (Ed.). *Natural antimicrobials for the minimal processing of Food.* Wood Head Publishing Limited, pp. 1–8.
Saks, Y., & Barkai-Golan (1995). Aloe Vera gel against plant pathogenic fungi. *Postharvest Biology Technology. 6*, 159–165.
Sanchez-Gonzales, L., Chafer, M., Gonzales-Martinez, C. Pastor, C., & Chiraltz, A. (2010). Interoperating Natural Antimicrobials into coatings for fruit preservation. *International Conference on Food Innovation.* Universidad Politec De Vale, pp. 1–4.
Sanzani, S.M., & Ippolito, A. (2011). State of the Art and future prospects of Alternative control means against Postharvest Blue mold of Apple: Exploiting the induction Resistance. In: Nooruddin, T. (Ed.). *Fungicides: The Beneficial and Harmful Aspects.* Intech, pp. 117–132.
Sanzani, S.M., Nigro, F., Mari, M., & Ippolito, A. (2009). Innovation in the control of postharvest diseases of fresh fruits and vegetables. *Arab Journal of Plant Protection. 27*(2), 240–244.
Schena, L., Nigro, F., & Ippolito, A. (2007). Natural antimicrobials to improve storage and shelf life of fresh fruits, vegetables and cut flowers. In: Roy, R.C., & Ward, O.P. (Eds.). *Microbial Biotechnology in Horticulture.* Vol. 2, pp. 259–303. Oxford and IBH Publishing, New Delhi, India.
Sharma, R.R., Singh, D., & Singh, R. (2009). Biological control of Postharvest Diseases of Fruits and Vegetables by Microbial Antagonists: A Review. *Biological Control, 50*(3), 205–221.
Shelf, I.A. (1983). Antimicrobial Effects of Spices. *Journal of Food Safety. 6*(1), 29–44.
Siddiqui, M.W. (2015). Postharvest Biology and Technology of Horticultural Crops: Principles and Practices for Quality Maintenance. CRC Press, Boca Raton, Florida, USA. pp. 550.

Siddiqui, M.W. (2016). Eco-Friendly Technology for Postharvest Produce Quality. Academic Press, Elsevier Science, USA, pp. 324
Siddiqui, M.W., Ayala-Zavala, J.F., & Hwang, C.A. (2016). Postharvest Management Approaches for Maintaining Quality of Fresh Produce. Springer, New York, pp. 222.
Suslow, T. (2000). Postharvest Handling for Organic Crops. University of California, DNR Publication 7254.
Tayel, A.A., El-Baz, A.F., & El-Hadary, M.H. (2009). Potential Applications of pomegranate peel extract for the control of citrus green mold. *J. Plant Dis. Prot. 116*(6), 252–256.
Thengavelu, R., Sundraraju, P., Sathiamoorthy, S. (2004). Management of Anthroenose Disease of Banana Using Plant Extracts. *J. Hort. Sci. Biotech., 79*, 664–668.
Valverde, J.M., Valero, D., Romero, D.M., Gullian, F., Castiello, S., & Serrano, M. (2005). Novel Edible Coating Based on Aloe Vera Gel to Maintain Table Grape Quality and Safety. *J. Agric. Food Chem., 53*(1), 7807–7813.
Walker, J.R.L. (1994). Antimicrobial Compounds in Food Plants. In: Dillon, V.M., & Board, R.G. (Eds.). *Natural Antimicrobial Systems and Food Preservation.* CAB International Wallingford, pp. 181–204.
Zhang, H.R., & Liu, W. (2011). Effect of chitin and its derivatives chitosan on postharvest decay of fruits: A review. *Int. J. Mol. Bio., 12*(1), 917–934.
Zhang, H., Wang, L., Dong, Y., & Zheng, X. (2009). Biocontrol of major postharvest pathogens on Apple using *Rhodotorula glutins* and its effects on postharvest quality parameters. *Biological Control, 48*(1), 79–83.

CHAPTER 10

ENHANCE: BREAKTHROUGH TECHNOLOGY TO PRESERVE AND ENHANCE FOOD

CHARLES L. WILSON

Founder/Chairman and CEO, World Food Preservation Center LLC, E-mail: worldfoodpreservationcenter@frontier.com

CONTENTS

10.1 Introduction 325
10.2 Hormesis: Well Established but Poorly Appreciated 326
10.3 Xenohormesis: Shared Chemicals for the
       Defense of Plants and Animals 327
10.4 Induced Plant Defensive Chemicals
       as Nutrients and Neutraceuticals 328
10.5 UV-C Light as a Hormetin 330
10.6 Summary: Xenohormesis to both Preserve
       and Enrich Our Food 331
Keywords 332
References 332

10.1 INTRODUCTION

Plants and animals have coevolved over a billion years. We are just starting to fully appreciate how they have evolved multiple ways to help one another survive adverse environmental stresses during this period. During their co-evolution, plants and animals have carried out "chemical conversations" on stressors in their environment and have adapted to their shared environment accordingly.

Plants being sessile cannot move to avoid adverse environmental conditions like animals. Consequently, plants have dedicated much of their secondary metabolism to their defense against adverse environmental insults. A wide array of compounds (alkaloids, cyanogentic glycosides, terpenoids, and phenoics) has been fashioned by plants for their defense. Many of these compounds are synthesized during stress. These compounds help plants to defend themselves against herbivores, pathogen, and adverse environmental conditions such as heat or drought (Siddiqui, 2015, 2016).

It has been shown that low dosages of plant or animal stressors that are normally toxic will often induce a beneficial or *hormetic* effect. In plants, the defensive compounds that are induced through *hormesis* for the plants defense are often beneficial nutritionally for animals. A new field of science has been established (*Xenohormesis*) based on the shared stress chemicals produced by plants in response to stressors.

The understanding and manipulation of Xenohormesis presents us an opportunity to enhance both the nutritional value and postharvest preservation of food. It has been shown that through hormetic treatments we can extend the shelf life of some foods while at the same time increasing their nutritional content.

10.2 HORMESIS: WELL ESTABLISHED BUT POORLY APPRECIATED

The phenomenon of hormesis came into prominence during the cold war. Russian scientists studied extensively the effects of radiation (Gamma, X-rays, and UV) on plants in conjunction with the development of atomic weapons (Lucky, 1980). In looking at the dosage effects of radiation on whole plants and seeds, Russian scientist repeatedly found a stimulatory effect by low dosages of irradiation. These results were initially attributed to experimental error but finally accepted as the concept of *hormesis* that states that low-dosages of a stressor, such as radiation that normally causes a detrimental effect to a living system can often cause a positive or beneficial effect.

As American scientists studied the effects of radiation dosages on living systems, they observed the low-dosage stimulatory effect as well. However, more for political than scientific reasons American scientists adopted a "Linear Non-Threshold Model" for the effects of radiation on living systems. They stated that low-dosage stimulatory effects were spurious and that all dosages of radiation were detrimental. Dr. Calabrese points out the damage that has been done to science by this lack of appreciation of hormesis in his article, "How a 'Big Lie' Launched the LNT Myth and the Great Fear of Radiation (Calabrese, 2011)." Dr. Lucky (2006) has similarly attacked the "Linear Non-Threshold Model."

What is striking about the phenomenon of hormesis is that this response to low dose stressors occurs throughout all living systems. It, therefore, must be a highly conserved and important genetic trait. Hormetic responses also occur broadly throughout all living systems including plants, animals, and microorganisms (Lucky, 1980). In our studies of the use of low-dose UV-C light to induce resistance to post-harvest diseases of fruits and vegetables we never failed to get a positive hormetic response in all the wide variety of commodities we studied (Wilson et al, 1994). This held true also for all the seeds that we studied (Brown et al., 2001).

A wide variety of agents (*hormetins*) can elicit the hormetic response in plants and animals including chemical (Belz and Piepho, 2012), biological (Nwachukwu, 2013), environmental (Kotak et al., 2007) and mechanical (Howitz and Sinclair, 2008) hormetins (Figure 10.1).

10.3 XENOHORMESIS: SHARED CHEMICALS FOR THE DEFENSE OF PLANTS AND ANIMALS

It has been shown that many of the compounds induced by hormetins in plants are beneficial nutrients for animals. This seemly unlikely association has recently found a logical explanation by Howitz and Sinclair (2008) at Harvard University in a field of science that they established and called *xenohormesis*. Hooper et al. (2010) have more recently

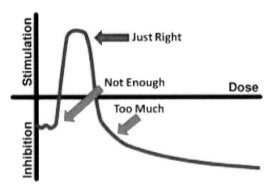

FIGURE 10.1 The hormetic dose response.

expanded the xenohormesis theory and suggested multiple applications as it applies to plants.

A basic tenant of the xenohormesis theory is that animals have learned to detect chemical changes in plants induced by adverse environmental conditions. This "chemical conversation" works toward the mutual benefit of both plants and animals. Animals may receive direct nutritional benefit from the "stress chemicals" produced by plants such as resveratrol (Adrian et al., 1997; Baur and Sinclair, 2009) or these compounds may "turn on" the animals own adverse environment defenses such as the SIRT1 gene (Howitz, 2003; Ingram et al., 2016; Knutson and Leeuwenburgh, 2008). SIRT1 is a gene found in humans and other mammals that helps to promote survival by protecting cells during times when food (and therefore energy) is scarce. Scientists have discovered similar genes in almost all species, including yeast, worms, and fruit flies. Herbivorous animals with their SIRT1 gene turned on presumably would put less pressure on plant populations by reduced feeding (Figure 10.2).

10.4 INDUCED PLANT DEFENSIVE CHEMICALS AS NUTRIENTS AND NEUTRACEUTICALS

Plants produce an array of phytochemicals in response to stress. Notable among these are antioxidants (Atkinson et al., 2005; Erkan et al., 2008),

ENHANCE: Breakthrough Technology to Preserve and Enhance Food

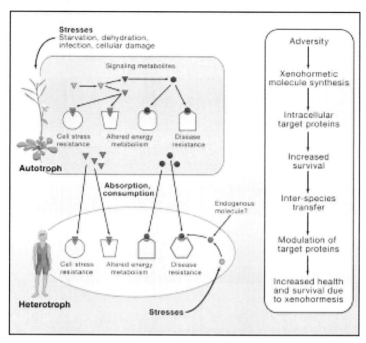

FIGURE 10.2 Human beneficial responses to hormetic induced phytochemicals in plants by stress (Howitz and Sinclair, 2008).

hydrolases (El Ghaouth et al., 2003), and phytoalexins (Adrian et al., 1997; Nwachukwu, 2013). Resveratrol is a phytoalexin produced by grapes, blueberries, and other fruit in response to stress (Wang et al., 2003). Phytoalexin are antimicrobial compounds that helps defend plant tissue against microbial invasion. Resveratrol has also been claimed to have extraordinary benefits for human health based on animal studies including: (a) Protecting the endothelial lining of arteries; (b) Reducing oxidative stress, which prevents premature aging of cells; (c) Blocking the production of noxious inflammatory agent; (d) Cellular support that improves mental function, and promotes oral/dental health; (e) Cancer suppression by preventing cancer cell replication and enhancing cancer cell death in a variety of laboratory cell culture studies; and (f) Improve muscle health, by reducing muscle wasting associated with diabetes and cancer. All these claims need to be validated in clinical studies, but they still show the remarkable potential of stress-induced chemicals in plants to impact on animal health.

The phytoalexins (glyceollins) elicited in soybean seeds by microorganisms, UV radiation, and physical injury have been found to have a variety of health benefits to humans including anti-tumor effects (Zhang et al., 2006).

10.5 UV-C LIGHT AS A HORMETIN

UV-C light has turned out to be an excellent hormetin that can elicit a number of improvements to harvested food. It is postulated that UV elicits beneficial effects by slight damage to the DNA of the plant tissue. The plant cell then responds with a variety of defensive responses including the production of defensive compounds (hydrolases, antioxidants, phytoalexins, flavonoids) and cell wall thickening (Charles et al., 2009; Siddiqui et al., 2016).

The pioneering work of the late Clauzell Steven at Tuskegee University and his associates (among whom I gratefully include myself) have shown the usefulness of low doses of UV-C light in reducing the postharvest decay of fruits and vegetables and extending their shelf life (Stevens et al., 1996, 1999, 2004, 2005). Stevens et al. (2005) demonstrated that the resistant response induced in harvested food was systemic. When just the ends of sweet potatoes were irradiated with UV-C light the induced resistance was actually greater than if all the surfaces of the sweet potato were irradiated.

Even with the great potential for the use of UV-C to enhance resistance in harvested food and increase its nutrient content there has been very limited commercialization of this technology. On-line applicators have been designed (Wilson et al., 1997). Valdebenito-Sanhueza in Brazil (Valdebenito-Sanhueza and Maia, 2001) has designed and used an on-line applicator of UV-C that was able to pay for itself after only 1.1 days operation on apples by the savings realized through disease control (Figure 10.3).

It has been demonstrated the pulsed treatment of mushrooms enhances their vitamin D content by as much as five times (Kalaras et al., 2012).

Dole has commercialized this technology both for whole mushrooms and in a mushroom powder (Figure 10.4).

10.6 SUMMARY: XENOHORMESIS TO BOTH PRESERVE AND ENRICH OUR FOOD

Hormesis and Xenohormesis provide us with breakthrough technology to enhance our food supply. World hunger is basically a nutritional hunger. If we can enhance the nutrimental content of our food, we have a way of attacking world hunger without increasing food production. This has appeal since concerns have been expressed over the environmental impact of expanded agricultural acreages to feed a rapidly expanding world population.

FIGURE 10.3 UV-C applicator for apples (Valdebenito-Sanhueza and Maia, 2001).

FIGURE 10.4 Commercial Dole mushroom with enhanced vitamin D content as a result of pulsed UV light treatments (Kalaras et al., 2012).

KEYWORDS

- ENHANCE concept
- Fruit and vegetable shelf life
- Hormesis
- Induced plant defense
- Irradiation treatment
- Postharvest quality
- Xenohormesis

REFERENCES

Adrian M., Jeander P., Veneau J., Westln L.A., & Bessis R. (1997). Biological activity of resveratrol, a stilbenic compound from grapevines, against Botrytis cinerea, the causal agent for gray mold. *J. Cem. Ecol. 23*, 1689–1702.

Atkinson, C.J., Nestby, R., Ford, Y.Y., & Dodds, P.A. (2005). Enhancing beneficial antioxidants in fruits: a plant physiological perspective. *Biofactors 23*, 229–234.

Baur, J.A., & Sinclair, D.A. (2009). Therapeutic potential of resveratrol: the in vivo evidence. *Nat. Rev Drug Siscov. 5*, 493–506.

Belz, R.G., & Piepho, H.-P. (2012). Modeling Effective Dosages in Hormetic Dose-Response Studies. *PLoS ONE 7*(3), e33432. doi: 10.1371/journal.pone.0033432.

Brown, J.E., Lu, T.Y., Stevens, C., Khan, V.A., Lu, J.Y., Wilson, C.L., Collins, D.J., Wilson, M.A., Igwegbe, E.C.K., Chalutz, E., & Droby, S. (2001). The effect of low dose ultraviolet light-C seed treatment on induced resistance in cabbage to black rot *(Xanthomonas campestris pv. campestris)*. *Crop Protection 20*, 873–883.

Calabrese, E.J. (2011). "How a 'Big Lie' Launched the LNT Myth and the Great Fear of Radiation." *21st Science and Technology*, 21–27.

Calabrese, E.J., & Blain, R.B. (2009). Hormesis and plant biology. *Environmental Pollution 157*, 42–48.

Charles, M.D., Tano, K., Asselin, A., & Arul, J. (2009). Physiological basis of UV-C induced resistance to Botrytis cinerea in tomato fruit. V. Constitutive defense enzymes and inducible pathogenesis related proteins. *Postharvest Biology and Technology 51*, 414–424.

El Ghaouth, A., Wilson C.L., & Callahan A.M. (2003). Induction of chitinase, β-1,3-glucanase, and phenylalanine ammonia lyase in peach fruit by UV-C treatment. *Phytopathology 93*, 349–355

Erkan, M., Wang, S.Y., & Wang, C.Y. (2008). Effect of UVC treatment on antioxidant capacity, antioxidant enzyme activity and decay in strawberry. *Postharvest Biology and Technology 48*, 163–171.

Hooper, P.L., Hooper, P.L., Tytell, M., & Vigh, L. (2010). Xenohormesis: health benefits from an eon of plant stress response evolution. *Cell Stress Chaperones 15*, 761–770.
Howitz, K.T., & Sinclair, D.A. (2008). Xenohormesis: sensing the chemical cues of other species. *Cell 133*, 387–391.
Howitz, K.T., Bitterman, K.J., Cohen, H.Y., Lamming, D.W., Lavu, S., Wood, J.G., Zipkin, R.E., Chung, P., Kisielewski, A., Zhang, L.L. et al. (2003). Small molecule activator of sirtuins extend Saccharoyces cerevisae lifespan. *Nature 425*, 191–196.
Ingram, D.K., Zhu, M. Mamczarz, J., Zou, S., Lane, M.A., Roth, G.S., & deCabo, R. (2006). Calorie restriction mimetrics: an emerging research field. *Aging Cell 5*, 97–108.
Kalaras, M.D., Beelman, R.B., & Elias, R.J. (2012). Effects of postharvest pulsed UV light treatment of white button mushrooms (*Agaricus bisporus*) on vitamin D2 content and quality attributes. *J. Agric. Food Chem. 60*(1), 220–225.
Knutson, M.D., & Leeuwenburgh, C. (2008). Resveratrol and novel potent activator of SIRT1, effects on aging and age-related diseases. *Natur. Rev. 66*, 591–596.
Kotak, S., Larkindale, J., Lee U, Koskull-Doring, P., Vierling, E., & Scharf, K.D. (2007). Complexity of the heat stress response in plants. *Curr. Opin. Plant. Biol. 10*, 310–316.
Lucky (2006). Radiation hormesis: the good, the bad, and the ugly. *Dose Response 4*, 169–190.
Lucky, T.D. (1980). *Hormesis with Ionizing Radiation*, CRC Press, Boca Raton.
Nwachukwu, I.D. (2013). The inducible soybean glyceollin phytoalexins with health-promoting properties. *Food Research International 54*, 1208–1216.
Siddiqui, M.W. (2015). *Postharvest Biology and Technology of Horticultural Crops: Principles and Practices for Quality Maintenance*. CRC Press, Boca Raton, Florida, USA. pp. 550.
Siddiqui, M.W. (2016). *Eco-Friendly Technology for Postharvest Produce Quality*. Academic Press, Elsevier Science, USA. pp. 324
Siddiqui, M.W., Ayala-Zavala, J.F., & Hwang, C.A. (2016). *Postharvest Management Approaches for Maintaining Quality of Fresh Produce*. Springer, New York. pp. 222.
Stevens, C., Khan, V.A., Kabwe, M.K., Wilson, C.L., Igwebe, E.C.K., Chalutz, E., & Droby, S. (2005). The effect of fruit orientation of postharvest commodities following low dose ultraviolet light-C treatment on host induced resistance to decay. *Crop Protection 24*, 756–759.
Stevens, C., Khan, V.A., Lu, J.Y., Wilson, C.L., Chalutz, E., Droby, S., Kabwe, M.K., Haung, Z., Adeyeye, O., Pusey, L.P., & Tang, A.Y.A. (1999). Induced resistance of sweet potato to Fusarium root rot by UV-C hormesis. *Crop Protection 18*, 463–470.
Stevens, C., Liu, J., Khan, V.A., Kabwe, M.K., Wilson, C.L., Igwegbe, E.C.K., Chalutz, E., & Droby, S. (2004). The effects of low-dose ultraviolet light-C treatment on polygalacturonase activity delay ripening and Rhizopus soft rot development of tomatoes. *Crop Protection 23*, 551–554.
Stevens, C., Wilson, C.L., Lu, J.Y., Khan, V.A., Chalutz, E., Droby, S., Kabwe, M. K, Haung, Z., Adeyeye, O., Pusey, P.L., Wisniewski, M.E., & West, M. (1996). Plant hormesis induced by ultraviolet light-C for controlling postharvest diseases of tree fruits. *Crop Protection 15*, 129–134.
Valdebenito-Sanhueza, R.M., & Maia (2001). Usage of UV-C light to protect Fuji apples from Penicillium expansum infection. *IOBC WPRS Bulletin 24*, 335–338.

Wang, C.Y., Chen, C., & Wang, S.Y. (2009). Changes of flavonoid content and antioxidant capacity in blueberries after UV-C illumination. *J. Food Chem. 117*, 426–431.

Wilson, C.L., El Ghauoth, A., Chalutz, E., Droby, S., Stevens, C., Lu, J.Y., Khan, V.A., & Arul, J. (1994). Potential of induced resistance to control postharvest diseases of fruits and vegetables. *Plant Dis. 78*, 837–844.

Wilson, C.L., El Ghauoth, A., Upchurch, B., Stevens, C., Khan, V., Droby, S., & Chalutz, E. (1997). Using an on-line UV-C apparatus to treat harvested fruit for controlling postharvest decay. *Horttechnology 7*, 278–282.

Zhang, B., Hettiarachchy, N., Chen, P., Horax, R., Cornelious, B., & Zhu, D. (2006). Influence of application of three different elicitors on soybean plants on the concentration of several isoflavones in soybean seeds. *J. Agric. Food Chem. 54*, 5548–5554.

INDEX

A

Abscisic acid, 102, 183, 205, 214
Achillea filipendulina, 115
Acidic water, 126
Aeromonashydrophila, 275
Agaricus bisporus, 152, 154, 157, 158, 162, 163, 171
Air distribution, 10, 12, 17
Air infiltration, 8
Air pre-cooling methods types, 9–15
  forced-air-cooling, 13
  operation, 14
  pallet racking, 13
  trapped tunnel pre-cooler system, 13
  vertical airflow, 13, 15
  natural convection air-cooling (room cooling method), 9–12
  modified room cooling method, 12
Air stream, 3, 22, 25, 137
Air velocity, 11, 53, 239
Air-cooling, 2, 7, 11, 13, 18, 19, 22, 27, 32, 33, 38, 45, 58, 123, 169
Air-freight terminal, 110
Ajuga reptans, 206
Aloe vera gel, 315
Alphonso, 183
Alstroemeria, 92, 96, 104, 124, 129, 132, 143
Alternaria alternata, 315
Amaranthus tricolor, 114
Aminoethoxyvinylglycine, 194, 195
Amrapali mango fruits, 199
Anemone coronaria, 215
Anthocyanin biosynthetic gene, 207
Anthurium, 93, 96, 98, 124, 128, 131–134, 139, 215
Antibacterial effect, 160
Anti-inflammatory, 159, 316
Antimicrobial
  activities, 316, 317
  properties, 316, 317
  sanitizing effect, 314
Antirrhinum, 107
Antirrhinum majus, 107, 115, 140
Anti-tumor effects, 330
Antiviral effect, 160
Antogonistic micro-organisms, 320, 319
Aranchnis, 128
Aranthera, 128
Arduino wireless sensor network, 55
Artichokes, 34
Ascorbic acid, 33, 71, 78, 170, 172, 198, 245
Asparagus, 34, 43
Aspergillus, 300, 303
Aspergillus flavus, 306
Astilbe hybrids, 114
Auxin efflux carriers, 105, 106, 189
Avocados, 34

B

Bacillus spp., 301
Bacterial blotch, 175
Bamboo baskets, 129, 174
Barn storage, 84
Bellis perennis, 114, 140
Bent neck in gerbera, 217
Biosensors, 250
Biosynthesis, 103, 105, 183, 184, 195, 214, 268, 279, 280
  enzymes, 100
  pathway, 100
Blackening symptoms, 108
Botrytis, 98, 142, 204, 218, 300, 303, 304
Botrytis blight, 204, 218

Botrytis cinerea, 106, 218, 300, 303, 315
Breathable materials, 245
Broccoli, 7, 34, 41–45, 251
   cooling methods, 45
   forced air-cooling, 45
   hydrocooling, 45
   package icing, 45
   room cooling, 45
Brown discoloration, 167
Brussels, 9, 22, 34, 44
Bud harvesting systems, 113, 115, 116, 256
Bunching, 117, 118
Byssochlammys, 299

C

Calcium carbonate, 164
Calcium ionophore, 98
Calendula, 113, 136
Calendula officinalis, 114, 140
Callistephus chinensis, 114, 140
Calocybe indica, 152, 154, 173
Campanula spp., 114
Candida and related genera, 298
Canning, 163, 171, 173, 266
   process, 171
      blanching, 171
      cooling, 171
      earning, 171
      filling, 171
      labeling and packaging, 171
      sterilization, 171
Cantaloupes, 34, 44
Capacity design, 18
Cape gooseberry, 201
Carbohydrate supply, 107, 109, 110, 126
Carbohydrates, 70, 107, 108, 110, 112, 137, 205, 215, 302
Carbon monoxide, 55
Cardboard (fiberboard), 242
Cardiovascular, 159
Cardoic tonic, 160
Careless handling and stacking, 243

Carnation, 92, 93, 96, 97, 100, 102, 105, 121, 124, 128, 129, 132, 133, 139, 142, 143, 204, 208, 209
Carotene, 200, 205
Carrots, 22, 34, 44, 266, 271, 317
Cassava, 69, 71–73, 75, 76
   Manihot esculenta Crantz, 70, 75
Cattleya, 113, 124, 131, 134
Cauliflower, 9, 22, 37, 156
Celery, 22, 34
Cellulose acetate, 131
Cellulose wood, 131
Centaurea spp., 114
Central equipment room, 31
Chemical conversation, 328
Chinaster, 208
Chlorophyll, 109, 205, 206, 214, 245, 268
Chrysanthemum, 92, 93, 120, 124, 125, 128, 132, 139, 143, 203, 204, 207–209, 215
Cinnamon, 317
Citric acid, 117, 125, 127, 128, 161, 170–172
Clamp storage, 84, 85
Clarkia unquiculata, 114, 140
Class-I, 122
Class-II, 122
Clostridium spp., 301
Cocoyams, 78
   Colocasia esculenta, 78
   Taro, 78
   Xanthosomas sagittifolium, 78
   Tannia, 78
Cold chain, 4, 5, 43, 44, 48, 49, 52, 53–55, 60, 232
Cold war, 326
Cold-storage, 174, 218
Color browning, 163
Compression test, 130
Consolida ambigua, 114
Control of the cold chain projects, 47
Controlled atmosphere, 51, 52, 97, 165, 240
   storage, 48, 141
   packaging, 165, 240

Convallaria majalis, 114, 140
Conveyer systems, 16
Cool storage, 4, 104, 108
Coreopsis grandiflora, 114, 140
Corrugated fiber board, 130, 242
Crop rotation (crop sequence), 304
Curing operation, 88
Cut flowers, 3, 7, 13, 22, 38, 52, 56, 94–99, 101, 104–108, 112, 116, 118, 122, 124, 126–128, 130, 131, 135, 136, 142, 144, 204, 212–219
   microbiology, 144
   storage, 135
Cyclamen, 105, 113, 139
Cyclopiazonic acid, 300
Cymbidium, 113, 124, 128, 134
Cytokinins, 103, 104

D

Daffodil, 102, 103, 142
Dahlia, 113, 203
Dahlia cvs., 114
Dashehari, 196, 200
Daylilies, 103
Decay organisms, 6
Decay-producing microorganisms, 3, 44
Dekkera/Brettanomyces, 298
Delphinium spp., 113, 114, 128, 134
Deterioration process, 2
Dhingri, 172
Dianthus barbatus, 115, 141
Diazo-cyclopentadiene, 101
Digitalis purpurea, 114, 140
Direct expansion, 3, 9, 11, 22, 24
   refrigeration system, 9
Direct use of slurry ice, 43
Directional mist eliminators, 22
Disease, 23, 74, 75, 79, 83, 95–99, 106, 110, 112, 136, 139, 218, 236, 253, 294, 300, 304, 305, 315, 316, 330
Disease resistance in rose, 218
Dormancy and spouting, 73, 88
Drop test, 130
Dry coil system, 2, 3
Dry storage, 136, 138, 139

Dry system, 24
Drying of mushrooms, 163, 168
Dynamic packaging systems, 247

E

Ear infrared, 248
Easiest icing method, 44
Ecopack, 18
Edible aroids (*Colocasia* Spp and *Xanthosomonas sagattifolium*), 70
Edible coatings, 281, 282, 318
Edible mushrooms, 155
Electronic Aroma Signature Pattern, 253
Electronic nose, 252
Emasculation, 187
Endogenous cytokinins, 103
Energy saving, 48, 56
ENHANCE concept, 332
Erwinia carotovora, 302, 303
Essential amino acid in edible mushrooms, 158
Ethanol sensors, 250
Ethephon, 107, 193, 196–198, 205
Ethylene, 3, 10, 43, 44, 92–94, 96, 98, 100–112, 115, 124, 125, 132, 135–137, 143, 169, 195, 236, 238, 240, 244–247, 268–270, 279, 280
   sensitive flowers care, 132
   synthesis, 10, 101, 102, 104
   tissue, 10
Eustoma grandiflorum, 114
Evaporative cooling, 2, 7, 169
Evaporator, 7, 8, 9, 11, 26, 49, 51, 52, 58
Exotic flowers special care, 131
External view of portable precooling, 29

F

Factors affecting postharvest life commercial flowers, 95
   chilling injury, 98
   cut flowers, 99
   desiccation, 99
   ethylene/hormones, 100
   genotype, 96

pre-harvest factors, 97
temperature, 97
poikilotherms, 97
water relations, 99
Fiberboard, 44, 130, 233, 235, 236
Field infections, 295
First-expired-first-out, 53
Floral preservative, 115, 117, 127, 128
Floricultural crops, 104
Florist shop, 108
Flow chart of processing, 161
Flower
  bud emergence, 204
  deterioration, 110
  harvesting, 204
  packaging, 144
  storage, 144
Food
  deterioration, 252, 295, 297
  spoilage, 302, 308
  processing methods, 266
    acidification, 266
    canning, 266
    dehydrating, 266
    fermentation, 266
    freezing, 266
    treatments with food additives, 266
Forced air, 2, 7, 12, 13, 15–19, 22, 23, 27, 30, 33, 45, 49, 51, 60, 97
Forced air precooling, 2, 17, 19, 23, 49, 51
  technique adapted to commodities, 2
Forced ventilation system, 22
Freesia hybrids, 114, 140
Fresh cut microbiology, 283
Fresh mushroom market, 174
Fresh produce, 3–7, 16–18, 37, 43, 45–47, 50–58, 166, 233, 236, 239, 240, 256, 257, 274–277, 283, 302
  cold chain, 4
  cut-flowers, 3
  fruits, 3
  vegetables, 3
Fresh-cut processing, 276

Fresh-cut products, 274–278
Fructose, 107, 205
Fruit splitting, 33
Fruits and vegetables, 5, 8, 29, 31, 34, 35, 42, 57, 58, 111, 136, 165, 232, 235, 236, 239–250, 253, 255–257, 266–271, 274–279, 295, 299, 300, 302, 312–314, 317–320, 327, 330
  shelf life, 332
Functions of packaging, 234, 257
Fungicides, 85, 106, 218, 304, 312, 313, 316
Fusarium moniliforme, 306
Fusarium spp., 300, 306
Future directives, 256

G

Gaillardia, 113, 136, 215
Gaillardia pulchella, 114, 140
Gaillardia x grandiflora, 114
Gas composition, 38, 240, 279
Gas indicators, 248, 252
Gas measurement systems, 252
Geotropic bending protection, 131
Geranylgeranyl pyrophosphate, 182
Gerbera, 93, 124, 128, 131, 134, 139, 210, 217
Gerbera hybrida, 209
Germicides, 125
Gibberellic acid, 105, 182, 185, 186, 188, 192, 194–196, 199, 203–208, 213–215, 219
Gibberellins, 105, 181–186, 201, 207, 210
  acidic, 182
  biosynthesis, 184
  commercial availability, 185
  diterepenes, 182
  ent-gibberellane, 182
  ent-kaurene, 182
  functions, 184
  growth retardants, 183
  cycocel, 183
  daminozide, 183
  paclobutrazol, 183

tetracyclic, 182
translocation, 184
Gladiolus, 92, 93, 107, 108, 113–115, 118–121, 124, 128, 131, 133, 136, 139–143, 207, 209, 215
Gloriosa superba, 114
Glucose, 107, 205
Glycoprotein, 159
Good agricultural practices, 277
Good manufacturing practices, 277
Grading, 117, 118, 120, 136, 137, 248
Gram-negative soil bacteria, 302
Grape vines, 185
Gravitropic response, 107
Green beans, 34
Green mold, 315
Greenhouse, 97, 101, 110, 115, 116, 135, 136, 213
Growth and tropic responses, 106
Gypsophila spp., 99, 115

H

Halogenated hydrocarbons, 39
Harvest, 2–5, 12, 33, 34, 38, 79, 80, 86, 88, 94–101, 108, 110–113, 117, 118, 123, 124, 132, 135, 157, 161, 163, 165, 169, 171, 174, 175, 181, 193, 194, 196, 198–201, 205, 215, 236, 240, 248, 253, 254, 268, 269, 292–295, 302, 303, 306, 312, 318
  bud harvesting, 113–116
  conditioning, 118
  handling, 116
  quality and grading, 118
  stages of harvest, 113
Harvesting indices, 88
Hazard analysis and critical control, 305
Helianthus, 113
Helianthus annuus, 114
Hemerocallis cvs., 114
Hibiscus sabdariffa, 206
High efficiency motors, 56
Himsagar mango, 199
Hippeastrum, 113
Histidine, 79

Homogeneity, 47
Hormesis, 326, 327, 332
Hormetic effect, 326
Horticultural produce, 181, 234, 236, 238, 300, 308, 316
Hydrating, 117, 119, 127
Hydrocooler belt, 31
Hydrocooling, 2, 7, 29–34, 38, 39, 45, 60
Hydrogen cyanide, 76
Hydro-vacuum methods, 39
Hydroxypyrene-1,3,6-trisulfonic acid, 100
Hypertension, 158, 160

I

Ice cooling, 7
Ice-bank cooling, 170
Icing systems, 46
Indirect use of slurry ice, 46
Induced plant defense, 332
Induced plant defensive chemicals as nutrients and neutraceuticals, 328
In-package-relative-humidity, 167
Insufficient water uptake, 109
Intelligent packaging, 247, 248, 257
Internet of Things, 55
Iris x hollandica, 114
Irish potato (*Solanum tubaeroson*), 70
Irradiation method, 86
Irradiation treatment, 163, 332

J

Jasmonic acid, 105
Jatropha curcas, 316
Juvenile stage, 201

K

Kinnow mandarin, 196
Kniphofia uvaria, 115
Kraft process, 243

L

Lactic acid bacteria, 302, 303

Larkspur, 118, 131, 136
Lathyrus odoratus, 115, 140
Leaf removal, 117
Leaf senescence, 104, 105, 219
Lettuce chicory potatoes, 22
Light intensity, 108, 109, 115, 127
Light-emitting diode, 59
Liliaceae, 104
Lilium, 104, 113, 115, 139, 140, 143
Limonium spp., 115
Linear nonthreshold model, 327
Listeria monocytogenes, 275
Litchi cultivar, 34
Low pressure storage (LPS), 142
Low-density polyethylene, 165
L-phenylalanine, 206
Lupin, 131, 136
Lupinus cvs. Russell, 114
Lysine, 79, 157

M

Magnetic resonance imaging, 248
Magnetic resonance relaxometry, 248
Maintenance, 2, 3, 52, 59, 142
Malate-dehydrogenase, 199, 201
Management of mycotoxins contamination, 305
MAP, 165–167, 245, 251, 257, 280
Marine containers, 54, 107
Mathematical modeling of vacuum cooling process, 40
Matthiola incana, 114, 140
Medicinal mushroom, 158
Mediocre package, 164
Methyl jasmonate, 106, 317
Microbiological safety, 271, 283
Microbiology, 270, 275
Mid infra-red, 248
Milky, 152, 162, 173
Milky mushroom, 154, 168, 173, 174
Minerals, 3, 71, 153, 155, 158
Minimal processing, 267–269, 320
Mobile pre-cooling facilities, 26
Modified atmosphere, 97, 139, 142, 165–167, 244, 245, 250, 278–282, 316

packaging, 165, 166, 244–247, 278–280
Moisture loss (dehydration), 239
Molded plastics, 243
Mucor, 299
Mushroom, 23, 37, 153–169, 171–175, 250, 298, 330, 331
packaging, 175
processing, 175
postharvest technology, 160
cucumbers, 22
Mycotoxins, 293, 299, 300, 305–307

N

N,N-dipropyl (1-cyclopropenylmethyl) amine, 101
Naphthyl phthalamic acid, 107
Narcissus cvs., 114
Natural antimicrobials, 313, 320
Natural coatings, 320
Neck length, 203, 209
Nitrites sodium benzoate, 316
Nitrosamines, 316
NMR spectroscopy, 248
Nuclear magnetic resonance, 248

O

Oncidium, 128, 134
On-off method, 51
Optimum precooling method, 6, 24
Orchid, 92, 93, 120, 134, 139
Orchid flowers, 100, 134
Organoleptic rating, 200
Oxygen transmission rate, 279
Oyster mushrooms, 166, 172

P

Packaging, 6, 17, 18, 23, 58, 87, 95, 117, 118, 129, 142, 162, 165, 166, 171, 174, 232–236, 240, 241, 244–248, 250, 253, 255–257, 266, 274, 278–282, 292, 318
functions, 233
material kinds, 241
different flowers, 132

carnation, 133
chrysanthemum, 132
rose, 132
Packaging requirements, 257
  fruits and vegetables, 236
Packing, 4, 17, 33, 86, 87, 108, 117, 119, 123, 124, 129–134, 161–166, 174, 243, 247, 255, 266, 315
  packaging, 164
  modified atmosphere packaging, 165
  modified humidity packaging, 166
Packing boxes, 130
Packing process, 4
Paddy straw mushroom, 154, 162, 174
Paper or plastic film, 243
Paphiopedilum, 124
Paraffin, 133
Parthenocarpic fruits, 185–187, 189, 202
Pathogens, 74, 106, 245, 250, 275, 277, 293–295, 304–307, 313–317
Pectin methyl esterase, 201
Peduncles, 134
Penicillium, 299, 300, 304
Penecillium digitatum, 315
Peonia cvs., 114
Peroxyacetic acid, 33
Petal cell elongation, 210
Petunias, 105
Phalaenopsis, 113, 128, 134
Phenylalanine ammonia lyase, 206, 270
Phloem, 108, 184
Phlox paniculata, 115, 140
Photosynthesis, 3, 108, 109
Physiological response, 283
Physiological weight loss, 192, 193
Physiology, 72, 73, 181, 268, 269
Phytoalexin, 329
Phytoalexins (glyceollins), 330
Pit storage, 84
Plastic sleeve, 131
Plate-type heat exchanger, 31
Platform storage method, 85
Pleurotus, 152–154, 157, 158, 172
Polianthes tuberosa, 115

Pollination, 100, 102, 104, 105, 111, 185
Poly vinyl chloride, 166
Polyaccetate films, 164
Polyethylene film, 138, 167
Polyethylene foil as protective cover, 131
Polygalacturonase, 201
Polypacks, 172, 174
Polyphenol oxidases, 163
Polypropylene, 133, 165, 257, 267
Polysaccharide, 156, 159, 281
Polysaccharide-K, 159
Polythene, 133–135, 164–166, 172–174, 243
Polythene sheet, 172
Pomegranate peel extract, 315
Postharvest,
  applications of,
    antogonistic micro-organisms in fruits and vegetables, 319
    natural coatings in fruits and vegetables, 318
  control, 106, 305, 306, 314
  cooling, 3, 5
  harvest handling, 88
  pathogens, 294
  pathology, 308
  quality, 33, 88, 96, 101, 175, 181, 215, 219, 240, 257, 315, 332
  shelf life, 219
  storage, 81, 253, 319, 320
  technology, 175
Potassium meta bisulphite, 170
Potato (*Solanum tuberosom*), 78
Precooling, 2–7, 15–18, 20, 22–29, 34–41, 43, 49–51, 56, 60, 117, 174, 257
  process, 2, 6, 17, 18, 25, 26, 35, 38, 41, 49, 51
  methods, 2
  air-cooling, 2
  evaporative cooling, 2
  hydrocooling, 2
  slurry ice, 2
  vacuum, 2

season, 6
systems, 3, 20, 38
Preharvest control, 305
Preharvest spray, 194, 199, 219
Preservation technologies, 163
Pre-storage, 4
Programmable logic controllers, 57
Propylene glycol, 43
Proteins, 71, 79, 156, 157, 189, 281, 297, 302
Proteoglycons, 159
Pseudomonas, 175, 302, 303, 319
Pulp-board punnets, 164, 166
Pulsing, 95, 124, 129, 213

Q

Quality preservation, 257, 283, 308
Quality standards, 53, 122, 247

R

Radiation preservation, 171
Radio frequency identification, 55, 248
Radiofrequency identification, 254
Rapid precooling, 5
Ready-to-eat, 256
Recommended storage conditions, 87
Reducing oxidative stress, 329
Refined smart cold chain system, 256
Refrigerated trucks, 47, 54, 130, 174
Refrigeration, 5–12, 18–25, 27, 43, 47–49, 54–59, 85, 115, 137, 169, 174, 240, 302
    capacity requirements, 19
    method, 85
    system, 8–12, 18, 23, 43, 47, 54, 57
Refrigerator, 9, 11, 169
Relative humidity, 2, 3, 11, 22, 25, 26, 46, 48–55, 57, 74, 82, 97, 131, 135, 168, 199, 239, 255, 268, 281
Respiration, 3, 5, 9, 35, 38, 50, 72, 73, 80, 83, 97, 107–110, 123, 128, 135–138, 163–169, 193, 214, 231, 236, 238–240, 244, 245, 267–269, 278–281, 318
Reverse effect, 201

Rhizopus, 299, 303, 304
Ripening/senescence stage, 292
Room cooling method, 7, 10, 13, 15
Root and tuber crops, 70–75, 78–88
    curing, 74, 75
    cell suberisation, 75
    formation of cork cambium, 75
    dormancy and spouting, 73
    harvesting, handling, transportation and marketing, 81
    handling of root and tubers, 82
    harvesting of root and tubers, 81
    marketing of root and tubers, 83
    transportation of root and tubers, 83
    major reasons for postharvest loss, 79–81
    mechanical damage, 79
    pathological factors, 81
    physiological factors, 80
    nutritional importance, 71
    physiology, 72
    postharvest losses, 79
    production statistics, 71
    storage method, 83
    common handling practices/conditions, 86
    improved/modern storage methods, 85
    recommended storage conditions, 87
    traditional method, 84
Rosa hybrida, 213, 214
Rudbeckia, 113

S

Saccharomyces, 298, 304
Safety measures, 308
Salmonella species, 275
Sanitation level, 6
Seed-borne fungal, 305
Seedlessness, 185, 187, 188
Senescence, 205, 207
    process, 2, 207
    regulation, 144, 219
Senescing florets, 107

Shelf life, 2, 7, 11, 32, 39, 46–48, 53, 54, 124, 161–167, 170, 171, 174, 175, 194, 198–201, 240, 245, 253–257, 267, 269, 276–279, 281, 283, 308, 315–318, 326, 330
Shewanella putrefaciens, 302
Shigella species, 275
Silver thiosulfate, 94, 101, 107, 111, 119
Simulated annealing (SA) technique, 41
  meta-heuristic technique, 41
Slurry ice, 2, 7, 42–47, 60
Smart cold chain system, 256
Smart packaging, 247, 257
Snapdragon, 131, 132, 136, 142
Society of American Florist, 120, 121
Sodium dichloroisocyanurate, 127
Sodium erythorbate, 164
Sodium hypochlorite, 33, 164
Solidago spp., 114
Spadix, 133, 134
Spoilage, 79, 86, 161–163, 167, 168, 172, 199, 200, 239, 250, 255, 270, 271, 275, 276, 293, 298–304, 307, 308, 312, 314, 317, 319
  molds, 299
  pathogens, 303, 308
Stalk length, 207, 208
Staphylococcus aureus, 275
Steeping preservation, 170
Storage, 167
  controlled atmospheric storage, 167
  life, 34, 78, 82, 87, 129, 142, 192, 198, 199, 238, 244
  methods, 84, 85, 135
  optimum storage conditions, 167
  packaging, 219
Strawberries, 20, 22, 24, 28, 32–34, 299
Strelitzia flower, 108
Styrofoam box, 130
Suberin, 75
Sucrose, 107, 124–129, 137, 205, 207, 213, 214
Supply chain, 52, 55, 232, 234, 248, 250, 254–256
  real-time tracking information, 55
Surfactants, 125

Sweet corn, 30, 34, 44
Sweet potato (*Ipomea batatas* L.), 70, 77
System classification, 20
  dry system, 24
  secondary coolant coil, 24

T

Tagetes erecta, 114, 140
Temperature and relative humidity control, 52
Thermodynamically equations, 40
Thermodynamically model, 40, 41
Thermophysical specifications, 19
Thidiazuron, 104
Time-temperature indicators, 248, 251
Titrable acidity, 196
Total electricity consumption, 58
Translocation, 184
Transportation, 4, 42, 43, 54, 80–83, 86, 87, 98, 101, 129, 130, 163, 174, 217, 233–236, 241, 243, 248, 249, 256
Tropaeolum majus, 114
Tuberose, 108, 206, 215
Tulipa cvs., 114, 141
Tulips, 107, 115
Tyrosine ammonia lyase, 206

U

Ultra-low-oxygen (ULO) storage, 48
Underground storage, 84
UV, 101, 243, 325–327, 330, 331

V

Vacuum, 2, 23, 30, 34–42, 51, 60, 166, 169, 170, 301
  cooler, 35
  cooling, 7, 23, 34–41, 169
  treatment of broccoli, 41
Vanda, 128, 134
Variable frequency drive, 26, 51
Variable speed drive, 52, 58
Vase life, 93–97, 99, 104, 108–110, 112, 113, 115, 116, 126–128, 135, 138, 144, 213–215, 217, 219

Vase solution, 99, 107, 113, 128, 144, 211
Vibration test, 130
Viola odorata, 115, 141
Viola x wittrockiana, 114
Vita Flora, 112
Vitamin content of edible mushrooms, 157
Volvariella volvacea, 152, 154, 157, 158, 174

W

Washing, 163–166, 248, 266, 274, 314, 317
Water loss during vacuum cooling determination, 39
Water resistant, 23, 130
Water spray systems, 39
Water tolerant packages, 44
Wet cooling system, 20
Wet deck system, 2, 20, 24
  low air temperature, 2
  relative humidity, 2
Wet packing of flowers, 131
Wetting agents, 125
White button mushroom, 154, 163
Wireless sensor network, 55, 56

World Health Organization, 266
WSN technology, 256

X

Xanthomonas campestris, 302
Xenohormesis, 325–328, 331, 332
X-rays, 326
Xylem tubes, 109

Y

Yam (*Dioscorea* Spp), 70, 76
  Bitter yam (*D. dumetorum* Pax), 77
  Chinese yam (*D. esculenta*), 77
  Water yam (*D. alata* L.), 76
  White yam (*Dioscorea rotundata* poir), 76
  Yellow yam (*D. cayenensis* Lam), 76
Yeast, 271, 291, 298, 304
Yersinia enterocolitica, 275

Z

Zinnia elegans, 115, 141
Zizyphus jujuba Mill, 191
Zygomycetes, 299
Zygosaccharomyces, 298, 304
  Debaryomyces hansenii, 298